Developments in Geotechnical Engineering 5

SOIL PROPERTIES AND BEHAVIOUR

Further titles in this series:

Developments in Geotechnical Engineering 5

SOIL PROPERTIES AND BEHAVIOUR

by

RAYMOND N. YONG
William Scott Professor of
Civil Engineering and Applied Mechanics,
McGill University, Montreal, Canada

and

BENNO P. WARKENTIN
Professor of Soil Science
Macdonald College of McGill University,
Montreal, Canada

ELSEVIER SCIENTIFIC PUBLISHING COMPANY
Amsterdam   Oxford   New York   1975

ELSEVIER SCIENTIFIC PUBLISHING COMPANY
335 Jan van Galenstraat
P.O. Box 211, Amsterdam, The Netherlands

AMERICAN ELSEVIER PUBLISHING COMPANY, INC.
52 Vanderbilt Avenue
New, New York 10017

Library of Congress Card Number: 73–85233

ISBN 0-444-41167-4

Printed in The Netherlands

PREFACE

Since publication of *Introduction to Soil Behaviour* by the authors a certain degree of formalization has been achieved in the study of soil behaviour. Whilst some aspects of soil behaviour are still in the stages of initial development, the subject of soil behaviour itself can now be considered to be a discipline standing on a base which draws from various allied fields. In this, it resembles most other disciplines. Considerable progress has been made in the development of various tools and techniques for evaluation of the specific interactions between soil particles and water, which allow the further development of fundamental concepts and mechanisms describing the behaviour of soil in response to external constraints.

The interaction of the various constituents in soil defines the integrity of the system. A proper modelling of the physics of the problem is required in order to provide adequate and meaningful analysis and prediction for application in the fields of geotechnical engineering where a knowledge of soil behaviour is needed to analyse or predict soil-response behaviour to external loads, of soil science and agronomy where soil water and the structure of soils are studied in relation to plant growth, and of agricultural engineering where response behaviour of surficial soils is required.

The material in the book is developed along the lines of the need for a fundamental appreciation of the physical constituents and their interactions which define structure of the soil-water system. Chapters 1 and 2 provide the background for an understanding of the nature of the mineral particles and the forces existing between the particles in the soil system. The structure and fabric of soil are developed in Chapter 3, and their relationship with water is examined in Chapter 4. The mechanisms involving water movement and soil performance are discussed in Chapters 5 and 6 in relation to the physics of soil-water movement and volume changes.

Application of the concepts developed for describing mechanisms of soil behaviour to the evaluation of mechanical properties of soil are provided in the subsequent chapters. No attempt is made in this book to provide design principles or specific parameters directly applicable for any one particular design problem. Rather, the concepts pertaining to the behaviour of soil, and the properties that are measured in relation to the response mechanisms evoked, are discussed. The central themes revolve around the need for an understanding of the physics of interaction, the various factors defining the integrity of the soil-water system, and the properties that are measured which may be conditioned by experimental constraints in the test procedures used. The need to distinguish between analytical and mechanistic parameters cannot be overstressed, since this bears directly on interpretation of the physical performance of the material. It is with this kind of concern that the authors have sought to present the material, with the interpretations of performance based upon the fundamental mechanisms of particle and fabric unit interaction. The role of soil water is emphasized since this is fundamental to an understanding of soil properties and behaviour.

The authors have benefited considerably from discussions with many of their colleagues at various research centres and universities. Few concepts explaining soil properties and behaviour are developed completely by one person. The concepts grow gradually in discussions and through studies of several research workers. The references cited in the book have been selected primarily with regard to ideas .shown and information used. References are also given to articles in which original work can be found, and to state-of-the-art texts and research papers. Many of these were used in the development of the ideas in this book.

The authors are indebted to their colleagues and to their present and former graduate students for helpful discussions of the contents of various chapters. S.E. McKyes, J.C. Osler and D.E. Sheeran provided assistance at various times in the development of the book. The suggestions and input provided by Professor G.G. Meyerhof are especially acknowledged. In addition, the support obtained by grants for research from both the National Research Council of Canada and the Defence Research Board of Canada are acknowledged with gratitude. Finally the authors wish to thank N. Ahmed for preparation of illustrations, and Mrs. M. Powell who typed the manuscript.

R.N. Yong
B.P. Warkentin

CONTENTS

Chapter 1

NATURE OF SOILS

1.1 INTRODUCTION

Soil constituents exist in solid, liquid and gas states. The solid phase is represented by mineral and some organic particles with size variation from submicroscopic to visually discrete. The liquid phase is a solution of various salts in water. The gaseous phase tends toward the composition of air but with a high content of water vapour. The variability of any or all of these three phases together with corresponding variation in specific interaction of the three phases provides for non-uniformity in soil composition, characteristics, and properties. These microscopic interactions and reactions make it difficult to predict precisely the response of soil to a set of external constraints. Macroscopically, one attempts to sum up all the microscopic interactions and reactions into one or two representative measurable parameters to describe soil behaviour. This is the requirement for the development of insight into the physical behaviour of soil.

Soil from a physical viewpoint

The most common physical properties of a body are probably those of weight and volume. The density of a soil is an important property. Another obvious physical property of the soil is the degree of wetness. Associated with this is water movement, swelling on wetting, and cracking on drying. The diffusion of air in the soil is regulated by the degree of wetness.

Soil consists of particles of different sizes, arranged in different ways, with voids or pore space between the particles. It has been described as a three phase (solid, liquid, gas) system in dynamic equilibrium; the equilibrium between these phases changes continuously. During a rainfall or during irrigation the liquid phase is increased and some of the gases are excluded. As a result of engineering construction, application of external stresses will cause particle movement.

Physically, the soil is made up of minerals of various sizes with some organic molecules strongly bonded to the minerals and some organic matter physically mixed in. The solid particles exist in a certain arrangement and the voids or pore spaces between these are partly filled with dilute solutions and partly with gases. The smallest particles, clay and silt, are not spread uniformly throughout the soil.

The clay particles are often arranged in parallel orientation and form coatings or clay skins around sand particles or soil fabric units, thus forming a boundary between sand and the voids or pore spaces. Many physical measurements are made on soil samples that are broken up or disturbed, i.e. not in their natural state. It should be remembered when the results of tests to determine these physical properties are evaluated that the particles in the soil have a definite arrangement upon which soil properties depend. The grain-size analysis, for example, gives only the proportion of different sized particles present, and not their arrangement with respect to each other.

Importance of physical properties

Physical properties of the soil often limit both agricultural and engineering use of soils. Since physical properties are much harder to alter than chemical properties, it is the physical properties which largely determine soil use.

The physical properties of importance in engineering are strength, compressibility, permeability, volume change, compactibility and frost susceptibility. Engineering structures may be designed with more confidence if one knows these properties and their changes with environment and time. Depth of soil, grain-size distribution, drainage, stability of soil crumbs, and water-holding capacity are among the common physical properties affecting cultivation of soil. These are also the properties which are considered in soil surveys, and land capability surveys, when soils are rated for crop production.

Physical properties of soils have been studied for many decades. The obvious differences in bearing capacity of soils and the necessity for tillage to prepare an adequate seed bed to allow germination of seed are two examples of practical reasons for this interest. In addition, there is the early interest of man in his physical surroundings. Physics, and the tools used in physics, were developed well before the tools used in chemistry. Thus many of the physical properties of soils, which are macroscopic, were realized before chemical and biological properties were considered.

1.2 ORIGIN OF SOIL

Soils are formed by the natural process of disintegration of rock and decomposition of organic matter. The minerals derived from parent rock material constitute the primary minerals of soils. The term "soils" bears different connotations depending upon the discipline in which the term is used. In its most general usage soil may be defined as finely-divided rock material. This can range from colloidal particles to boulders up to a few feet in diameter.

Weathering or disintegration of rock can be by physical or chemical processes. In physical weathering there is generally no alteration of the chemical or mineralogical composition of the rock material. This process involves the crushing of rock into smaller sizes. In chemical weathering, the decomposition of rock is by chemical changes with alteration of the minerals in the rock.

The forces associated with weathering and erosion constitute the chief agencies for production of soil from rock. Of these, water has by far the greatest effect. Other factors also contribute such as temperature, pressure from glaciers or other sources, wind, bacteria, and human activity. In the Mississippi River system for example, where the drainage area covers approximately one and a quarter million square miles, over 600,000,000 tons of soil are torn from the drainage area and carried downstream to the mouth of the river in the Gulf of Mexico in one calendar year.

The atmosphere absorbs at least 100,000 cubic miles of moisture a year, most of which returns to the earth in the form of rain, snow and other forms of moisture deposition. At least one third of this falls on land. A fair proportion of this will be used to erode the material on the surface.

After the breakdown of rock by forces associated with glaciers, temperature, rivers, wave action and other means, further particle breakdown may occur either chemically or physically. Water seeping through fissures in limestone, for example, can disintegrate limestone if the water contains acids. Weak carbonic acid may be formed by water passing through the atmosphere collecting carbon dioxide. Also in passing through vegetative cover, the pH of water may reach as low as 4. The percolation of weak carbonic acid through the limestone causes erosive action disintegrating and breaking it down into smaller particles.

The derivation of rock fragments either physically or chemically is the first requirement for formation of soil. Rock fragments are broken down due to the action of glaciers or the forces exerted by flowing rivers and waves from the ocean. With time and continued action, the fragments are ground down to smaller and smaller particles. The product of the erosive forces whether by wind, water or glacier, find themselves deposited in lower regions such as deltas, valleys and plains.

Weathering is at its highest intensity where interfaces between the atmosphere, hydrosphere, biosphere and lithosphere overlap, e.g., the upper soil zone. The upper part of the regolith contains minerals which have been decomposed to less complex compounds. Some chemical weathering and other types of weathering also occur at the surface during erosion, transportation, etc.

The rate of weathering is influenced by many factors such as: size of rock particles, permeability of rock mass, position of groundwater table, topography, temperature, composition and amount of the groundwater, oxygen and other gases in the system, organic matter, wetting and drying, relative surface area of rock exposed, relative solubilities of the original rocks and the weathered materials, environmental changes, etc.

Weathering is a spontaneous reaction involving geologic material and energy. It is a change in the direction of a decrease in the free energy of the system. Thus, it is possible to predict the thermodynamic susceptibility of minerals to weathering. The assemblages of minerals in soils and sedimentary rocks are a reflection of temporary equilibrium stages in the dynamic process of change from source material to final end-product.

The weathering sequence involved in the development of clay minerals from parent rocks may be represented in terms of a reaction series portraying sedimentary rocks and soils as equilibrium stages in a continuous progression from their ultimate source (common igneous rocks such as basalt and granite) to a final end-product such as laterite.

1.3 SOIL CLASSIFICATION

There are many methods available for classification of soils. The choice of method depends upon the specific use intended for the soil. Geological classifications of soils differ from those used in soil engineering and from those used in agriculture. Geologic classifications are mostly genetic but partly descriptive, mostly in terms of surficial deposits. The major sub-divisions in the geologic classification of surficial deposits are:

(a) Transported: (1) fluvial deposits; (2) alluvial deposits, such as alluvial fans and deltas; (3) aeolian deposits; (4) glacial fluvial deposits; (5) glacial deposits; (6) volcanic deposits; (7) marine deposits; (8) mass-wasting products, such as mud flows, slide rock, talus, rock glaciers, slope wash, etc.

(b) Residual: (1) Soils — zonal soils, azonal soils, intrazonal soils; (2) marine and lacustrine deposits; (3) organic-mineral complexes such as muskeg.

The pedologist, or soil scientist working in agriculture, regards soil not in terms of the complete unconsolidated material but in terms only of the weathered and altered upper part of the unconsolidated material. The term "horizon" is a general term used to define a stratum in the soil. A soil profile consists of a succession of horizons or strata. The surface horizons are formed by interaction of rock material with organic material and with the atmosphere. The "A" horizon denotes the top stratum that is altered by natural processes such as leaching and washing away of the material. The "B" horizon which lies below the "A" horizon represents the stratum of accumulation or deposition. The material that is leached or washed down from the "A" horizon accumulates in the "B" horizon. The "C" horizon lying below the "B" horizon contains soil that is unaltered by weathering subsequent to deposition or formation. The development of A and B horizons in the soil profile is of primary importance. Soil formation or the development of the soil profile is determined by climate, vegetation, parent material, topography and time. Classification depends upon the complete description of the soil profile and relies to a large extent on profile development. Because of the complete description, this system serves a very useful purpose.

The other groups of soil classification methods of interest to us in the study of soil behaviour are the classifications of soils for engineering purposes. These classification methods rely primarily on the grain size and its gradation within the soil mass, its consistency, and probably its relation to frost penetration or its reaction to frost effects. These methods are outlined in various books on soil engineering.

While it is possible to infer from particle-size distribution an estimate of the water-holding capacity of soils, the strength and compressibility, etc., there is no substitute for actual measurement and evaluation of the properties of the soil. It is not possible, however, to include these actual measurements in classification methods, since they involve tedious laboratory studies. The purpose of classification is not to become involved in a long laboratory study programme but to provide a general description of the soil in a terminology for classification which is understood by all concerned. The present available classification systems for engineering purposes are meant specifically (and have been developed in general) for use in engineering projects involving highways and airport pavements. They concern themselves primarily with soils limited to the first few feet from the surface. Consequently, the properties measured and the descriptions of the soils are usually only those concerned with particle-size distribution and consistency. With a knowledge of clay mineralogy and the characteristics of clay—water interaction, it is possible to infer from a particle-size distribution curve the various properties of interest to the engineer. However, it is not generally possible for one to obtain a knowledge of the clay minerals present in the general classification of the system. Hence, for example, given a general grain-size distribution curve, it is frequently necessary to obtain consistency limits to give an idea of the clay—water interaction.

1.4 PROCEDURE FOR THE MINERALOGICAL ANALYSIS OF SOILS

It has become customary to recognize two particle categories in soils:
(1) Granular particles (gravel, sand and silt).
(2) Colloidal particles (clay-size particles).
The boundary between these is taken to be at 4 microns by geologists using the

Wentworth scale, or 2 microns by engineers. This division is convenient also for mineralogical soil analysis, and the list of methods is divided accordingly.

(A) *Grain-size analysis and fractionation.*
    1. Sieving methods.
    2. Sedimentation methods (wet analysis).
    3. Infra-sizer method (air elutriation).

(B) *Methods for coarse particles.*
    1. Chemical analysis.
    2. Microscope studies.
        (a) Binocular microscope (> 0.25 mm).
        (b) Polarizing microscope (> 5.0 microns).
            Refractive indices by immersion.
            Determination of minerals by their optical properties.
        (c) Staining methods.
        (d) Particle shape, size, sphericity, angularity, statistical counts of particles, composition and texture studies.
    3. Specific gravity determinations.
    4. Heavy-liquid analysis.
    5. Electromagnetic separation.

(C) *Methods for clay-size fraction.*
    1. X-ray diffraction and fluorescence.
    2. Differential thermal analysis.
    3. Optical (microscope) study of aggregates.
    4. Electron microscopy.
    5. Chemical methods.
        (a) Bulk chemical analysis.
        (b) Chemical treatment for removal of specific substances.
        (c) Organic content analysis.
        (d) Cation-exchange capacity determination.
        (e) Potash determination.
        (f) Glycol or glycerol retention.
    6. Surface area determination.
    7. Infrared methods.

Certain methods used for characterizing the clay-size fraction are also applicable to the coarser fraction, particularly the X-ray diffraction and differential thermal analysis methods. The methods for examination of clay-size fractions are given in Chapters 2 and 3.

TABLE 1.1

Summarized properties of clay-mineral groups (From Soderman and Quigley, 1965)*

Clay type	Symbol[1]	Negative charge per unit cell	Cation exchange capacity[2] (m.eq./100 g)	Glycol retention (mg/g)	Basal spacing glycol (Å)	Basal spacing dry (Å)	Liquid limit W_L	Plasticity index I_P	Activity a_c	Compression index C_c	Friction angle Drained
Kaolin	strong H + bond	−0.01	3	20	7	7	50	20	0.2 av	0.2	20–30°
Illite	strong K + bond	−1.0	25	80	10	10	100 to 120	50 to 65	0.6 av	0.6–1	20–25°
Chlorite	strong bond Al, Mg or Fe hydroxide sheet	−0.5 to −1.0	10–40	30	14	14	———— Probably like illite ————				
Vermiculite	weak Mg bond	−0.5 to −0.7	150	150	14	10	———— Probably between illite and montmorillonite ————				
Montmorillonoid	very weak bond	−0.3	100	260	16 to 18	10	150 to 700	100 to 650	1–6	1–3	12–20°

1 ⟶ silica sheet; ☐ hydroxide sheet.

2 m.eq./100g = milliequivalents/100 grams of dry soil.

*Reproduced by permission of Nat. Res. Council of Canada, *Can. Geotech. J.*, 1965, p. 168.

Most of the methods used in mineral analysis of the coarser grain fraction of soils are the standard chemical and geological methods. Mechanical analyses and sedimentation methods are discussed in the standard laboratory texts on soil mechanics. The infra-sizer such as that designed by Hultain, is useful for separating the finer fractions using air elutriation rather than liquid sedimentation, as the basis for separation. It is more rapid and convenient than the liquid separation methods.

Microscope studies and identification of minerals are discussed in standard texts on optical mineralogy and petrography. Of the methods listed, microscope techniques and electromagnetic separation are the quickest methods for mineralogical analysis of granular soils. Magnetic or electromagnetic separators use the small difference in the magnetic susceptibility of minerals. Rapid separation of feldspars, carbonates or any other mineral, is possible. Weighing the various separated mineral fractions gives a quick quantitative mineral analysis of the soil.

With the microscope, especially the polarizing microscope, a visual examination of the particles, together with observation of their birefringence and determination of their optical sign, is usually sufficient to identify most minerals. It is not usually important to determine the exact composition of the minerals in soil analysis. A general grouping into quartz, feldspars, micas, pyroxenes, amphiboles, and the accessory minerals such as zircon, iron oxides, sphene and apatite is sufficient. Should a more detailed analysis be desirable, the refractive indices of the minerals by the immersion method, as well as their other optical properties, may be determined.

Different clay minerals result in different soil properties, so a mineral analysis is often necessary to an understanding of soil behaviour. This is discussed further in Chapter 2. Table 1.1 summarizes some of the different properties of common clay minerals.

1.5 PARTICLE-SIZE COMPOSITION OF SOILS

The particle-size distribution of a soil influences chemical, physical and biological properties of soils. Separates consisting of larger particles, the sands and gravels, form the skeleton of the soil and determine many of its mechanical properties. The finest separate, consisting of clay particles, has a large surface area and determines most of the chemical and physical-chemical properties of the soils. Particle-size distribution in soils influences type and weight of cultivation machines required, the susceptibility of soils to erosion, the water-holding capacity and hence the water supply to plants, and spacing of irrigation and drainage ditches and types of drainage and irrigation systems required. It also influences the strength and compressibility of soils, both of which are important in the consideration of bearing

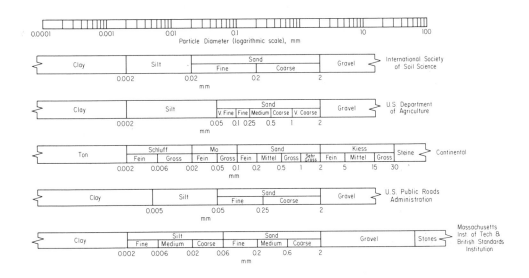

Fig.1.1. Principle particle-size scales. (From Anonymous, 1954: Soil Mechanics for Road Engineers, Chapters 3 and 4.)

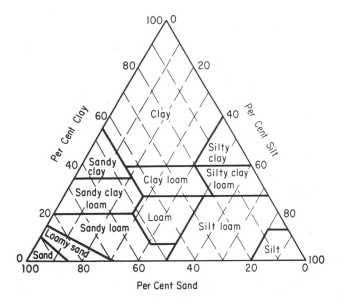

Fig.1.2. Chart showing per cent clay, silt, and sand in the soil textural classes of U.S. Department of Agriculture.

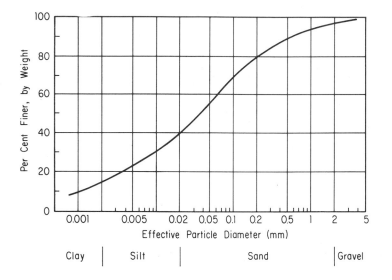

Fig.1.3. Particle-size distribution curve.

and stability for engineering purposes. Particle-size distribution is also important in the determination of soil temperature, aeration, and frost susceptibility.

Various systems have been used to group soils according to the proportion of different sizes of the constituent particles. The size limits used for sand, silt, and clay vary slightly in different systems. The Atterberg or International system, the M.I.T., International Society of Soil Science, U.S. Bureau of Public Roads, and the U.S.D.A. systems are all used (Fig. 1.1). Fig. 1.2 is a triangle used in soil science giving textural classes corresponding to different proportions of sand, silt and clay. Sometimes a size-distribution curve (Fig. 1.3) is used to show the results of particle-size analysis. This does not require an arbitrary division of particles into separates based upon size, but for many purposes it is easier to work with the numbers obtained by considering sizes such as sand, silt and clay.

Measurement of particle-size distribution

The procedure for determining the particle-size composition of soils involves three steps: (1) obtaining a representative sample; (2) dispersing the aggregates of the soil sample into their ultimate particles or grains; and (3) measuring the weight of the particles which occur within a certain size range.

Since the composition and properties of soils in situ vary from metre to metre or even from centimetre to centimetre it is often difficult to obtain a representative sample. The method commonly used is to take small samples from various parts of a field and make a large composite sample. Sampling procedure, which is based upon statistical considerations of the variability of the soil and the kind of analysis desired, will not be discussed in detail here. However, it must be kept in mind that the analysis will be useful only insofar as it represents the soil in the field.

Dispersion of the sample consists in removing cementing materials which are responsible for aggregating particles, and in deflocculating the sample so that these particles fall individually. The main cementing material to be removed from surface soil samples of the temperate region is organic matter. In soils where carbonate is present it will also act as a cementing material between particles. Iron and aluminum oxides are the common cementing materials in tropical soils. These oxides often make up an appreciable proportion of the total soil, therefore it is sometimes difficult to know how drastic the removal procedure should be. The same is true of soils where carbonate forms an appreciable proportion of the total soil mass. The only valid procedure for such soils is to remove the oxides and carbonates, determine the amounts present, and take these amounts into account in the calculation.

Organic matter is destroyed by oxidation, usually with hydrogen peroxide. Carbonates can be dissolved in acid solutions. Procedures are also available for reducing and dissolving iron oxides.

The sample must be deflocculated or peptized after the cementing materials are removed. Forces of attraction are always present between clay particles (Chapter 2) making them tend to flocculate or floc together and settle as large units. The attractive forces cannot be easily decreased, but can be overcome by increasing the force of repulsion (Chapter 2) between clay particles. The method of increasing interparticle repulsion consists in having a monovalent exchangeable ion on the clay and a low salt concentration in the solution. This is usually done by adding sodium ions and washing the soil suspension free of soluble salt. The sodium source used in most particle size determinations is sodium metaphosphate. As well as supplying sodium, the metaphosphate complexes or ties up any remaining calcium or magnesium. After the cementing materials have been removed and the soil has been chemically peptized, mechanical shaking or boiling is usually sufficient to separate all the particles. A dilute suspension of the soil is then made, and the particle-size determination carried out. The coarser size fractions (sands) are separated out by sieving, and finer materials (silt and clay) are separated by allowing them to settle in water.

Principles of sedimentation analysis

The basis of most methods of particle-size analysis is that particles falling through a medium of lower density will become separated as to weight because the larger or heavier particles fall faster. Therefore an "effective settling size" can be measured.

A particle falling through a liquid is acted upon by two forces, the force of gravity pulling it down and the buoyant force of the liquid resisting the fall. In a vacuum where the buoyant force is absent a particle would continue to accelerate at 980 cm/sec^2 and a separation of particle sizes would be impossible. However, in a liquid, a constant velocity known as the terminal velocity is soon reached and the particle falls thereafter with this constant velocity. It can be shown from a mathematical consideration of the force of gravity and the magnitude of the buoyant force of water that soil particles falling through water reach a terminal velocity within a few seconds. For the determination of particle size we then need the relationship between size of particle and the constant terminal velocity of fall.

The downward force is simply the force of gravity on a body, mg, and the buoyant upward force on a sphere, given by Stoke's law is:

$$F = 6\pi r \eta v \tag{1.1}$$

where: r = radius of falling sphere, η = viscosity of the liquid, and v = terminal velocity of fall. Since the velocity is constant, the acceleration and hence the net force are zero. Therefore, the upward and downward forces balance. By equating these tow forces, we can derive the relationship for the velocity of fall of a sphere:

$$6\pi r \eta v = mg = \gamma_{eff} \, Vg = (\gamma_s - \gamma_l) \, Vg$$

where: γ_{eff} = effective density of particle, γ_s = density of solid, γ_l = density of liquid, and V = volume of particle. Since $V = \frac{4}{3} \pi r^3$ for a sphere:

$$6\pi r \eta v = (\gamma_s - \gamma_l) \frac{4}{3} \pi r^3 g$$

$$v = \frac{2g}{9\eta} (\gamma_s - \gamma_l) r^2 = Kr^2 \tag{1.2}$$

where K = constant depending upon temperature. Since $v = h/t$, where h = depth at which suspension is sampled and t = time:

$$r^2 = h/Kt \tag{1.3}$$

which gives the relationship between depth of sampling, sampling time, and effective radius of particles.

Several assumptions are made in the derivation of Stoke's law:

(1) Particles must be sufficiently large so that Brownian movement will not

influence their rate of fall, that is, the particles must be much larger than water molecules.

(2) The particles must fall independently of each other. This requires suspension concentration of less than 5%.

(3) There must be no slipping between the particle and liquid. Since soil particles are hydrated, this condition is fulfilled.

(4) The velocity of fall must be sufficiently low so that the viscosity of the liquid is the only resistance to fall.

(5) Particles must be rigid and smooth. Soil particles are rigid but are definitely not smooth spheres.

Stokes' equation for the buoyant upward force holds for laminar flow. This is assumption 4 listed above. If a critical particle velocity is exceeded, turbulent flow is found and the buoyant force becomes more complicated. This would be the case for large sand particles, but in a particle-size analysis sand is separated out first, so the assumption of laminar flow holds.

Stokes' law was worked out for spheres settling in a buoyant fluid (assumption 5). The equations have also been solved for plate-shaped particles; and the constant is no longer equal to 6π. But since clay particles are irregular plates, this refinement is of little value. Hence, we use the constant for spheres and speak of an "effective settling diameter" or of a "Stokes' diameter" as the size of particle measured.

Interpretation of accumulation curves

Particles become separated during sedimentation because heavier particles fall faster, but the distance between all particles of the same weight (size) remains constant. Therefore, the concentration of all particles in a plane remains constant until they have all settled below this plane. As one particle moves down, another moves into the plane to take its place. This is illustrated in Fig. 1.4.

Particles settle independently, at a velocity depending upon their size. After a given time, t_1, at a given height, h, all particles with a velocity larger than h/t_1 will have fallen below the plane, but the concentration of particles with a velocity smaller than h/t_1 will still be the same in the plane h as it was originally (Fig. 1.5). If a pan is placed at h, all the larger particles will have collected plus some of the smaller ones which started from points below the top of the suspension. These smaller particles will have been collecting at the same rate since time $t = 0$, because the distance between them remains constant. This rate is the slope of the weight vs. time curve and can be taken as $\Delta W/\Delta t$. The weight of smaller particles collected during time, t_1, is, therefore, $t_1 (\Delta W/\Delta t)$ (Fig. 1.6). When this weight is subtracted from the total weight, the weight of the larger particles with settling velocity

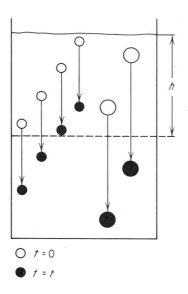

○ $t = 0$
● $t = t$

Fig.1.4. Settling of particles in a resisting fluid. (t = time.)

greater than h/t_1 is obtained. This may be performed at any time, t, to obtain the
fraction desired, since h is constant and h/t fixes v which is related to r by the
settling formula eq. 1.2.

The accumulation curve for a soil would be a smooth curve because soils

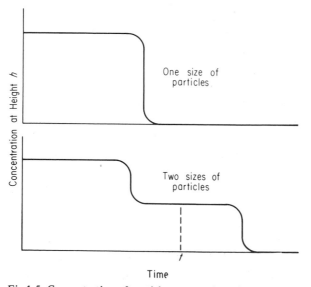

Fig.1.5. Concentration of particles at one plane during settling.

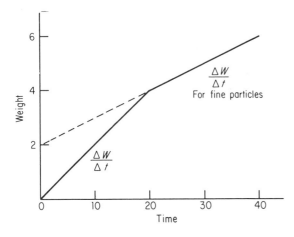

Fig.1.6. Weight accumulation at height h for sample consisting of 2 g of particles settling in 20 min. plus 4 g of particles settling in 40 min.

contain particles of all sizes. This is most easily analyzed graphically (Fig. 1.7). Instead of using a balance pan at a level h to collect the material, readings of density are taken with a hydrometer or a sample is withdrawn in a pipette at time t. The concentration of particles in the plane h is then determined from these measurements.

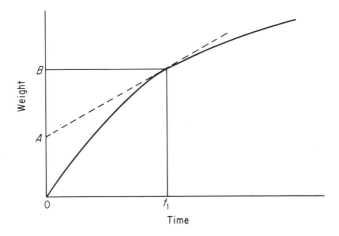

Fig.1.7. Weight–accumulation curve for soil.

1.6 BASIC SOIL PROPERTIES

Weight and volume

The basic soil properties may be easily defined by visualizing the three component phases in the soil — namely solid, liquid and gaseous. Consider the schematic diagram in Fig. 1.8. By separating the phases into three distinct parts, we may obtain relationships to define the basic soil properties.

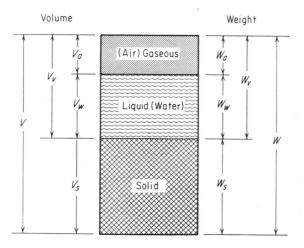

Fig.1.8. Idealized three-phase soil system. V_a = volume of air; V_w = volume of water; V_s = volume of solids; $W_s = V_s \gamma_s$; $W_w = V_w \gamma_w$; $V_v = V_a + V_w$ = volume of voids; W_a = weight of air = 0; W_s = weight of solids; W_w = weight of water; W = total weight; γ_s = density of solids; γ_w = density of water.

Void ratio e is defined as the ratio of volume of voids V_v to volume of solids V_s:

$$e = V_v / V_s \tag{1.4}$$

The ratio of *volume of voids* V_v to total volume V is defined as the *porosity, n*:

$$n = V_v / V \tag{1.5}$$

But $V = V_v + V_s = (1 + e)V_s$. Hence:

$$n = \frac{V_v}{V_s(1 + e)} = \frac{e}{1 + e} \tag{1.6}$$

The *water content*, ω, is given as $W_w/W_s \cdot 100\%$. For a fully saturated soil-water system, since all the voids will be completely filled with water:

$$V_v \gamma_w = W_w \tag{1.7}$$

where W_w = weight of water, and γ_w = density of water. For partial saturation:

$$V_w \gamma_w = W_w \text{ or } (V_v - V_a)\gamma_w = W_w$$

Hence the relationship for S_r the degree of saturation is given as:

$$S_r = \frac{(V_v - V_a)\gamma_w}{V_v \gamma_w} \cdot 100\%$$

Hence:

$$S_r = \frac{V_v - V_a}{V_v} \cdot 100\% = \frac{V_w}{V_v} \cdot 100\% \tag{1.8}$$

The unit weight γ, of the saturated soil-water system is defined as W/V, i.e. total weight divided by total volume. It is frequently more desirable to express unit weight relationships in terms of a *dry density* or *bulk density* γ_d, i.e. $\gamma_d = W_s/V$. Since:

$$\gamma = \frac{W}{V} = \frac{W_w + W_s}{V}$$

$$\gamma = (1 + \omega)\frac{W_s}{V}$$

But since $W_s/V = \gamma_d$, we have $\gamma = (1 + \omega)\gamma_d$ or:

$$\gamma_d = \frac{\gamma}{1 + \omega} \tag{1.9}$$

In a submerged state, because of the buoyancy effect of water, the submerged or *buoyant density* γ' is given by $\gamma - \gamma_w$.

Consistency limits

The definitions of the various states of consistency (Fig. 1.9) depend upon

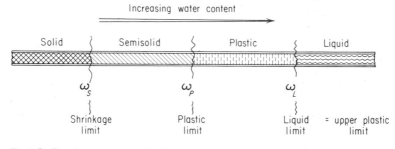

Fig. 1.9. Consistency states of soil.

laboratory tests for establishing the limits of consistency. These limits are by no means absolute and are sensitive to several environmental and operative factors.

The generalized relationship between water content and volume as a proportion of oven-dry volume is shown graphically in Fig. 1.10.

Points *1, 2* and *3* in Fig. 1.10 represent the liquid limit ω_L, plastic limit ω_P and shrinkage limit ω_S respectively. In terms of a water content—volume change definition, only the shrinkage limit can be rigorously defined. The *shrinkage limit* ω_S defines the water content at which volume change is no longer linearly proportional to change in water content. The relationship shown in Fig. 1.10 is

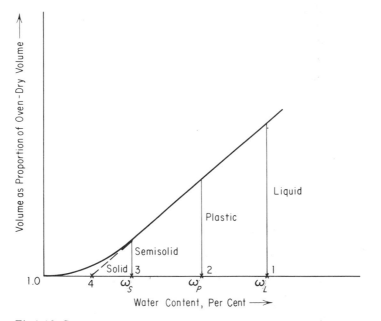

Fig. 1.10. Consistency states during drying of an initially saturated clay.

linear until the shrinkage limit is reached. Beyond this stage, any decrease in water content will not cause a proportional decrease in volume. If the straight line is projected until it intersects the abcissa (water-content line), the point *4* as shown in Fig. 1.10 is reached. This represents the total shrinkage of the soil sample.

The actual measurement of both the liquid and plastic limits depends on standardized laboratory tests. The *liquid limit* ω_L is that water content at which a groove cut in a standard liquid limit device is closed after 25 taps. The *plastic limit* ω_p is that water content at which a $\frac{1}{8}$ inch diameter thread of soil begins to crack and crumble under continued rolling by hand. The standardized laboratory procedures may be found in the American Society for Testing Materials (ASTM) specifications. The nature of plasticity of soils and clays is discussed in Chapter 2.

1.7 SUMMARY

One of the principal objectives for classification of soil is to provide a common language for use in the communication of soil types between interested parties. What to one party may be a fine silt, to another may be a clay. Based upon particle-size distribution, the separation between sand, silt and clay depends on the choice of limiting sizes. Fig. 1.2 gives a comparison between five different systems.

Methods based on particle-size distribution are adequate for classification of coarse-grained soils. For finer grained soils such as clays and the active silts, these methods must be augmented with other pieces of information. As a progression from the particle-size scale, the textural method may be used (Fig. 1.3). Its value, however, is still limited to coarse-grained soils. Identification of soil for purposes of classification, taking into account the behavior of the soil, provides a better means of evaluating both coarse and fine-grained soils.

Chapter 2

CLAY MINERALS IN SOILS

2.1 INTRODUCTION

The constituent particles of soils are grouped on the basis of size into clay, silt, sand, gravel, etc. Each size fraction contributes different properties to the soil. The methods of mechanics satisfactorily describe the behaviour of soils made up of the larger particles; little regard need be paid to mineralogical nature unless it is such that the particles break up or crush under load. Most of the problem soils in engineering practice are clay soils, and for the description of their behaviour the methods of mechanics are less successful. Two clay soils with the same grain-size distribution can have very different colloidal properties, depending upon their mineralogical nature. These varying characteristics will lead to large differences in engineering behaviour. In order to interpret these differences it is necessary to study some of the distinctive properties of clays.

Particles with a diameter smaller than 0.001 mm but larger than molecular size (10^{-6} mm) are classified as colloids, or are said to be in the colloidal state. Clay particles fall within this range. Colloidal properties such as plasticity and adsorption of molecules arise from the large surface area associated with a small mass. Surface forces become dominant and the influence of gravitational forces is small. These surface forces of interparticle attraction and repulsion will be discussed in this chapter, because they determine the behaviour of clay soils.

2.2 NATURE OF CLAY MINERALS

Definition

The term "clay" has different meanings. It refers to that size fraction of soils with particles of less than 0.002 mm effective diameter as measured from their settling velocity in water. It also refers to certain minerals, called clay minerals, which are the result of chemical weathering of rocks and which are usually not present as larger particles. Where a distinction between these two uses is not obvious from the context, the terms "clay size" and "clay minerals" should be used.

In a less frequent usage, clay is a general term used for earth deposits which have plastic properties. This imprecise meaning is no longer used.

The clay-size fractions of most soils contain clay minerals. However, there are soils in which clay-size particles are either noncrystalline, amorphous materials or primary minerals such as quartz, mica, feldspar, etc. Physical disintegration of rocks by ice produces clay-size particles consisting of primary minerals.

Amorphous materials are alumino-silicate materials which are amorphous to X-rays, i.e. do not show an X-ray diffraction pattern. There is no regularity in positions of the atoms to give a simple repeating structure. Primary minerals are the minerals in igneous rocks. They are larger and more regular than clay minerals formed from the primary minerals.

In early studies of soils, clay-size material was thought to consist of amorphous aluminum and iron silicates. This material had the properties of plasticity, water adsorption and ion exchange associated with the clay fraction. Since the optical methods used in identifying larger minerals could not resolve particles of this size, no further identification was possible. The name kaolin was applied to clays or used synonymously with "clay"; differing properties of clays were ascribed to different amounts of impurities mixed with the kaolin.

By 1920 several workers had evidence that clays consisted of definite crystalline minerals, and in 1923 the first X-ray diffraction patterns of clay deposits confirmed this. The clay fraction of soils was investigated later, and also found to contain clay minerals. In the 1930's most of the crystalline structures of clays were worked out on the basis of their X-ray diffraction patterns. Pauling worked out the structure of micas, Gruner the kaolinite structure, and Hofmann with others the montmorillonite structure. Since then much work has been done on identification of clay minerals in soils and on relating soil properties to the constituent minerals.

Structure

The various schemes proposed for classifying clay minerals according to their structure differ in detail, but the general outlines are agreed upon. Clay mineralogy is an actively developing study, and as more information becomes available on clay-mineral structures these schemes will be modified. The simple classification presented in Table 2.1 will serve for the discussion of the properties of representative clays and will also allow the reader to place in it other minerals which he may encounter.

The crystal structure of clay minerals will be discussed only briefly as an introduction to the discussion of surface properties. For a detailed discussion of clay minerals the reader is advised to consult some of the standard references such as Grim (1954), Mackenzie (1957), Marshall (1964) and Van Olphen (1963).

TABLE 2.1

Classification of layer-lattice clay minerals[1]
(Based upon Mackenzie, 1957, p.310)

Composition of layers	Group	Minerals	Population of octahedral sheet	Isomorphous substitutions
1:1 (one silica to one alumina sheet) 7.2 Å thick	kaolinites	kaolinite dickite	dioctahedral 2/3 of positions filled with Al	–
		halloysite		–
	serpentine minerals	chrysotile	trioctahedral	Mg for Al
		cronstedtite	Mg or Fe in all positions	Fe for Al
2:2 (14 Å thick)	chlorites	chlorite	dioctahedral, trioctahedral or mixed	Al for Si Al for Mg
2:1 (two silica to one alumina sheet) 10 Å thick	micas	illite glauconite }	usually dioctahedral	Al for Si Al for Si Fe and Mg for Al
	montmorillonites	montmorillonite beidellite nontronite	dioctahedral	Mg for Al Al for Si Fe for Al
		hectorite saponite	trioctahedral	Li for Mg Al for Si
	vermiculites	vermiculite	usually trioctahedral	Al for Si
1:1, 2:1 or mixed	interstratified minerals	bravaisite		–

[1] In addition to these layer-lattice minerals, the X-ray-amorphous alumino-silicate, allophane, and the chain silicates, palygorskite (attapulgite) and sepiolite are usually included with clay minerals.

Clay minerals are alumino-silicates, i.e. oxides of aluminium and silicon with smaller amounts of metal ions substituted within the crystal. The aluminium-oxygen combinations are the basic structural units which are bonded together in such a way that sheets of each one result. The stacking of these sheets into layers, the bonding between layers, and the substitution of other ions for aluminium and silicon account for the different minerals. This substitution occurs for ions of approximately the same size, and is therefore called isomorphous substitution.

Sizes of the unit cells of clay minerals are given either in Ångstrom units, Å, where Å = 10^{-10} m, or in kX units defined in terms of calcite spacings. 1Å = 1.00202 kX. In this chapter the more common Å units will be used.

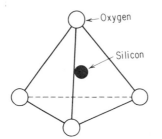

Fig.2.1. Schematic diagram showing single silica tetrahedron.

The silica unit consists of a silicon atom, Si, surrounded by four oxygen atoms, O, equidistant from the silicon (see Fig.2.1). These oxygens are arranged at the corners of a tetrahedron with each of the three oxygens at the base of the tetrahedron shared by two silicons of adjacent units. This sharing forms a sheet with the tetrahedral units packed in such a way that the sheet has hexagonal holes. Since the silicon is smaller than the oxygens, the sheet can be visualized as two layers of oxygen atoms with silicon atoms fitting in the holes between. This sheet has a thickness of 4.93Å in clay minerals. The O—O distance is 2.55 Å, leaving a hole within the tetrahedron of 0.55 Å radius into which the silicon of 0.5 Å radius can fit without distortion. Silicon has a positive valence of 4 and oxygen a negative valence of 2. With each silicon having one oxygen atom and sharing three other

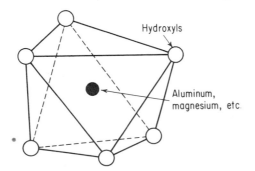

Fig.2.2. Schematic diagram showing single aluminium or magnesium octahedral unit.

oxygens, this unit has a negative charge of 1. When the top oxygen takes on a
hydrogen with a positive valence of 1 to become hydroxyl, OH, the unit is neutral.
Oxygen and hydroxyl have about the same radius in clay minerals.

The alumina unit is an aluminium atom, Al, equidistant from six oxygens or
hydroxyls in octahedral coordination as shown in Fig.2.2. Each oxygen is shared by
two aluminium ions, forming sheets of two layers of oxygen (or hydroxyl) in close
packing, but only two thirds of the possible octahedral centres are occupied by
aluminium. This sheet is 5.05 Å thick in clay minerals. When all the oxygens are
hydroxyls, this is called gibbsite, with a chemical formula $Al_2(OH)_6$. If magnesium,
Mg, is present in place of aluminium, all the octahedral positions are filled and the
mineral is called brucite with a chemical formula $Mg_3(OH)_6$. The radius of
aluminium is 0.55 Å and of magnesium 0.65 Å. The O−O distance in octahedral
coordination is 2.6 Å and the OH−OH is 2.94 Å, leaving an octahedral space of
0.61 Å radius. Clay minerals in which two thirds of the octahedral positions are
filled are called dioctahedral. When all the positions are filled, they are termed
trioctahedral.

The best way to gain an understanding of the crystal lattice structures of clay
minerals is by working with models. These can be obtained commercially or can
easily be made, as illustrated in the various books dealing with this topic, from a
variety of materials. Balls about the size of ping-pong balls can serve for oxygen and
hydroxyl, with other atoms to scale.

Clay minerals are classified on the basis of the component layers as shown in
Table 2.1. The layers are shown schematically in Fig.2.3. The most common clay
minerals will be discussed in the following section.

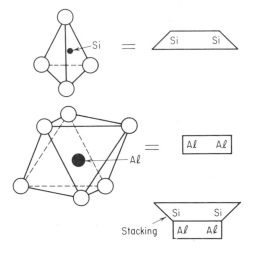

Fig.2.3. Schematic representation of silica tetrahedron and alumina octahedron.

2.3 CLAY MINERALS COMMON IN SOILS

Kaolinite

The kaolinite crystal consists of repeating layers, each layer consisting of a silica sheet and an alumina sheet sharing a layer of oxygen atoms between them. This is shown schematically in Fig.2.4. Each layer is three oxygen atoms thick. The layers are held together by hydrogen bonding (see Appendix 1) between hydroxyls from the alumina sheet on one face and oxygens from the silica sheet on the opposite face of the layer. These forces are relatively strong, preventing hydration between layers and allowing many layers to build up. A typical kaolinite crystal may be 70–100 layers thick.

Kaolinites are most easily identified from their distinctive X-ray diffraction pattern shown in Fig.2.5. The thickness of the layers, 7.13 Å, is the distance between diffracting planes. The peak at 3.56 Å is a second-order reflection of this spacing, and 2.37 Å is a reflection from another plane. Kaolinites are found in soils that have undergone considerable weathering in warm, moist climates. They have a low liquid limit and a low activity. The quantity of kaolinite in a soil is most readily determined from differential thermal analysis.

Halloysite differs from other minerals in the kaolin group in that it can occur in hydrated form, with a layer of water molecules between the layers of the kaolinite crystal. The hydrated form found in nature loses the water layer readily on heating and does not rehydrate when put in water. The dehydrated form is sometimes called metahalloysite, although preferred usage is halloysite with a specification of the state of hydration. It is most easily recognized by its tubular shape in electron micrographs, and from its X-ray diffraction pattern.

Minerals with a kaolinite structure but with all octahedral positions filled

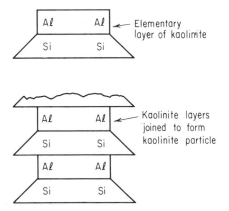

Fig.2.4. Schematic representation of typical kaolinite structure.

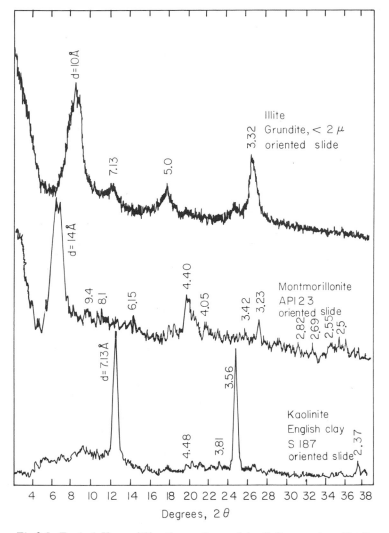

Fig.2.5. Typical X-ray diffraction pattern of kaolinite, montmorillonite and illite minerals (oriented particles using CuKα radiation).

(trioctahedral) are not common clay minerals. They occur in the disintegration products of serpentine rocks and are included in Table 2.1 for the sake of completeness. These minerals are sometimes classed with chlorites.

Chlorite

The repeating layers of the chlorite minerals are composed of a silica sheet, an alumina sheet, a second silica sheet and another alumina or brucite sheet. On the

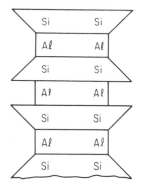

Fig.2.6. Schematic representation of typical chlorite structure.

basis of the structure of mica, discussed in the next section, chlorite can be described as being made of mica layers held together by brucite or gibbsite sheets (Fig.2.6). The repeating layer has a thickness of 14 Å, and the chlorite mineral is recognized by an X-ray diffraction pattern with this line and its higher orders (Fig.2.7). The first-order reflection at 14 Å is often weak, and since the higher orders coincide with the X-ray pattern for kaolinite, identification may be difficult. The brucite or gibbsite layer of chlorite can be dissolved out in warm acid, so to aid identification the sample is treated in acid. If the lines disappear, the mineral is chlorite; if they remain at the same intensity it is kaolinite.

Substitution of other cations for silicon and aluminium occurs in chlorite. In the silica sheet there may be substitution of trivalent aluminium for tetravalent silicon, giving a negative charge. Part of this charge is balanced by substitution of aluminium for divalent magnesium of the brucite layer, giving a positive charge. The remainder gives a net negative charge to the mineral which is the cation exchange capacity.

Chlorite is found in metamorphic rocks, e.g., chloritic schist, and in soils derived from them. It is also a product of weathering in soils. The chlorite found in soils may have a partly random stacking of layers and some hydration between layers. It then has an X-ray diffraction pattern which is less regular than that of the unweathered mineral.

Clay mica (illite)

The mica minerals have repeating layers consisting of an alumina sheet between two silica sheets with oxygens shared to give a unit four oxygen atoms thick. This layer is 10 Å thick. The layers are bonded together by potassium ions which are just the right size to fit into the hexagonal holes of the silica sheet

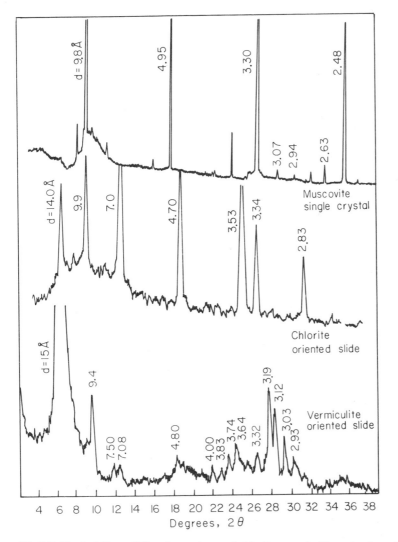

Fig.2.7. Typical X-ray diffraction pattern of chlorite, vermiculite and mica minerals (oriented particles using CuKα radiation). Traces of other minerals present with chlorite and vermiculite.

(Fig.2.8). The potassium ions exist in 12-coordination, bonding six oxygens from one silica sheet to the adjacent six oxygens of the silica sheet of the next layer. The negative charge to balance the potassium cations arises from substitution of aluminium for silicon in the silica sheet.

The clay micas, illite and glauconite, differ from the minerals such as muscovite and biotite in having a smaller particle size, less substitution of aluminium for silicon, less potassium, and a more random stacking of layers with consequent weaker bonding between them. The loss of potassium is made up by exchangeable

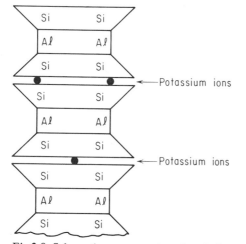

Fig. 2.8. Schematic representation of typical mica structure.

ions such as calcium, magnesium and hydrogen in amounts to balance the charge.

Clay micas are recognized chiefly by their X-ray diffraction pattern, shown in Fig. 2.5, with a strong 10 Å line and the higher order reflections. Soils containing illite have properties intermediate between kaolinite with a low activity and montmorillonite with a high activity. Illite occurs widely in sedimentary rocks in temperate and in arid regions.

Montmorillonite

The montmorillonite minerals have the same layers as the micas discussed in the previous section. However, isomorphous substitution occurs mainly in the alumina sheet, with magnesium or iron substituting for aluminium in the dioctahedral minerals. The different montmorillonite minerals have different substitutions. There are no potassium ions to bond the layers together, and water enters easily between layers, as illustrated in Fig. 2.9. A typical X-ray diffraction pattern of montmorillonite is shown in Fig. 2.5.

The distance of separation of the layers on hydration can be controlled if certain organic liquids rather than water are used. The diagnostic X-ray spacings of montmorillonite saturated with glycerol will show spacing of layers of 17.7 Å, of which 10 Å is the thickness of the layer and 7.7 Å is glycerol.

The montmorillonite clays have a high activity and high liquid limit. The most obvious characteristic is the swelling of the clay to several times its dry volume when placed in contact with water. This swelling is due to water adsorbed between layers, pushing the layers apart. The layers become separated by several

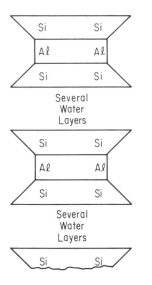

Fig.2.9. Schematic representation of typical montmorillonite structure.

water layers, often about 9 Å thick. A group of such layers is called a tactoid, with distance between tactoids being several tens of Ångstrom units. The tactoids of sodium-saturated montmorillonite can be dispersed into individual layers with each layer more than 10 Å from the next. A particle of such a dispersed montmorillonite is only one layer, or 10 Å thick.

Montmorillonites occur in sediments of semiarid regions and are the main minerals in bentonite rock. They are formed when volcanic ash weathers in marine water or under conditions of restricted drainage. They are also found during weathering in soils.

Vermiculite

Vermiculite, shown in Fig.2.10, also has the mica structure, and resembles montmorillonite except that absorption of water between layers is usually limited to two thicknesses of water molecules and the water enters between a smaller proportion of the layers. There is substitution of aluminium for silicon in the silica sheet, leaving a high net negative charge. The dominant exchangeable cation in natural vermiculites is magnesium.

The X-ray diffraction pattern (Fig.2.7) shows a first-order reflection at about 15 Å, which is the thickness of the layer plus two layers of water molecules. This water can be driven off by heating; if the temperature exceeds 700°C, the mineral surfaces will not rehydrate. The X-ray pattern will then resemble that of the micas.

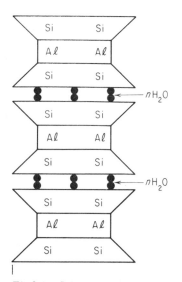

Fig.2.10. Schematic representation of typical vermiculite structure.

Vermiculite has the highest cation exchange capacity of all the clay minerals. This is often a clue to its identification.

Vermiculite is a weathering product of mica and occurs as an accessory mineral in many clay soils. The typical commercial form of expanded vermiculite used for packing is made by rapidly heating the mineral so that steam forces the layers apart.

Interstratified minerals

Clay minerals found in soils often do not have a regular repetition of the same layers, i.e. they are not one of the single minerals discussed in previous sections. They often contain interstratified layers of two minerals. Since the layers of all 2:1 minerals are similar, this might be expected. Bravaisite is the name given to interstratified montmorillonite and clay mica. This may be a fairly regular alternate stacking of mica and montmorillonite layers or a stacking with a random occurrence of the two layers. Interstratified vermiculite and chlorite is also common. Hydrobiotite is a mixed mica-vermiculite. Such minerals are difficult to identify, but are very common in soils. They have properties intermediate between those of the two components.

Allophane

Allophane does not have a sufficiently regular structure to show X-ray diffraction peaks. It consists of random arrangements of silica tetrahedra and metallic ions, usually aluminium, in octahedral coordination.

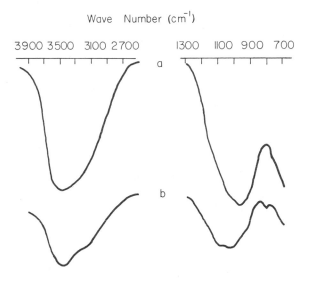

Fig.2.11. Infra-red absorption pattern of two typical allophane samples from Japan (a) and the West Indies (b).

Allophane is often identified in soils by the absence of X-ray diffraction lines. The infrared spectra are useful in identifying allophane (Fig.2.11). Allophane is sufficiently distinctive that it is referred to as a clay mineral, even though it does not have the same uniformity of composition as the minerals described above. Some authors refer to any X-ray-amorphous alumino-silicates as allophane, others restrict the name to a range of composition of about equal molar proportions of aluminium and silica. The cation-exchange capacity of allophane is high, and the amount increases sharply with increase in pH.

Allophane is often found associated with halloysite, and is a weathering product of volcanic ash under conditions of good drainage. Soils containing allophane have a high activity and high sensitivity. Their unusual plastic properties are discussed in Section 2.8. They often flocculate during particle-size analysis under treatments which will disperse other soils.

Attapulgite

The crystal structure of attapulgite does not have repeating layers. It is a chain silicate in which the units are continuous in only one direction. Attapulgite is a double chain of silica tetrahedra with two of the four oxygens shared. The chains are linked through oxygens. The octahedrally coordinated ions are linked continuously only in one direction. This structure has open tubes, which contain water molecules.

Since the mineral structure is continuous in only one direction, attapulgite occurs as rod- or lathe-shaped particles. These can be recognized in an electron micrograph. Attapulgite occurs occasionally in desert soils.

Clay-mineral mixtures

The properties of specific minerals have been discussed in previous sections. Soils usually contain a mixture of clay minerals and amorphous iron, silicon and aluminium oxides. Some primary minerals such as quartz and feldspar may also be present. Identification of the components is sometimes difficult because one mineral may interfere with the identification of another. The behaviour of soils containing a mixture of minerals is influenced by properties of the different minerals, but the extreme properties of the pure minerals are not encountered.

Primary minerals and amorphous materials predominate in the clay-size fraction of some soils. Primary minerals are usually easy to identify. It is difficult to determine the amount of amorphous material, so that its significance has probably not been fully realized, especially when it occurs as one component of a mixture. However, there is considerable interest now in amorphous material, and its properties and occurrence in soils are being studied. It is more readily soluble than the crystalline components and can often be separated in this way.

Weathering of clay minerals

Clay minerals are not stable, static entities in soils. Weathering gradually changes them by dissolving some minerals and reprecipitating others. Intensity of weathering increases with increasing temperature, and is higher in moist than in dry soils.

Some minerals have a more stable structure than other minerals, and on this basis weathering reactions have been worked out. A weathering scheme worked out by Jackson (1964) is as follows, where the numbers in parenthesis are weathering indices with more stable minerals having higher numbers.

Mica   →   Vermiculite (8)   →   Montmorillonite (9)   ⇌   Pedogenic
Biotite (4)   2:1 to 2:2
Muscovite (7)   swelling 18 Å
Illite (7)   intergrade (9)

(Fe, Mg, Al)   Secondary   Pedogenic   Al   Kaolinite
Chlorite   →   Chlorite → 2:1 to 2:2 14 Å → Chlorite → and Halloy → Gibbsite
   (4)   (8)   intergrade (9)   (9)   site (10)   (11)

Weathering attacks the outer surface of clay minerals first. Therefore, the outer coating may have a different composition than the inner part. This has led to the concept of "rind" and "core" of clays in soils. Many methods of clay identification measure the outer surface of minerals. On the other hand, cleaning the surface with drastic chemical treatment to measure properties of the "core" can give misleading information on the reactions of a mineral in the soil. A pattern for weathering of mica to a vermiculite mineral in Fig.2.12, shows the change which would occur at an edge and which would constitute the "rind" with properties different from the "core".

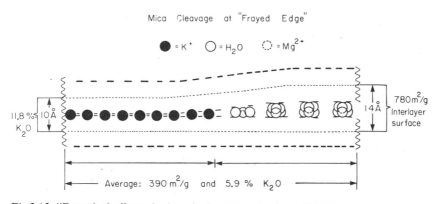

Fig.2.12. "Frayed-edge" weathering of mica. (From Jackson, 1963.)

Occurrence of clay minerals in soils

The regions in which the various clay minerals typically occur are listed in Table 2.2. Since clay minerals are the products of chemical weathering of rocks, both the climate, which determines weathering, and the parent rock, influence the type of minerals found. These two factors are stressed in Table 2.2. A knowledge of local geology together with some analyses for identification of clays will give the background needed to interpret geotechnical properties of soils.

TABLE 2.2

Summary of occurrence of clay minerals in soils

Clay mineral group	Occurrence
Kaolinite	Highly weathered soils with good drainage. Generally in older soils. Common in tropical and subtropical areas.
Chlorite	Areas of metamorphic parent rock. Common in marine sediments and sedimentary rocks. Not normally present in dominant proportions.
Clay Mica	In soils derived from weathering of sedimentary rocks. Dominant mineral in slate and shale.
Montmorillonite	Results from weathering of volcanic rocks or ash under poor drainage. Common in sediments of arid areas and often mixed with clay mica.
Allophane	Results from weathering of volcanic ash under adequate drainage. Common in areas with recent volcanic activity such as Japan, New Zealand and western South America.

2.4 IDENTIFICATION OF CLAY MINERALS

No one method is satisfactory for identification of a variety of minerals in soils. This is partly because in a mixture one mineral interferes with the measurement of another, and partly because there is a range in composition and in crystal structure of clays from different sources. Usually several methods are required, especially if quantitative estimates are required. X-ray diffraction is the most useful method of identification, often aided by treating the sample, e.g., with glycerol to expand montmorillonite or with potassium saturation and heating to collapse vermiculite. Details of these methods can be found in books on clay mineralogy and soil science.

After obtaining the X-ray diffraction pattern, it may be necessary to obtain the differential thermal analysis (DTA) curve for the sample, or to measure the infrared absorption spectrum (IR spectra). An electron micrograph may be helpful. Further information on the sample can then be obtained from chemical methods such as cation-exchange capacity or chemical analysis for the ions present. A scheme used for detailed analysis of a sample is shown in Table 2.3.

TABLE 2.3

Scheme for analysis of minerals in a soil sample

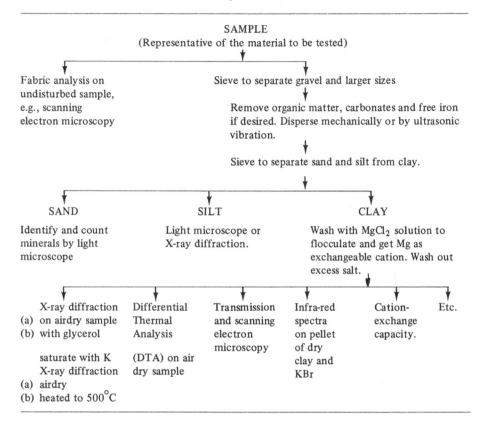

X-ray diffraction

The most widely used method of identification of clay minerals is from an X-ray diffraction pattern of a powdered sample of the clay-size fractions of a soil. The minerals present can usually be identified from diffraction lines, although it may be difficult to specify the proportions of each mineral in a mixture. Interpretation of the diffraction patterns is based upon the crystal structure of a clay.

The unit cell of a crystal is the smallest repeating unit always having the same structure. The crystal can, therefore, be considered to be made up of identical unit cells. The lengths of the sides of the unit cell in the X-, Y-, Z-coordinate system are *a, b* and *c,* and the angles between the sides are α, β and γ. Different clays have unit cells of different dimensions and different angles between the faces.

In the unit cells of clay minerals *a* is about 5 Å and *b* about 9 Å. These values vary slightly with the clay mineral. The large variation is in the *c* length, which is the length in the direction of the thickness of the layers. These lengths, often referred to as the *c*-axis spacings, are characteristic of the mineral and are the main "*d*" values used in identification of clays from their X-ray diffraction patterns (see Table 2.4).

The principle of X-ray diffraction by planes of atoms in a crystal is shown in Fig.2.13.

TABLE 2.4

X-ray diffraction spacings obtained from [001] planes of layer-silicate minerals for different sample treatment
(From Whittig, 1965. *Am. Soc. Agron. Monogr.*, 9, p.672)

Diffraction spacing (Å)	Mineral (or Minerals) indicated
	Mg-saturated, air-dried
14 −15	montmorillonite, vermiculite, chlorite
9.9 −10.1	mica (illite), halloysite
7.2 − 7.5	metahalloysite
7.15	kaolinite, chlorite (2nd order maximum)
	Mg-saturated, glycerol-solvated
17.7 −18.0	montmorillonite
14 −15	vermiculite, chlorite
10.8	halloysite
9.9 −10.1	mica (illite)
7.2 − 7.5	metahalloysite
7.15	kaolinite, chlorite (2nd order maximum)
	K-saturated, air-dried
14 −15	chlorite, vermiculite (with interlayer aluminium)
12.4 −12.8	montmorillonite
9.9 −10.1	mica (illite), halloysite, vermiculite (contracted)
7.2 − 7.5	metahalloysite
7.15	kaolinite, chlorite (2nd order maximum)
	K-saturated, heated (500°C)
14	chlorite
9.9 −10.1	mica, vermiculite (contracted), montmorillonite (contracted)
7.15	chlorite (2nd order maximum)

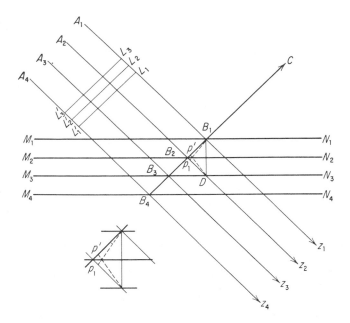

Fig.2.13. Representation of a set of planes in a crystal and a beam of parallel X-rays.

The lines M_1N_1, M_2N_2, ..., represent a set of densely populated planes in a crystal. If A_1Z_1, A_2Z_2, ..., constitute a beam of parallel X-rays, and L_1L_1', L_2L_2', ..., are successive wave fronts, then θ = angle $A_1B_1M_1 = CB_1N_1 = \theta'$. Since the atoms are arranged in planes, they act as a diffraction grating for X-rays.

When a wave front arrives at the position B_1p_1, the scattered ray in the direction B_2C arrives at position p', where $B_2p_1 = B_2p'$. Since p' is the foot of the perpendicular projection from D upon B_2C, and p_1 is the perpendicular projection from B_1 on to B_2Z_2, then from geometry, the angle p_1B_1D = angle $p'DB_1 = \theta$. Hence the wavelet scattered from B_2 by the wave front arriving at B_1 and B_2 lags behind the one scattered from B_1 by a path difference of $p'B_1$.

Since $B_1D = 2d$, the path difference $p'B_1$ is $2d \sin \theta$. The same reasoning may be used for B_2B_3, B_3B_4, etc. If this difference is $n\lambda$ where n is an integer defining the order of reflection, we will obtain a diffracted beam in direction B_1C when:

$$n\lambda = 2d \sin \theta \qquad\qquad (2.1)$$

This is Bragg's equation, which relates the distance between planes d to the monochromatic X-ray wavelength λ. This equation is obtained from the condition that the X-rays reflected from the planes of atoms in a crystal are in phase.

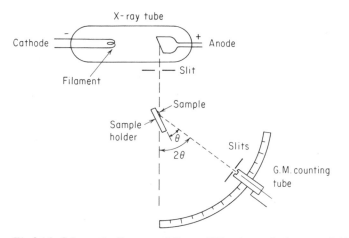

Fig.2.14. Schematic diagram of X-ray diffraction unit for crystal identification. The sample holder rotates through θ degrees while the Geiger-Muller counting tube rotates through 2θ degrees. The X-rays are diffracted by the crystalline sample and when θ is such that the refracted rays are in phase, a higher intensity is recorded by the G.M. tube.

A diffraction spectrometer is shown schematically in Fig.2.14. A motor rotates the specimen and counting tube to vary the angle of incidence of the X-rays. The intensity of X-rays reflected from the sample and entering the Geiger-Müller counter is recorded as a function of the angle θ. When θ corresponds to a distance d between planes of the clay crystal required to bring the diffracted rays into phase (eq.2.1) they are reinforced and a peak in X-ray intensity will be measured by the counter. The number of counts per second is proportional to the intensity of the

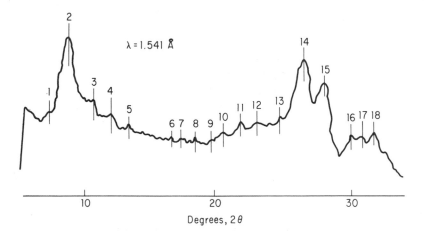

Fig. 2.15. X-ray diffraction pattern for "Seven Islands" clay.

X-rays reflected into the counter. This signal is amplified and fed into a recorder to obtain a diffraction tracing such as shown in Fig.2.15. This X-ray diffraction pattern was obtained from a powdered clay sample and shows the peaks which are used for calculations of d values. The abscissa represents the value of 2θ (the angle of the counter) in degrees, while the ordinate shows the intensity of the reflection. The wavelength of the incident radiation was 1.541 Å.

The spacing d of the reflecting planes of atoms is evaluated from Bragg's equation. It is evident, from this equation, that for any one diffraction angle there may be several corresponding lattice spacings, depending upon the order of the diffraction n. Thus, for instance, peak *14* (Fig.2.15) may be caused by a first-order reflection from lattice planes spaced at 3.36 Å, or by a second-order reflection from a 6.7-Å lattice spacing, or from a third-order reflection from a 10.1-Å lattice spacing, or from a fourth-order reflection from a 13.5-Å lattice spacing, etc. These values are obtained from the Bragg equation (eq.2.1) where the diffraction order n is given consecutive values of 1, 2, 3, etc.

The identification of minerals from such a diffraction tracing would proceed as follows. The θ values at the peaks are measured and converted to d values using

TABLE 2.5

Diffraction peaks and lattice spacings from X-ray tracing of Seven Islands powder clay sample shown in Fig.2.15

Peak No.	Lattice spacings: d corresponding to different orders of diffraction (Å)		
	$n = 1$	$n = 2$	$n = 3$
1	14.2		
2	10.1		
3	8.4		
4	7.1	14.2	
5	6.6	13.1	
6	5.2	10.5	
7	5.0	10.1	15.1
8	4.78	9.6	14.3
9	4.62	9.2	13.9
10	4.26	8.5	12.8
11	4.03	8.1	12.1
12	3.78	7.6	11.2
13	3.55	7.1	10.6
14	3.36	6.7	10.1
15	3.21	6.4	9.6
16	3.00	6.0	9.0
17	2.94	5.9	8.8
18	2.83	5.7	8.5

eq.2.1 and *n* values of 1, 2 and 3, as shown in Table 2.5. From Tables 2.4, 2.6 and the more extensive values given in reference books, the peaks are identified as resulting from certain minerals. The higher orders, $n = 2$ and 3, are examined to see whether the peaks are higher orders.

Identification of complex mixtures of minerals requires experience and a knowledge of the relative intensity of different peaks. When a peak is assigned to one mineral, other *d* values should be examined to ensure that the other peaks for the mineral are also present. In this way it is usually possible to distinguish higher order reflections of one mineral from peaks due to another mineral.

Five minerals can be identified in Fig.2.15. Peaks *1, 4* and probably *13* are due to chlorite. Experience with these clays indicates that the second order is as strong as the first, which is the reason for assigning peak *4* to chlorite. Peaks *2* and *7* are due to mica. Peak *3* is probably from amphibole. Peaks *10* and *14* are the typical quartz lines and *11, 12, 15* and *18* are identified with feldspar. Mica is the dominant mineral. The peaks for quartz and feldspar are also high, but experience shows that small amounts of these minerals give strong reflections.

Differential thermal analysis

Differential thermal analysis (DTA) determines the temperature at which changes occur in a mineral when it is heated continuously to a high temperature. The intensity of the change is proportional to the amount of the mineral present.

TABLE 2.6

Diagnostic X-ray and differential thermal analysis (DTA) values for minerals found in soils

Mineral	Strongest X-ray peaks, *d* (Å)	Strongest D.T.A. peaks	
		endothermic, *T* (°C)	exothermic, *T* (°C)
Kaolinite	7.15; 3.57; 2.33	550−600	950−980
Chlorite	14; 7.1; 4.7	600−650	840−900
Clay mica	10.0; 4.5; 3.3	550−600	about 900
Montmorillonite			
air dry	14−15; 4.5	150−200	−
glycerol-saturated	17.7; 8.9		
Vermiculite	14−15; 3.5	150−200	−
Allophane	−	about 150	about 950
Quartz	3.55; 4.21	−	−
Feldspar	3.2; 4.0; 1.8	−	−

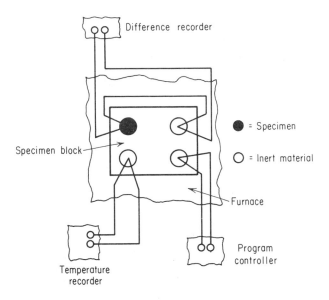

Difference recorder

Specimen block

● = Specimen

○ = Inert material

Furnace

Temperature
recorder

Program
controller

Fig.2.16. DTA apparatus with associated recording and control mechanisms.

Clays lose water or go through phase changes which either give off or require heat, at specific temperatures. The temperatures at which these reactions occur are characteristic of the mineral and can, therefore, be used for identification. The loss of water molecules causes an endothermic reaction, in which heat is taken up by the sample. In an exothermic reaction, usually a phase change in the structure of the mineral, heat is given off. Diagnostic DTA temperatures are given for several clay minerals in Table 2.6.

For the DTA measurement, a sample of clay and a sample of inert material are slowly heated in a furnace. The difference in temperature as well as the temperature of the ceramic are measured with thermocouples, as shown in Fig.2.16. The reference inert material must experience no thermal reaction within the test temperature range. Generally, calcined aluminium oxide or ceramic are used as the reference material. When a temperature is reached at which the clay loses water by vaporization, the sample temperature will drop below that of the inert material. This temperature difference is recorded as an endothermic peak. Similarly a phase change will produce a temporary temperature difference between the sample and the ceramic, often an exothermic reaction. The e.m.f. generated by the thermo-couple is amplified and fed into a recorder. Characteristic endothermic and exothermic peaks are shown for several clays in Fig.2.17.

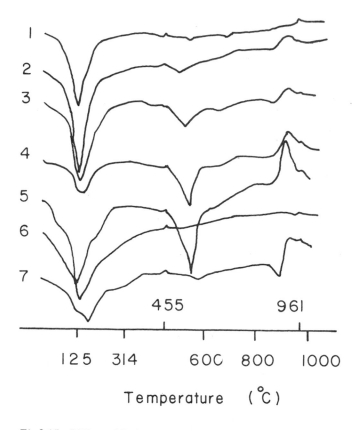

Fig.2.17. Differential thermal curves of soil clays. *1* = almost pure montmorillonite; *2* = montmorillonite with 5% kaolinite, *3* = montmorillonite with 10% kaolinite; *4* = montmorillonite and hydrated halloysite, 1:1; *5* = montmorillonite and kaolinite, 2:1; *6* = soil allophane; and *7* = montmorillonite and vermiculite, 1:1. (From Barshad, 1965).

Infrared spectroscopy

Clay minerals have absorption bands in the infrared region of the energy spectrum. Infrared vibration frequencies are in the range of molecular bond vibration frequencies. Many of these bonds are not specific to one mineral, because they are due to interatomic bonds common to many minerals (Table 2.7). However, allophane can usually be recognized from its IR spectrum with broad absorption bands between 3,400–3,500 cm^{-1} and between 1,000 and 1,100 cm^{-1} (Fig.2.11). The mechanism of absorption of many compounds onto the surface of clay minerals can often be elucidated from IR spectra. The method is being increasingly used in studies of clay minerals.

TABLE 2.7

Infrared absorption bands for some soil minerals
(From White, 1971. *Soil Sci.*, 112, p. 24, by permission of the Williams and Wilkins Co., Baltimore)

Mineral	Frequency (cm⁻¹)																			
Kaolinite	3,695	3,670	3,650	3,620						1,108	1,038	1,012	940	915		700		540		472
Halloysite	3,695			3,620		3,400*			1,640**	1,100	1,040	1,020		918		695		545		474
Montmorillonite				3,620		3,400*			1,640***	1,100	1,040	1,020		915				520		470
Nontronite				3,560		3,400*			1,640***	1,130	1,050			827				490		430
Muscovite				3,628						1,120		1,020		928	828	750		535	480	
Biotite			3,658		3,550						1,000					760	690	465		445
Vermiculite					3,550	3,380*			1,640**			985			812		670		480	
Chlorite Al-rich				3,620	3,520						1,004		940		825		692	528		475
Gibbsite				3,620	3,525	3,445	3,395	3,340		1,102	1,030	975			800	745	670	560		540
Quartz										1,172	1,084				800	780	697	512		462
Microcline										1,110	1,030	1,000				769	727	647		
Calcite									1,435					877		712				

* OH-stretching frequency for water molecules.
** OH-bending frequency for water molecules.

Electron microscopy

Both transmission and scanning electron microscopy are used in studies of clay minerals to learn about shape and arrangement (fabric) of clay particles. Micrographs of several clay minerals are shown in Chapter.3 in the discussion of soil fabric.

2.5 SURFACE AREA OF CLAYS

A schematic comparison of the size, shape, and surface area of several clay minerals is shown in Fig.2.18. These clay particles are plate-shaped or tabular because the layer-lattice structure results in strong bonding along two axes but weak bonding between layers. The clay-particle thickness depends upon the magnitude of the forces of attraction between the layers. The variation in specific surface area is primarily due to different thicknesses of the tabular particles. Montmorillonite may occur, in ultimate dispersion, as particles one layer (10 Å) thick; usually, it is only several layers thick.

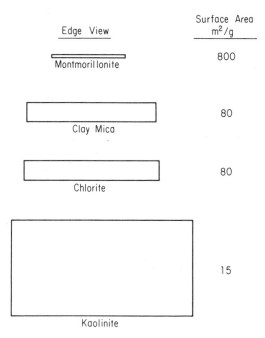

Fig. 2.18. Edge-view sketches to show relative size and shape of clay particles; dimension not shown is equal to length.

Variation in the other two dimensions of clay particles is related to the degree of crystallinity of the clay minerals. A well-crystallized kaolinite has large particles; if it is poorly crystalline the particles may be no larger than those of montmorillonite.

Size and shape are most directly determined with an electron microscope, although birefringence and light-scattering measurements can give some of the information. A magnification of 100,000 can be achieved in electron micrographs of clay particles. A thickness of 10 Å can be seen at this magnification, and becomes a length of 0.1 mm. Fine details of this magnitude are usually observed after subjecting the particles to a process called shadow casting, where a metal vapour is shot at the particles at a certain angle to leave behind the particle a shadow free of metal atoms. This area has a lower absorption of electrons and hence the contrast in the picture is increased.

Kaolinites show the most uniform crystals, often hexagonal plates with a typical diameter of 0.3–4 μ and a thickness of 0.05–2 μ (see Chapter 3). Halloysite shows plates with curled edges, or tubes of 0.04–0.2 μ typical diameter with a wall thickness of 0.02 μ. It is difficult to obtain good pictures of montmorillonite; the particles are thin plates typically around 30 Å thick and 0.1–1 μ in diameter. Illite particles are plates with a typical thickness of 300 Å.

The activity of a clay and its various colloidal properties such as adsorption of water and formation of gels at low concentrations, increase with increasing specific surface area. This area could be calculated if the size and shape distributions of the particles in a sample were known, but this method is too tedious to be practical. Surface area is therefore determined by measuring the amount of a liquid or gas required to cover the surface. Water vapour, nitrogen and organic liquids such as ethylene glycol are used. A method based on the weight of ethylene glycol adsorbed as a monomolecular layer on the clay surface is widely used. It is rapid and does not require special equipment, although the results are not as accurate as methods using adsorption of gases at low pressure.

Most of the differences between clay minerals in properties such as water retention, plasticity, or cohesion can be explained by the different amounts of surface. This explains the high swelling and high liquid limit of montmorillonite. As a first approximation, surface area is a good guide to predicting magnitude of surface properties. Surface area is then a fundamental property of clays which one would wish to measure. However, if surface properties are calculated per unit of surface area as shown in Table 2.8 for three clays, kaolinite is shown to have the most "active" surface. This may be due to its high surface density of charge (Table 2.9, p. 50).

Surface area is not a common measurement for geotechnical studies of clays, largely because it is time-consuming. The liquid limit is closely related to surface

TABLE 2.8

Comparison of activity of representative clay-mineral types based on plasticity index (P.I.)

Clay mineral	P.I. / % Clay	P.I. / Surface Area
Montmorillonite	0.89	0.14
Allophane	0.85	0.12
Kaolinite	0.49	0.19

area and may be a more useful measurement because it "measures" the nature of the surface as well as the area. For example, Farrar and Coleman (1967) found that liquid limit and surface area were related by: Liquid limit = 0.67 (surface area) + 0.93 with an "r" value of 0.91. The correlation coefficient "r" is a measure of the degree of correlation, with $r = 1$ being perfect agreement. The slope of the liquid limit-surface area plot depends upon the nature of the surface.

2.6 WATER AND ION ADSORPTION AT CLAY SURFACES

Hydration of clays

Clay particles in soils are always hydrated, i.e. surrounded by layers of water molecules called adsorbed water. These water molecules should be considered part of the clay surface when the behaviour of clay soils is considered. Plasticity, compaction, interparticle bonding and water movement in soils are all influenced by the water layers. The properties of clays change as the thickness of this hydration shell changes, and consequently the engineering characteristics of soils change. These changes will be discussed in more detail in appropriate chapters. In this section the properties of the hydration shell will be investigated briefly.

The forces holding water molecules to the clay surface arise both from the water and from the clay. Water is a dipolar molecule, with a separation of centres of positive and negative charge (see Appendix 1). This means that water will be attracted by the charges on the clay surface. Further, the hydrogen ions of water will lead to hydrogen bonding of water molecules to the exposed oxygen atoms of the clay mineral surface. Hence the clay contributes both the negative charge and the oxygen or hydroxyl surface to attract water molecules. Cations in water are always hydrated. Therefore the exchangeable cations held near the negatively charged surface hold some of the water at the surface as water of hydration of ions. Since cations are hydrated to different degrees, the hydration of the surface would vary depending upon the cation present.

It is incorrect to picture adsorbed water as dipolar water molecules lined up uniformly at the surface with positive ends at the clay and negative ends away from it. The surface density of negative charges on the clay is too small to hold a complete water layer in this way, and the water molecule is not a rodlike dipole. The main force bonding water to the surface is due to the hydrogen bond. This is a very important bond in many natural materials and is discussed in Appendix 1. The first layer of water molecules is held by hydrogen bonding to the clay surface. The second water layer is held to the first, again by hydrogen bonding, but the force becomes weaker with distance as the orienting influence of the surface on the water molecules decreases. Each successive layer is held less strongly, and the bonding quickly decreases to that of free water.

The properties of this water close to the clay surface differ from those of free water. The density of adsorbed water is higher than that of free water. Values up to 1.4 g/cm^3 have been measured for the first layer of water molecules. The density decreases as further layers are added, dropping to 0.97 g/cm^3 at about four water layers, and then increases to 1.0 g/cm^3 for free water. The viscosity of this water, as measured by the diffusion of ions near the surface, is greater than that of free water. In the first water layers it may be a hundred times greater. The dielectric constant decreases with closer approach to the surface, and may fall to one-tenth that of free water.

Water adsorbed on the clay mineral surfaces can best be visualized as composed of water molecules which are relatively free to move in the two directions parallel to the clay surface but are restricted in their movement perpendicular to, or away from, the surface. Movement parallel to the surface is a transfer from one bonded position to another. The thermodynamic properties of this water are not the same as those of ice, where movement of water molecules is restricted in all three directions.

There is lack of agreement as to the thickness of the hydration layer. The forces holding water become gradually weaker further away from the surface, so there is no sharp division between water of hydration and free water. With increasing distance from the surface, the water molecules gradually become more free to move in a direction perpendicular to the surface. This rate of decrease is a function of exchangeable cations and of the surface. It is agreed that the first two or three layers of water molecules are strongly bonded to the surface, and that properties of adsorbed water differ from those of free water for several more water layers.

Electric charge

Substitution of one ion for another in the clay crystal lattice and

TABLE 2.9

Charge on clay minerals

Clay mineral	Range of charge (me/100g)	Reciprocal of average surface density of charge (Å^2 / electronic charge)	Source of charge
Kaolinite	5–15	25	broken bonds at edges, some ionization of hydroxyl
Clay mica and chlorite	20–40	50	ion substitution in lattice, some broken bonds at edges
Montmorillonite	80–100	100	ion substitution, small contribution from broken bonds
Vermiculite	100–150	75	ion substitution in lattice, small contribution from broken bonds
Allophane	40–70	120	broken bonds, aluminium in tetrahedral coordination

imperfections at the surface, especially at the edges, lead to negative electric charges on clay particles. Cations from the pore water are attracted to the particles (and anions repelled) to maintain electroneutrality. These are the exchangeable cations and their number is the cation-exchange capacity or the amount of negative charge per unit weight or per unit surface area of the clay. This is usually expressed as milliequivalents per gram or per 100 g, me/g or me/100 g. The amount of charge is given for several clays in Table 2.9. The energy with which ions are held at the surface varies with the nature of the charge.

Charge arising in the clay particle from isomorphous substitution of an ion by another of nearly equal size but lower valence has been discussed under structure of the minerals. This occurs during crystallization or formation of the mineral. If the substituting ion has a lower positive valence than the substituted ion, then the lattice is left with a net negative charge. The main substitutions found are aluminium for silicon in the silica sheet, and ions such as magnesium, iron or lithium substituting for aluminium in the alumina sheet. These substitutions account for most of the charge in 2:1 and 2:2 minerals, but only a minor part in the 1:1 kaolinites.

A second source of electric charge on clay particles is unsatisfied valence charges at the edges of the particles. These are referred to as broken-bond charges. The clay crystal lattice is continuous in two directions, but at the edges there must be broken bonds between oxygen and silicon and between oxygen and aluminium.

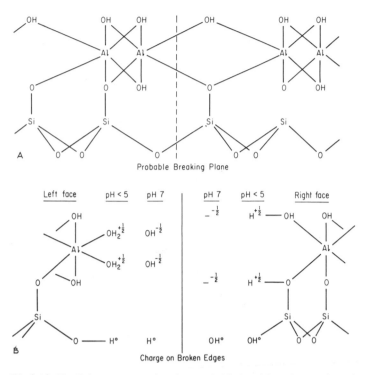

Fig.2.19. Kaolinite structure showing probable breaking plane and mechanism for edge charge by picking up hydrogen or hydroxyl from solution to give positive charge at low pH and negative charge at higher pH.

The most probable breaking plane is shown in Fig.2.19A. The amount of this charge per unit weight of clay increases with a decrease in particle size, because the proportion of edge area to total area is increased. Charge due to broken bonds would also arise if pieces were broken out of the flat surfaces.

These broken bonds attract hydrogen or hydroxyl ions from the pore water, as shown in Fig.2.19. The hydrogen can be exchanged for other cations. The term hydrogen ion is used in this discussion, although H^+ as such does not exist in solution. It is usually referred to as the hydronium ion, $H^+ \cdot H_2O$ or H_3O^+. The ease with which the hydrogen can be exchanged (the negative of its bonding energy) increases as the pH of the pore water increases, i.e. as the hydrogen ion concentration of the pore water decreases. Therefore, the charge due to broken bonds increases as pH increases. At low pH some of these sites on kaolinite may become positively charged by attracting extra hydrogen ions. The mechanism proposed for this is shown in Fig.2.19B. The peculiar flocculation and structure properties of kaolinite below pH 4 are due to flocculation by attraction of positive and negative charges.

In surface horizons of soils, and in organic deposits, the organic matter also contributes to the electric charge or the cation exchange capacity. It can account for one third of the exchange capacity of a surface soil. The negative charge on organic matter arises from ionization of hydrogen from carboxyl and phenolic hydroxyl groups.

The kind and number of exchangeable cations have an important influence on the behaviour of soils. Monovalent cations such as sodium increase the activity of the clay, its swelling, etc.

Exchangeable cations

Exchangeable cations are the positively charged ions from salts in the pore water which are attracted to the surface of clay particles to balance the negative charge. They are termed exchangeable because one cation can be readily replaced by another of equal valence, or by two of one-half the valence of the original one. For example, if a clay containing sodium as exchangeable cation is washed with a solution of calcium-chloride, each calcium ion will replace two sodium ions, and the sodium can be washed out in the solution. This process is called cation exchange. It can be written formally as:

$$Na_2 \text{ clay } + CaCl_2 \rightleftharpoons Ca \text{ clay } + 2NaCl$$

The quantity of exchangeable cations held is the cation-exchange capacity for that clay, and is equal to the amount of negative charge.

The predominant exchangeable cations in soils are calcium and magnesium. Potassium and sodium are found in smaller amounts. Aluminium and hydrogen are the predominant exchangeable ions in acid soils. Geological environment and subsequent leaching determine which exchangeable ions will be present. Clays deposited in seawater will have predominantly magnesium and sodium. Calcareous soils will contain mostly calcium. Extensive leaching removes the cations which form bases (calcium, sodium, etc.), leaving a clay with the acidic cations, aluminium and hydrogen.

The relative energy with which different cations are held at the clay surface can be found by measuring their relative ease of replacement or exchange by a chosen cation at a chosen concentration. These measurements show that a small amount of calcium easily replaces exchangeable sodium, but the same amount of sodium does not replace much exchangeable calcium. The valence of the cation has the dominant influence on ease of replacement. The higher the valence the greater the replacing power, or the harder to replace if the cation of higher valence is at the surface. For ions of the same valence, increasing ion size gives greater replacing power. In addition, there are specific effects due to either the ion or the clay mineral surface which influence replacing power. Potassium, a monovalent cation, has a high replacing power. It is held strongly because it fits into the hexagonal holes at the clay surface. This was discussed in the section on structure of the mica minerals. The result is that potassium will replace calcium much more easily than will sodium.

The hydrogen ion was shown in the older literature to have a high replacing power. These measurements were made with acid clays. It was later found that acid clays had mostly aluminium rather than hydrogen as exchangeable ion. The measurements had actually been made with trivalent aluminium, which would be expected, on the basis of valence, to possess a high replacing power. Clays with only hydrogen as exchangeable ion can be prepared in the laboratory, but they are unstable. The clay edges break down, liberating aluminium, which replaces hydrogen. The actual replacing power of hydrogen is about equal to that of sodium.

The cations can thus be arranged in a series on the basis of their replacing power. While the position in this series will depend upon the kind of clay and upon the ion which is being replaced, the cations will appear approximately as follows, arranged in order of increasing replacing power:

$$Li^+ < Na^+ < H^+ < K^+ < NH_4^+ \ll Mg^{2+} < Ca^{2+} \ll Al^{3+}$$

The number of exchangeable ions replaced obviously depends upon the concentration of ions in the replacing solution. If a clay containing sodium cations is placed in a solution of calcium ions, exchange will take place until at equilibrium a certain percentage of the exchangeable ions will still be sodium and the remainder will be calcium. Similarly both sodium and calcium will be present in the pore water or the solution. This exchange of sodium by calcium would occur if lime were added to the soil. The proportion of each exchangeable cation to the total cation-exchange capacity, as the outside ion concentration varies, is given by exchange-equilibrium equations. Several equations have been derived with different assumptions about the nature of the exchange process. The most simple and useful equation is that first used by Gapon:

$$\frac{M_e^{+m}}{N_e^{+n}} = k \frac{\sqrt[m]{M_0^{+m}}}{\sqrt[n]{N_0^{+n}}}$$

where M and N are cations of valence m and n respectively, subscript "e" refers to exchangeable and subscript "o" to ions in the outside solution, k is a constant depending upon specific cation adsorption effects and upon the clay surface, decreasing as the surface density of charge increases.

For the exchange of monovalent sodium and divalent calcium this equation would become:

$$\frac{Na_e}{Ca_e} = k \, \frac{Na_o}{\sqrt{Ca_o}} \tag{2.2}$$

For example, if an illite clay smaple with a k value of 0.4 (litre/mole)$^{\frac{1}{2}}$ was placed in a solution containing 0.02 M CaCl$_2$ and 0.02 M NaCl, the ratio of exchangeable sodium to calcium would be:

$$\frac{Na_e}{Ca_e} = 0.4 \, \frac{0.02}{\sqrt{0.02}} = \frac{0.008}{0.14} = 0.057$$

If the exchange capacity of the clay was 40 me/100 g, this would mean 38 me of Ca and 2 me of Na.

This calculation shows the valence effect. From the same concentration of calcium and sodium in solution, nineteen times as much of the divalent ion is present in exchangeable form. As the solution concentration increases, this ratio becomes smaller. At salt concentrations usually found in pore water, the ion with the higher valence is adsorbed predominantly.

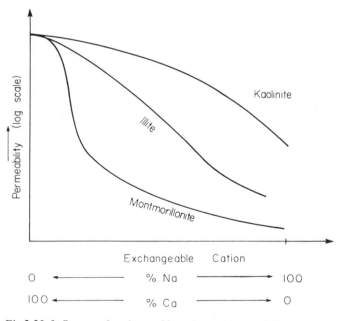

Fig. 2.20. Influence of exchangeable cations on permeability of different clays.

The effect of exchangeable cation on behaviour of clay soils can be best illustrated by examining how a soil property changes as the proportion of exchangeable cations changes. This is illustrated in Fig.2.20. The effect of a given ratio of cations also depends upon the clay type, being more marked for montmorillonite. This is because Na-montmorillonite has a well-developed diffuse ion layer, as discussed in the next section.

Diffuse ion-layers

The exchangeable cations are not all held in a layer right at the clay surface, but are present at some average distance from the surface. The lower the valence, the greater is the average distance from the surface. The electrical force between negatively charged surface and positively charged ions attracts the cations to the surface, but their thermal energy makes them diffuse away from this space with a high ion concentration. The balance of Coulomb electrical attraction and thermal diffusion leads to a diffuse layer of cations, with the concentration highest at the surface and gradually decreasing with distance from the surface. The interaction of diffuse ion-layers of adjacent particles gives an explanation for the properties of swelling, plasticity, and water retention of clays. These are discussed in relevant chapters.

Theoretical expressions have been derived for this distribution of cations in a diffuse layer. Often the assumptions made in the derivation do not fit clay—cation—water systems, so that quantitative calculation of soil properties has not been possible. Laboratory samples of sodium-saturated montmorillonite can be prepared to conform with the assumptions required, and for these samples quantitative agreement with theory has been obtained for cation and anion distribution, and for swelling pressure.

The main assumptions in the derivation of the theoretical distribution of ions in a diffuse layer are that the clay particle can be considered a simple charged plate for which the electric field is described by the Poisson equation, and that the distribution of ions in this field is described by the Boltzmann equation.

The theoretical distribution of ions at a negatively charged surface as a result of thermal and Coulomb forces was worked out by Gouy and later by Chapman. It often bears their names. The derivation is summarized in Appendix 2. The resulting equation for cation is:

$$n_+ = n_o \; \coth \frac{x}{2}\left(\sqrt{\frac{8\pi^2 z^2 c_o N}{\varepsilon kT}}\right)^2 = n_o \, (\coth 0.16z \sqrt{c_o}x)^2 \qquad (2.3)$$

where: n_+ = number of cations per unit volume at any distance x from the surface;

n_o = number of cations per unit volume in the pore water away from the influence of the surface; z = valence of cations; c_o = concentration of cations in moles/litre away from the influence of the surface; x = distance from surface in Ångstrom. The constants are defined in Appendix 2.

The concentration of ions at any distance from the surface can be calculated from this equation. For example, in 0.001 M salt with monovalent ions, the value of n_+ at x = 50 Å is:

$$n_+ = 0.001 \; (\coth 0.16 \times 1 \times \sqrt{0.001} \times 50)^2 = 0.001 \; (\coth 0.25)^2$$
$$= 0.001 \; (4.08)^2 = 0.016 \, M$$

At this distance from the surface the concentration of cations is already 16 times as great as in the pore water. The extent of the diffuse ion-layer, which can be taken as the distance to which n_+ is still appreciably larger than n_o, depends upon concentration and valence. This can be seen from eq.2.3. The lower the concentration and the valence, the larger is the ratio n_+/n_o. Monovalent ions at low concentration would give the most extended diffuse ion-layer. Increasing either the valence or the salt concentration in the pore water would reduce it. For a divalent ion at 0.001 M concentration, the value of n_+ at x = 50 Å is 0.004 M. This is illustrated in Fig.2.21. This diffuse ion-layer may extend far enough from the clay-particle surface so that much of the water in the micropores of a soil is within its influence.

The ion distribution for interacting clay particles rather than the ion distribution around a single particle is of interest in interpreting soil behaviour. The investigation of how the diffuse ion-layer affects particle interaction requires a calculation of the ion distribution when the particles are sufficiently close to cause the individual diffuse ion-layers to overlap. This distribution can be calculated in the same way as for the isolated particle in Appendix 2, but the resulting equation cannot be solved explicitly. Good approximations derived for the two limiting conditions can, however, be made. For a small amount of overlap, or interaction at relatively large interparticle distances, the resulting electric potential is the algebraic sum of the potentials from single layers (see Appendix 2). At close spacing of particles, the anions can be neglected. This calculation was made by Langmuir and is given in Appendix 2. The resulting equation is:

$$y_c = 2 \log_e \frac{\pi}{0.32 z \sqrt{c_o} x} \tag{2.4}$$

where y_c = electric potential in the line midway between parallel interacting particles.

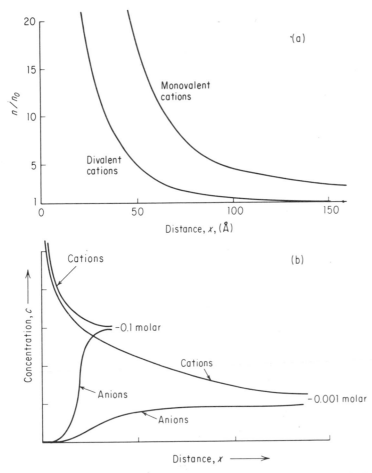

Fig.2.21. Theoretical distribution of ions at a charged surface. a. Influence of valence on thickness of diffuse ion-layer. b. Influence of salt concentration on cation and anion distribution.

Since:

$$y_c = \log_e \frac{n_c}{n_o}$$

$$n_c = n_o \left[\frac{\pi}{0.32z\sqrt{c_o}\,x} \right]^2 \qquad (2.5)$$

where n_c = number of ions per unit volume at the midplane between particles. This equation is solved for potential rather than ion concentration because potential will be used later in calculations of swelling.

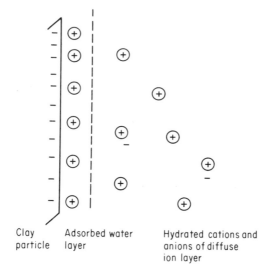

Clay   Adsorbed water   Hydrated cations and
particle   layer   anions of diffuse
                                       ion layer

Fig.2.22. Schematic diagram of clay particle with adsorbed water layer and some cations at the surface, and the remainder of the exchangeable cations in the diffuse ion-layer.

These calculations of electric potential and resulting ion distribution are not quantitatively applicable to most clay soils. The equations are included here because they describe some of the principles of particle interaction, and give the influence of factors such as valence and salt concentration. An improvement is made by applying the Stern correction, by which some of the ions are held in a layer near the surface and the remainder are in the diffuse layer. Most of the calcium ions are held in the Stern layer, while sodium ions occur mostly in the diffuse layer.

The clay particle with its layer of adsorbed water and diffuse layer of exchangeable cations can be visualized as illustrated in Fig.2.22. This model will be used, where relevant, to discuss behaviour of clay soils. The interaction of these particles, or the forces acting between particles, is outlined in the next section.

2.7 INTERACTION OF CLAY PARTICLES

Clay particles interact through the layers of adsorbed water, through the diffuse layers of exchangeable cations and in some cases through direct particle contact. Interparticle forces of attraction and repulsion will be discussed here as a basis for later explanations of soil behaviour.

Repulsion

Repulsion results from interpenetration of ion layers of adjacent particles, and from adsorption of water on surfaces of adjacent particles. Repulsion is manifested in swelling of clay on wetting and in rebound on unloading in a consolidation test. The small part of the rebound on unloading caused by unbending and elastic rebound of particles is due not to interparticle repulsion but to particle contact.

Water molecules adsorbed on the surface will force adjacent particles apart. This accounts for swelling at low water contents where adsorbed water is strongly held. The repulsion arising from adsorbed water has not been quantitatively predicted.

When two clay particles are less than 15 Å apart, the exchangeable ions are uniformly distributed in the interparticle space and do not separate into two diffuse layers, one associated with each surface. Under these conditions there is a net attraction between particles. However, when the interparticle distance exceeds about 15 Å, diffuse ion-layers form, with a resulting net repulsion. This repulsion can again be visualized as being due to water attracted between the particles forcing them apart. In this case the water moves due to osmotic activity of the ions between particles rather than due to properties of the surface.

The concentration of ions is higher in the plane midway between two parallel particles than in the outside solution. Water moves in response to this concentration gradient. The concentration difference depends upon the distance between particles and upon how far the diffuse ion-layers extend, i.e. upon the valence and concentration of ions as discussed in the previous section. Repulsion will be greatest with monovalent exchangeable ions and with distilled water as pore water. Interparticle repulsion is most directly manifested in swelling and in swelling pressure, and is discussed in detail in Chapter 6.

All clays show some swelling on wetting. High-swelling clays are distinguished because the swelling continues to very high water contents. These are clays with high surface area, e.g., montmorillonite, and with monovalent exchangeable ions such as sodium. This high swelling is due to repulsion resulting from diffuse ion-layer interpenetration. The small amount of swelling found for most clays is mostly due to adsorbed water. These two effects cannot be easily separated, and probably work together in most clays.

Attraction

Several different forces must be considered in describing attraction between clay particles. First, there is the attraction between molecules and atoms described by the London—Van der Waals theory (Appendix 1). These are short-range forces inversely proportional to the seventh power of the distance between atoms. They therefore decrease very rapidly with increasing distance of separation. The magnitude of the force depends upon the properties of the surfaces. For two quartz bodies a force of 0.002 dyne has been measured for a gap distance of 1,000 Å, and the force increased rapidly as the distance was reduced. These forces are present in clays where the particles have been brought close together by drying or by consolidation.

At interparticle distances of less than 15 Å there is a net force of attraction between clay particles when their exchangeable cations are in the interparticle space. This was discussed in the previous section. The attraction can be calculated on the basis of image forces.

Under certain conditions electrical forces inversely proportional to the square of the distance (Coulomb forces) act between clays. Kaolinite develops positive edge charges below pH 5, and the attraction of positive to negative charges results in interparticle attraction. This may occur to a limited extent in other clays.

The most important forces holding particles and secondary soil structural units together at field water content are bonds due to nonclay material, either inorganic or organic, bonding to surfaces of more than one clay particle. Iron oxide, aluminium oxide, and carbonates are the most important of the inorganic bonding materials. They are bonded to the clay particle either by chemical bonds such as those existing in crystals or by the weaker, intermolecular Van der Waals bonds. Since aluminium, iron and oxygen are component atoms of the clay crystal, chemical bonds can be easily formed when the oxides precipitate between particles. Calcium or magnesium carbonates also precipitate between particles, forming bonds from one particle to another.

Organic matter contained in surface soils forms interparticle bonds. The organic molecules are held at the clay surface, usually by hydrogen bonding. Sometimes electrical bonds are formed between the negative charge on the clay and a positive charge on the organic matter, e.g., protein, or between negative organic acids and positive charge on clay edges. Organic matter also accounts for much of the bonding of clays to larger particles such as sand, as well as bonding between sand particles.

If a soil contains both water and air, particles with an adsorbed water layer around them can be held together by surface tension forces of the curved air—water interface. This occurs in unsaturated sands and in soils with sand size particles, but is less important in clays. Clays remain water saturated to relatively low water contents. The total volume decreases on drying, giving the observed shrinkage, but no air enters the sample and therefore no air—water interfaces exist.

The magnitudes of the forces of attraction and repulsion in clay soils vary, but the maximum attraction is lower than the maximum repulsion. The forces of attraction can be manifested only if the conditions do not favour repulsion. In the high-swelling clays, repulsion is dominant.

Particle arrangement

The interparticle forces of attraction and repulsion determine the clay particle arrangement or fabric. In most soils, sand and silt-size particles are also present to influence the arrangement. These larger particles either form a skeleton with clay coating them or occurring in the pores, or occur in a clay matrix. Particle arrangement and fabric are discussed in detail in Chapter 3.

Flocculation and dispersion

Flocculation and dispersion characteristics of clays are usually measured in dilute suspensions (about 5% solids), such as those used for particle-size analysis. After the addition of dispersing agents, each clay particle of the dispersed suspension settles independently. If salt is added, the particles flocculate or clump together and settle as large units. These changes from dispersed to flocculated are manifestations of changes in interparticle forces.

The forces of attraction between particles in suspension cannot be easily varied, and can be assumed in first approximation to be constant. Bonds due to inorganic and organic impurities have already been broken during preparation of the suspension and are not considered here. The forces of repulsion can, however, be easily varied. Dispersion of a suspension consists in creating conditions for maximum repulsion, i.e. a low salt concentration in the pore water, monovalent exchangeable ions, high pH to prevent positive charges, and high water content to increase interparticle distance. When these forces of repulsion are decreased — e.g., by adding salt — the forces of interparticle attraction can act and the clay flocculates. A flocculated clay can be redispersed easily by shaking, showing that the forces of attraction are not strong enough to bond clay particles together against the mechanical forces of shaking. However, the clay flocculates on standing after shaking is stopped. Removing salt from a flocculated clay will again cause dispersion. Flocculation and dispersion of clay in suspension is, therefore, reversible. In this it differs from many other inorganic colloids, e.g., the classical silver iodide colloid, which flocculates irreversibly, and which cannot be redispersed by shaking.

2.8 PLASTICITY

The rheological behaviour of a soil changes with a change in water content. This property is especially marked for clay soils, and influences the uses to which clays can be put. At high water contents, soils are suspensions, with the flow properties of liquids. As the water content is gradually reduced, the flow properties of clay soils change to the non-Newtonian flow of pastelike materials. As the water content is decreased further, the clay soil becomes sticky. On further drying the stickiness disappears, and the clay can be moulded. At this water content it is plastic. As the water content decreases still further the plasticity is lost; the soil becomes harder to work and at low water content takes on the properties of a solid.

The physical condition of a soil at a given water content is called its consistency. Consistency is the resistance to flow of the soil and therefore an indication of its rheological behaviour. Consistency is obviously related to the force of attraction between individual particles or aggregates of these particles. It is easier to feel consistency than to describe it quantitatively. Different soils have different consistency at different water contents, and the specification of this condition gives some information about the type of soil material.

Liquid and plastic limits

Plasticity may be defined as the ability of a material to change shape continuously under the influence of an applied stress, and to retain the new shape on removal of the stress. This distinguishes it from an elastic material which regains its original shape on removal of a stress, and from a liquid which does not retain its own shape under stress. The type of flow curve obtained for a plastic material is shown in Fig.2.23.

Only the smaller particles of soils, clays and to some extent silt, exhibit plastic behaviour. A soil exhibits plastic behaviour over a range of water content from a lower limit to an upper limit. The plastic limit is that water content below which the soil is not plastic when it is worked, and crumbles on application of pressure. At the liquid limit the change is from plastic to flow behaviour. Most of the present measurements of plasticity and plasticity limits are based on the early work of Atterberg and are sometimes called Atterberg limits.

Measurements of plasticity are made on remoulded soil samples. For this reason plastic limits have not been widely used in classification of agricultural soils, where the structure or aggregation of the soil is important because it determines the large voids necessary for plant growth. Plastic limits, in conjunction with particle-size distribution, are widely used in evaluation and classification of soils for engineering purposes, e.g., classification systems such as the Revised Civil

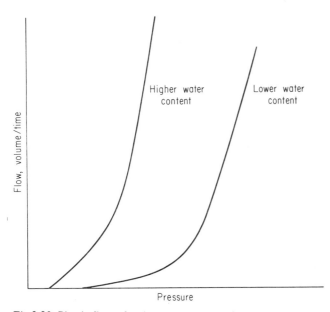

Fig.2.23. Plastie flow of a clay-water paste under pressure through an orifice.

Aeronautics Administration system which uses the plasticity index ($\omega_L - \omega_P$), and the revised Bureau of Public Roads system. The liquid (ω_L) and plastic (ω_P) limits depend upon both the type and amount of clay present in the soil. The difference between the two limits, the plasticity index, depends to a first approximation only on the amount of clay present. This is true except where the clays have unusual properties. Compressibility of soil increases markedly with increasing water content at the plastic limit, whereas strength of the soil decreases under the same conditions. With increasing plasticity index the strength of a soil increases.

The lower plastic limit of a soil is measured by rolling a soil sample into a thread. The water content is adjusted by trial and error to where the soil will roll into threads about 1/8 inch in diameter before it crumbles. If the water content is too low, crumbling occurs at larger diameters; if it is too high crumbling will not occur until the diameter is below 1/8 inch. The plastic limit is both a measure of cohesion of the soil particles and a measure of resistance to cracking when the sample is worked. A plastic material will not crack, while a solid will. Therefore, this point is taken as the lower limit of plasticity of the soil.

The upper limit is generally determined with a mechanical device which jars the sample in a reproducible way. The soil is thoroughly mixed with water and placed in the bottom of a cup. A groove is made in the soil with a grooving tool, and the cup is then dropped repeatedly from a standard height until the soil flows together in the groove. The liquid limit is defined as the water content at which 25

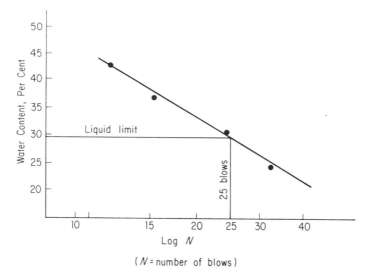

Fig.2.24. Typical semi-log plot for liquid limit determination.

of the standardized blows are required to close the groove over a standardized distance. Below this water content the soil is plastic, and above it begins to exhibit flow properties.

In the actual laboratory-determination, a series of trials at varying water contents is made to establish the water content at the 25 blow point. The trial results plotted on a semilog plot (Fig.2.24) should fall on a straight line. With a proper choice of initial water contents, the 25 blow point can be straddled in the trials and the liquid limit established by interpolation on the semilog plot. The slope of the line depends upon the nature of the clay minerals and upon the salt concentrations.

The plasticity index is defined as the liquid limit minus the plastic limit. For nonplastic soils it may be impossible to measure either the plastic limit or even the liquid limit, or the plastic limit may exceed the liquid limit. In each of these cases the soil is considered to be nonplastic. This occurs with soils having small amounts of fine particles. The activity of a soil is defined as the ratio of the plasticity index to the per cent of clay-size particles.

The explanation of the plastic properties of clay soils must be sought in the interaction of the soil particles. The interparticle forces involved depend upon size, shape and type of clay particles present. When a soil is deformed plastically, particles move relative to each other, taking up new equilibrium positions.

The liquid limit depends only on the fine particles present. The liquid limit of a mixture of clay and sand decreases as the per cent clay in the mixture decreases,

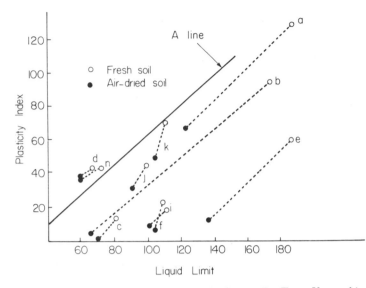

Fig.2.25. Plasticity of wet and air-dry allophane soils. (From Yamazaki and Takenaka, 1965.)

but the liquid limit per unit weight of clay remains essentially constant. The water content at the liquid limit may be lowered if the soil is air dried or oven dried before the determination is made. This is especially true for clays containing organic matter or allophane, which tend to become irreversibly dehydrated. Generally an increase in organic matter content will increase the liquid limit of a soil. Plastic limits for different clays are shown in Table 2.10.

Allophane-containing soils have plastic properties which set them apart from the other clays (Fig.2.25). The high plastic limit, and a resulting low plasticity index and the large change on drying are diagnostic properties for allophane. This is important because the standard methods for clay-mineral identification are unsatisfactory for allophane. Physical properties which are easily measured, such as plasticity, can then be used for identification. Highly allophanic soils become nonplastic on air-drying, or if slightly less allophanic, on ovendrying. They then have a gritty consistency, and the plastic limit cannot be measured.

Interpretation of plastic limit and liquid limit

In a plastic soil the particles must be able to move or slide past one another to take up new positions, and then retain these new equilibrium positions. The cohesion between particles, or units of particles, must be sufficiently low to allow this movement and yet sufficiently high to allow the particles to maintain the new

TABLE 2.10

Plastic limits for clays

Description of sample	Plastic limit	Liquid limit	Reference
Kaolinite – Na	26	52	White (1949)
Kaolinite – Ca	36	73	White (1949)
Illite – Na	34	61	White (1949)
Illite – Ca	40	90	White (1949)
Montmorillonite – Na	97	700	White (1949)
Montmorillonite – Ca	63	177	White (1949)
Allophane – undried	136	231	Birrell (1952)[1]
Allophane – airdried	78	85	Birrell (1952)[1]
Na – montmorillonite			
– water	–	950	Warkentin (1961)
– 0.01N NaCl	–	870	Warkentin (1961)
– 1.0N NaCl	–	350	Warkentin (1961)
Ca – montmorillonite			
– water	–	360	Warkentin (1961)
– 1.0N CaCl$_2$	–	310	Warkentin (1961)
Kaolinite, pH 4			
– water	–	54	Warkentin (1961)
– 0.01N CaCl$_2$	–	46	Warkentin (1961)
– 1.0N CaCl$_2$	–	39	Warkentin (1961)
Na – kaolinite, pH 10			
– water	–	36	Warkentin (1961)
– 0.01N NaCl	–	34	Warkentin (1961)
– 1.0N NaCl	–	40	Warkentin (1961)
Attapulgite	145	171	
Halloysite	65	94	

[1] See Grim (1954).

moulded position. The lower plastic limit is the lower water content at which these properties are exhibited.

As seen in Table 2.10, the plastic limit varies much less than the liquid limit. It increases as the surface area of the clay increases, but not in direct proportion. The surface area of montmorillonite is 40 times as great as that of kaolinite, but the plastic limit is only 2–3 times as large. The plastic limit can, therefore, not be related simply to a thickness of water films around the particles. Sufficient water is required to wet all the surfaces and to fill the small pores where water is held at very high suction values. At this water content, the particles will slide past one another on application of force but there is still sufficient cohesion to allow them to retain a shape. If the soil is not water-saturated, air—water interfaces contribute to cohesion. Varying the exchangeable cations and changing the salt concentration changes the particle arrangement and hence the size and distribution of pores. This changes the water content at the plastic limit.

The variation in liquid limit of different clays is greater than that of the plastic limit. The influence of exchangeable cations and salt concentration is also greater. Interparticle forces have a more prominent role in determining the liquid limit. The distance between particles, or between structural units of particles, is such that the forces of interaction between the clay particles become sufficiently weak to allow easy movement of particles relative to each other. The soils are water, saturated at this point, and cohesion between particles in remoulded samples is small. In high-swelling clays such as montmorillonite, the dominant interparticle force is one of repulsion. This force of repulsion determines the distances between particles. Therefore an increase in salt concentration or substitution of divalent for monovalent exchangeable cations, which decreases the repulsion, decreases the liquid limit (Fig.2.26).

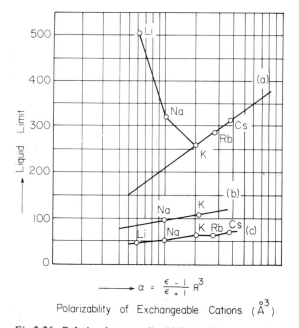

Fig.2.26. Relation between liquid limit of montmorillonitic and illitic clays and polarizability of the exchangeable cations. a = montmorillonite; b = 90% illite, c = 40% illite. (From Rosenqvist, 1962.)

In low-swelling clays, such as kaolinite, the influence of exchangeable ions and salt concentration is in the opposite direction. Interparticle forces again determine particle arrangement. Edge-to-face flocculated clay has a random particle arrangement with a consequent decrease in the liquid limit. A dispersed kaolinite or amount of water trapped within the structural units. With increasing salt concentration the particle arrangement changes from edge-to-face to a more parallel arrangement with a subsequent decrease in the liquid limit. A dispersed kaolinite or one with monovalent exchangeable cations has a lower liquid limit because the repulsion and hence the distance between particles is small.

2.9 RHEOTROPY

Many clay soils exhibit the property of rheotropy at water contents above the liquid limit, and also to a lesser degree at water contents in the plastic range. This is the change to a more fluid consistency on stirring or disturbance; when the disturbance has ceased the system reverts to its less fluid or more rigid condition. This is often called thixotropy, although the strict definition of thixotropy is a reversible, isothermal sol–gel transformation. A sol, by definition, has no yield value, while a gel has rigidity. The change in clay-water systems is generally from a system with higher yield value to one with a lower yield value. In the engineering sense, one may consider a sol as a colloidal dispersion. This restricts sols to liquidlike behaviour. When hardening of the sol occurs, a gel is formed. This requires a change of state from a liquidlike substance (sol) to a semisolid (gel).

There are several ways of measuring the rheotropy of clay soils. It can be measured in certain types of shear apparatus, for example the vane shear, or at higher water contents with a viscometer. A simple test consists in applying force by inverting the test tube containing the clay. If a rheotropic sample has been undisturbed for some time it will retain its shape when the tube is inverted. However, if it is stirred or shaken, it will flow as soon as the tube is inverted.

A loss in shear strength of clay soils on remoulding is usually observed. If such a soil is tested at increasing times after remoulding, an increase in strength with time is generally measured. This is illustrated in Fig.2.27. The entire undisturbed strength may not be regained. This strength regain has been called "thixotropic regain" but is more properly termed age-hardening or rest-hardening. Following Boswell (1961) rest-hardening will be used here. Sensitivity is the ratio of undisturbed to remoulded shear strength.

The property of rest-hardening has been explained either by changes in particle rearrangement and interparticle forces, or by changes in adsorbed water. On stirring, the particles and fabric units (Chapter 3) are rearranged and the bonds

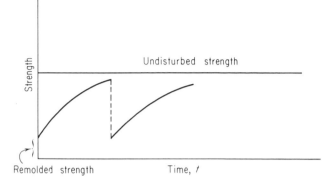

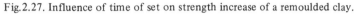

Fig.2.27. Influence of time of set on strength increase of a remoulded clay.

between particles and units are broken. Also the structure of the adsorbed water is broken up, and the clay mass will be more susceptible to deformation under self weight. After deformation, the clay fabric will seek a status of minimum energy with maximum attraction between particles and fabric units. The adsorbed water also regains its quasi-crystalline form to give the system sufficient rigidity to have a yield value.

There are several factors which contribute to the regain of part or all of the strength. These are original structure, activity of the clay minerals, and degree of disturbance.

Rest-hardening and structure

The loss in strength of a clay with random fabric is generally less than that of a flocculated clay at the same void ratio and water content. Since this is attributable to the strength of the bonds developed in the undisturbed state, it follows that bonds which give rise to a "structural" strength found in the flocculated structure would have greater strength (Chapter 10). The dispersion of particles resulting from remoulding destroys initial "structural" strength. This loss is greater for flocculated fabrics, since random fabrics possess a measure of dispersion to begin with. It follows therefore that for the same soil type the degree of strength regain will be more for a random structure than for a flocculated structure, although quantitatively the regain might be greater for the latter. In Fig.2.28 this is shown in terms of percentage regain and actual regain.

The strength regain by rest-hardening following remoulding depends upon the degree to which particle and fabric unit reorientation are allowed. In the presence of strong interparticle forces of repulsion, the likelihood of regain of flocculated

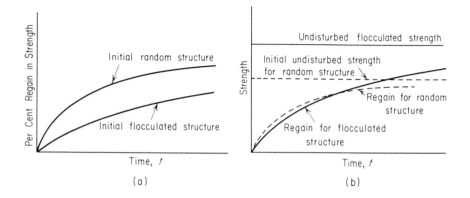

Fig.2.28. a. Per cent regain in strength for remoulded samples. b. Total strength regain with time of set.

fabric is rather small, since the repulsion will not allow the dispersed particles and units to come sufficiently close to establish the edge-to-surface or edge-to-edge arrangements of a flocculated fabric (Chapter 3). If, on the other hand, the interparticle forces of repulsion are not sufficiently large to overcome the electrical forces of attraction, then the possibility of reformation into a metastable flocculated state becomes quite feasible and likely.

2.10 SUMMARY

The crystal structure and appearance of some common clay minerals are discussed briefly. Some methods for identification which are discussed are X-ray diffraction patterns, cation-exchange capacity, infrared absorption, differential thermal analysis and microscopy.

Surface properties, both extent and nature of surface, distinguish clay-size grains from larger soil grains. The water bonded to these surfaces, and the ions held by the electrical charges at the clay surface, give rise to the physical-chemical forces of attraction and repulsion between clay particles.

Plasticity is a diagnostic characteristic of clay, and is discussed on the basis of interparticle forces in clays.

Chapter 3

SOIL FABRIC AND STRUCTURE

3.1 INTRODUCTION

In the typical class of soil-engineering problems, one is concerned with: (a) the stability of soil as a supporting material; (b) the overall rheologic performance of soil; and (c) the general transmissibility characteristics of the material in its natural or altered (compacted) state.

For an analysis or evaluation of stability of a soil mass, the governing equations which describe the problem will consist of the field relations, the constitutive equation, and the continuity relationship. Whilst the field relation examines the physics of the problem and provides one with the condition of equilibrium or the state of the system, the property inter-relationships are established by the constitutive relationship. Thus for example, in a stability problem one is generally concerned with the stress—deformation—time (rheologic) behaviour of the material. Continuity relationships, on the other hand, provide the separate links in terms of material performance characteristics. In order to analyze and predict response soil behaviour, it is necessary to understand how the individual microscopic constituent elements (mineral particles) interact.

Where agreement between the mathematical description or analytical model and the physical behaviour of the system is reached, an accurate prediction or assessment of the behaviour of the soil-water system can be obtained. From the viewpoint of evaluation of behaviour, test results and performance characteristics can be completely interpreted with a proper knowledge of the fabric and structure of the soil material.

3.2 STRUCTURE AND FABRIC

We define soil structure as that property of soil which provides the integrity of the system and which is responsible for response to externally applied and internally induced sets of forces and fluxes. Soil structure, as a property, includes the gradation and arrangement of soil particles, porosity and pore-size distribution, bonding agents and the specific interactions developed between particles through associated electrical forces. In essence, soil structure provides a physical description of the various constituents of the soil-water—air system and the interactions occuring within the system which provide for the integrity of the overall soil-water system.

A major component of soil structure is the geometrical inter-relationships established by individual soil particles. In granular soils, the arrangement of individual particles is generally referred to as "packing" of particles. The corresponding term in clays is "fabric". Fabric denotes the geometrical arrangement of the constituent mineral particles, including void spaces, which can be observed visually or directly using optical and electron microscope techniques. In addition, particular clay-fabric patterns can be deduced through indirect means, using certain physical properties of the mineral constituents, e.g., intensity of X-ray diffraction as a function of particle orientation. Soil fabric constitutes an integral and vital component of soil structure. An assessment of soil fabric is required for a proper evaluation of soil structure, in addition to physical performance measurements which would provide for a more complete evaluation of the response behaviour of the soil-water—air system.

3.3 GRANULAR SOIL PACKING

In this section we will consider the arrangement of granular particles — where the influence of surface forces is negligible. We have pointed out previously that the arrangement of granular particles is better considered as "packing" of soil particles rather than "fabric".

The packing of grains of soil or other particulate media is very strongly influenced by particle shape and size distribution. A knowledge of optimum packing conditions is desirable since soil-engineering problems concerned with stability, such as subgrade fill and embankments, require optimum density of the granular material for development of maximum resistance to shear displacement. The same requirement applies to aggregates for concrete and bituminous mixtures. This requires a particular range of grain sizes to fill the voids created by packing particles which are either angular (e.g., crushed stone products) or rounded (e.g., river-washed gravels and sands). Fig. 3.1 shows the required particle-size distribution for optimum packing based upon the condition of four large rounded particles as initial constraints. While this is an ideal situation, it demonstrates the need for an ideal distribution of grain sizes to achieve optimum density.

Ideal grain-size distributions generally do not exist in natural soils. Fig. 3.1 shows the required grain-size distribution to fit the void spaces, and establishes one end of the scale. The other end of the packing spectrum is obtained by selecting particles of the same size. This means taking only the four large particles shown in

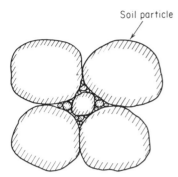

Soil particle

Fig.3.1. Ideal particle-size distribution for optimum packing.

Fig. 3.1 without having smaller particles filling the voids. It is possible to obtain various kinds of packing with uniform spheres. Maximum porosity for regular packing is obtained with a cubic packing (open packing) and minimum porosity exists in the tetrahedral rhombic packing (close packing). These are shown in Fig. 3.2.

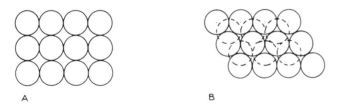

A   B

Fig.3.2. Ideal packing of spheres. A. Simple cubic packing. B. Rhombic packing.

The five modes of regular packing that may be achieved by spheres of equal radius R are (a) simple cubic, (b) cubic tetrahedral, (c) tetragonal spheroidal, (d) pyramidal, and (e) tetrahedral.

Defining the coordination number as the number of spheres in contact with any given sphere, and the density of packing D_p as the ratio of the volume of space occupied by solid matter to total volume, the porosity–density relationships for the five modes of packing may be arrived at (Table 3.1). The arrangement for the modes of packing is given in Fig. 3.3.

The number of adjacent particles in contact with any particle is related to porosity. The distribution of coordination number for several porosities, using stacked lead shot in a large cylindrical vessel to simulate well-packed aggregate is shown in Fig. 3.4. The relationship between average number of contacts per sphere and porosity n may be obtained by assuming that statistically the grouping of the

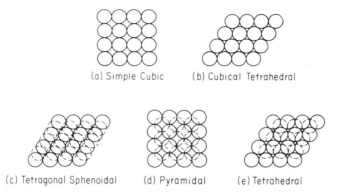

(a) Simple Cubic   (b) Cubical Tetrahedral

(c) Tetragonal Sphenoidal   (d) Pyramidal   (e) Tetrahedral

Fig.3.3. Models of regular packing of equal spheres. (From Deresiewicz, 1958, fig. 1, by permission of Academic Press Inc.)

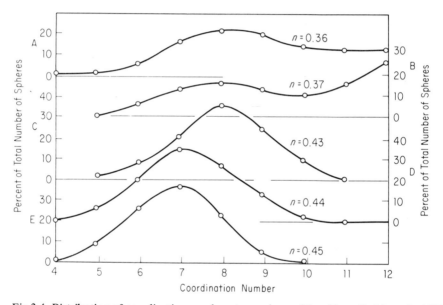

Fig. 3.4. Distribution of coordination number at several porosities. (From Smith et al., 1929.)

spheres as a whole may be regarded as groups of closepacked spheres in cubic array. Using m to represent the fraction of closepacked spheres, we have:

$$n = mn_{c.p.} + (1 - m)n_{cub} \tag{3.1}$$

where the subscripts "c.p." and "cub" represent closepacked and cubic respectively. On a unit volume basis, the average number of contacts N per sphere will be obtained as:

$$N = \frac{m\dfrac{N_{c.p.}}{V_{c.p.}} + (1-m)\dfrac{N_{cub}}{V_{cub}}}{\dfrac{m}{V_{c.p.}} + \dfrac{(1-m)}{V_{cub}}} \qquad (3.2)$$

where $V_{c.p.}$ and V_{cub} represent the volume of closepacked and cubic array of spheres in terms of R, the radius of the spheres. The values for $N_{c.p.}, N_{cub}, V_{c.p.}$ and V_{cub} may be obtained from Table 3.1. In this instance, $N_{c.p.}$ and $V_{c.p.}$ may be taken as either the pyramidal or tetrahedral type of packing.

TABLE 3.1

Packing of spheres
(From Deresiewicz, 1958, table 1)

Type of packing	Coordination number	Spacing of layers	Volume of unit prism	Density	Porosity (%)
Simple cubic	6	$2R$	$8R^3$	$\pi/6$ (0.5236)	47.64
Cubic–tetrahedral	8	$2R$	$4\sqrt{3}R^3$	$\pi/3\sqrt{3}$ (0.6046)	39.54
Tetragonal–sphenoidal	10	$R\sqrt{3}$	$6R^3$	$2\pi/9$ (0.6981)	30.19
Pyramidal	12	$R\sqrt{2}$	$4\sqrt{2}R^3$	$\pi/3\sqrt{2}$ (0.7405)	25.95
Tetrahedral	12	$2R\sqrt{2/3}$	$4\sqrt{2}R^3$	$\pi/3\sqrt{2}$	25.95

Thus from eq. 3.1 and 3.2 :

$$N = \frac{1}{\sqrt{2}-1}\left[6(2\sqrt{2}-1)-\frac{\pi\sqrt{2}}{1-n}\right]$$

which reduces to:

$$N = 26.4858 - \frac{10.7262}{1-n} \qquad (3.3)$$

For particles of nonuniform size, porosity may be reduced if interstitial particles are available to fill the void spaces created by the larger particles as in Fig.

3.1. The combinations of packing that one can obtain are a function of gradation of particles, particle shape, texture and manner of placement.

The packing of unequal spheres, which represents real systems more closely, may be studied mathematically. In doing so it is necessary to prescribe a set of limiting conditions. The study by Wise (1952), for example, requires that one large sphere A be taken such that all other spheres of smaller sizes may be placed on its surface. The first two spheres must not only contact each other, but also touch A. Each new subsequent sphere to be added must touch, in addition to A, at least two others which are in contact with each other (Fig. 3.5). The network of triangles formed by joining the centres of spheres surrounding one sphere (shown in Fig. 3.5) should provide for triangles on different planes. A polyhedron is obtained if the vertices are joined to the centre of sphere A, since these triangles are in effect the faces of a polyhedron whose vertices are the centres of the spheres. Thus a set of tetrahedra associated with any given sphere may be obtained.

Wise (1952) expressed the properties of packing for the four radii in each tetrahedron in terms of a probability distribution function. In the case where the logarithms of the radii conform to a normal distribution of standard deviation $s = 0.4$, a mean radius of 1.08 and estimated mean density of packing of 0.8 is obtained. Fig. 3.6 shows the distribution of spheres and the mean coordination number for a sphere with radius R.

In engineering practice and especially for stability considerations, optimum packing of granular particles is needed. Not only is shear strength increased through more particle contact and lateral support, but volume change (subsidence and compression) through both particle rearrangement and compression is reduced

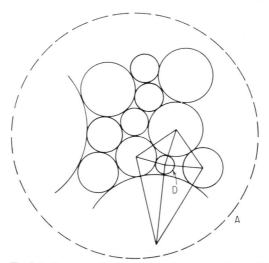

Fig.3.5. Dense random packing of unequal spheres. (From Deresiewicz, 1958, fig. 6, by permission of Academic Press Inc.)

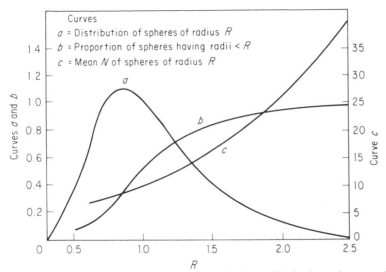

Fig.3.6. Dense random packing of spheres having radii obeying a log-normal distribution of standard deviation = 0.4 (From Deresiewicz, 1958, fig. 7, by permission of Academic Press Inc.)

because of optimum packing. With loose packing there is a tendency for particles to readjust to new positions of equilibrium under external stress, to attain a minimum potential energy level. This readjustment may be detrimental since it results in volume change. To avoid problems arising from loose packing of subsurface granular material, it is necessary to compact the material prior to construction on the material. For shallow deposits of loose material, a vibratory compactor may be used successfully. For deep deposits of loose material it is necessary to resort to piling to achieve denser packing. Piledriving serves to vibrate the immediate vicinity of the pile, in addition to the use of the support itself.

3.4 CLAY SOIL FABRIC

Typical electron photomicrographs of a remoulded clay are shown in Fig. 3.7 and 3.8. Whilst single clay particles can be easily distinguished, we note that by and large, the clay particles tend to form identifiable groups or group units. The geometrical arrangement of the group units, identified as *fabric units,* constitutes the *first-order* recognition pattern. The geometrical arrangement of single particles within each fabric unit in turn constitutes the *second-order* fabric recognition pattern.

The total arrangement of all particles, fabric units and voids comprises the total fabric of clay soils. Whilst individual fabric unit classification can be achieved

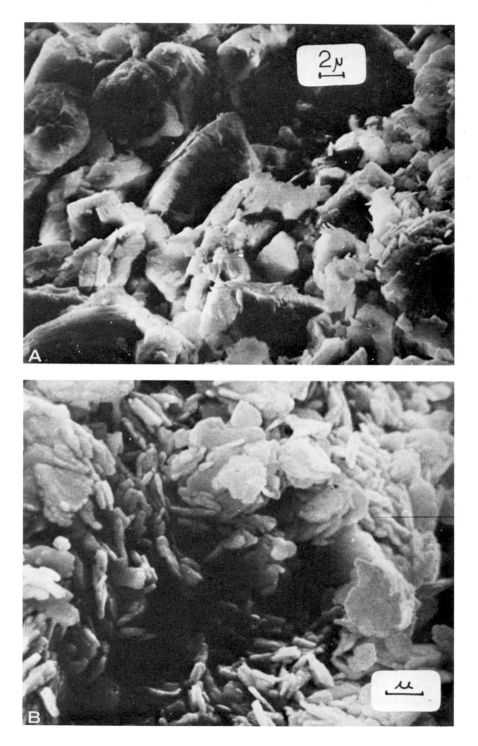

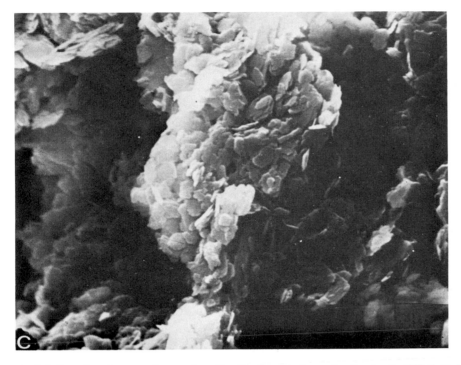

Fig.3.7. Scanning electron micrographs of kaolin clays. A. Coarse-fraction kaolin (MP clay). B, C. Medium size particle fraction kaolin (English clay). Note ped size fabric unit in (C) and macro pores. Pore characteristics between ped units are different from pore characteristics in ped unit itself.

through examination of photomicrographs, overall fabric classification can be difficult to obtain because of the interactions of the various fabric units. Fig. 3.7 and 3.8 and a study of many other photomicrographs of various natural and remoulded soils show that single particle action is not the general rule in clay soils. Response behaviour in clay soils is through fabric unit interaction, and through the various bonding mechanisms between units.

Fabric classification

There are three *levels* of first-order fabric recognition. These are categorized on the basis of degree of magnification required for a proper observation of fabric patterns or soil particles:

(a) *Macroscopic.* The fabric units are distinguishable with the naked eye. Granular particles, for example, will in general constitute single fabric units. In clay soils, fabric units that can be identified visually with the naked eye, will generally

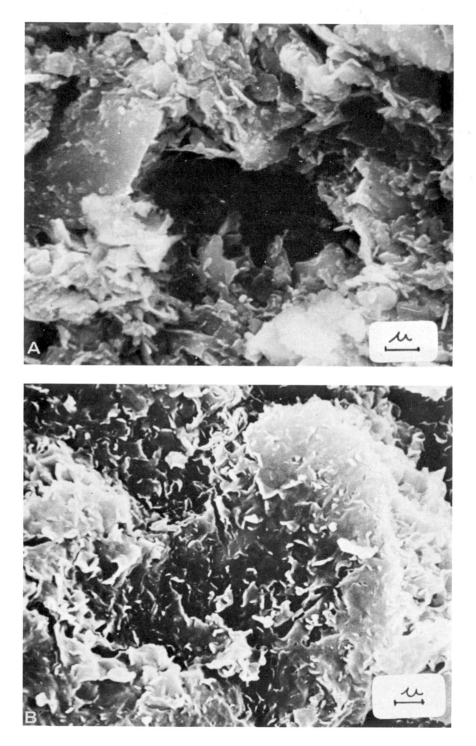

Fig.3.8 Scanning electron micrographs of clays. A. Natural sensitive clay (Champlain Clay from Eastern Canada). B. Montmorillonite, APl 23. C. Illite, AP1 H-36.

consist of an aggregation of clay particles. These units are defined as *peds*. Each ped consists of an aggregation of particles. Other terms used are "crumbs" and "aggregates". However, as a general term, peds is recommended.

(b) *Microscopic*. The fabric units are visually observed under the light microscope. In the case of clays, single particles will not be distinguishable at this level of viewing. The fabric units identified in the microscopic range consist of several particles or groups of particles and are defined as *clusters*. Whilst the term "floc" has also been used, the preferred definition is clusters. The composition of a cluster is somewhat similar to that of a ped except in regard to size. Clusters can be combined to form peds.

(c) *Ultra-microscopic*. The fabric units are visually observed in the ultra-microscopic level using electron microscopy (e.g., either transmission or scanning electron microscopy). Single or individual clay particles can be distinguished at this level. The small fabric units observed and distinguished at this level of viewing are *domains* or single clay (or other fine-grained) particles. Domain units consist of two or more particles acting as a unit. In general, particles within domains are in parallel array. Several domains could combine to form a cluster. Other terms used are tactoids (for montmorillonites) and packets.

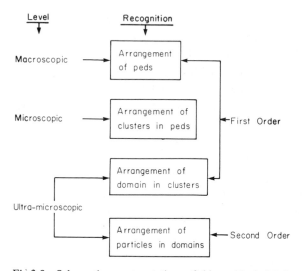

Fig.3.9. Schematic representation of hierarchical fabric recognition system. The level of recognition corresponds to the technique used to establish the recognition pattern and is thus defined accordingly.

Second-order fabric recognition relates to a description of the geometrical arrangement of particles within the fabric units. Since clay particles can generally be recognized only through the use of electron microscopy, it is apparent that the fabric units examined are domains and clusters. A description of the arrangement of domains within clusters would constitute a first-order fabric description. Fig. 3.9 gives a simplified schematic description of the hierarchical system of fabric recognition.

In the scanning electron photomicrographs of natural and remoulded clays (Fig. 3.7, 3.8 and 3.10–3.13) we can identify the many kinds of fabric units defined above.

Particle arrangement in fabric units

Since the presence of single particles as individual entities in clay soil behaviour is not a common occurrence, we will examine here the possible arrangement of groups of particles comprising individual fabric units, i.e., second-order characterization.

We have seen in Chapter 2 that clay particles interact through (a) the layers of adsorbed water, (b) the diffuse ion-layers, and (c) through mineral contact in certain particular cases. In placement or deposition of clay particles, interparticle forces of attraction and repulsion play a dominant part in particle arrangement. The presence and proportion of larger-sized particles will also influence final arrangement of the soil particles. The most important consideration in the structure and fabric of clay soils is the nature and magnitude of forces originating from the soil particles and fabric units, and between soil and water. The term "surface forces" will be used to include both types of forces. Since many of the soil structure and fabric theories which have been advanced to explain particle arrangement are incomplete, it is useful to examine clay soil fabric on the basis of surface forces and particles (including fabric units) interacting in the presence of these forces.

Typical patterns (Smart, 1969; Barden and Sides, 1971; Yong, 1971) observed are (a) random, (b) cardhouse, (c) dispersed, (d) oriented, and (e) flocculated. These patterns are shown in Fig. 3.10. Other variations undoubtedly exist and can be classified. However, for realistic application in the evaluation of fabric contribution to soil structure, it is necessary to provide the distinct pattern guidelines as identification units.

The definition of a flocculated fabric implies a close aggregation of particles and fabric units without any specific orientation. The term, random, applied to soil fabric indicates an aggregation of fabric units in random orientation, but not in close contact.

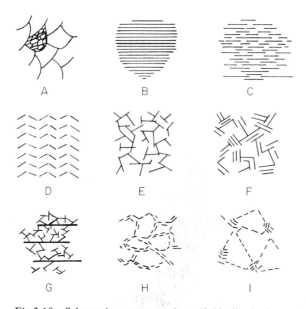

Fig.3.10. Schematic representation of idealized clay-particle arrangements. (From Smart, 1969.) A = kidney structure, sketch; B = packet, cross-section, each stroke represents a platy particle; C = domain, cross-section; D = herring bone structure, cross-section; E = cardhouse structure, two-dimensional analogue; F = salt-flocculated structure, analogue; G = Brownian structure analogue; H = flocculent structure, cross-section; I = Pusch's structure, cross-section.

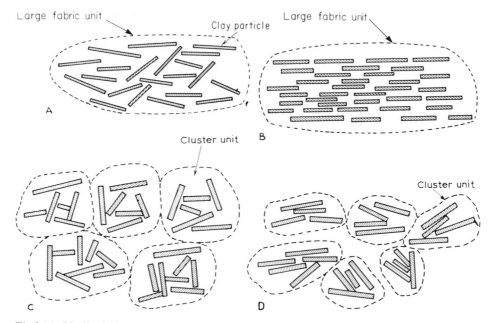

Fig.3.11. Idealized clay structures. A. Partially-oriented structure. B. Fully-oriented structure. C. Cluster units in random orientation. D. Packet clusters.

A dispersed fabric indicates some degree of orientation of the fabric units. The extent of orientation of the units will include both first- and second-order characterization, i.e., orientation of particles in a fabric unit indicates second-order characterization, whilst orientation of fabric units, i.e., domains or clusters, denotes the first-order characterization.

A decrease in interparticle repulsion can result in flocculation of clay particles and fabric units. Soil fabric of high swelling soils in the presence of high salt concentration which reduces interparticle repulsion would be a flocculated type shown in Fig. 3.11.

In some low swelling soils such as kaolinite, distinction must be made between flocculation due to development of edge charge at low pH and flocculation due to high salt concentration. At high salt concentration, face-to-face flocculation occurs, and produces partial orientation of particles in the fabric units and also partial orientation of the fabric units themselves. This may be identified as dispersed particles and dispersed overall fabric (see Fig. 3.11B). The forces holding particles in an edge-to-surface and edge-to-edge configuration can include not only the combination of a positive edge charge to a negative surface charge, but also the bonding derived from cementation, organic matter, and surface tension forces.

We have described the fundamental clay fabric structural models generally in terms of interparticle forces of attraction and repulsion. Cementation bonds which require time for precipitation to induce cementation, will also exist, e.g., bonded sensitive clays of Eastern Canada (La Rochelle and Lefebvre, 1971) can occur due to the presence of nonclay minerals and organic matter.

Interparticle forces of repulsion arise because of the interaction between interpenetrating diffuse ion layers from adjacent particles, from adsorbed water layers, and from interaction between like charges.

Attractive forces are of two kinds: (a) London-Van der Waals, and (b) Coulombic. The London-Van der Waals forces are short-range forces which vary inversely as the seventh power of the distance between atoms, and inversely as some higher power for larger units. Coulombic forces are inversely proportional to the square of the distance between particles.

Cementation bonds may be due to inorganic bonding materials such as carbonates or oxides precipitated between particles. In natural soils these may be quite common. Bonding from organic matter can arise when an organic molecule is tied to two or more particles. This may be Coulombic bonding between the negative charge on a clay-particle surface and the positive charge on the organic matter, or London-Van der Waals bonding between atoms of organic and inorganic constituents.

First-and second-order fabric characterization

The two orders of fabric characterization relate directly to the level of fabric identification and the overall macroscopic response behaviour of the soil-water system.

Since mechanical and physical soil-testing procedures provide only macroscopic measurements, it is evident that the primary assessment of fabric relates only to first-order characterization. From the characterization given and from the examples shown in Fig. 3.10, we demonstrate that the first-order arrangement describes the geometry of the fabric units, e.g., oriented or random arrangement of fabric units, whilst the second-order characterization deals with the arrangement of particles within the units. This second-order characterization which in general is not directly measured in most soil-testing procedures will nevertheless play an important part in the manifestation of response soil behaviour (see Chapters 7 and 10).

Fig. 3.12 illustrates the characterization procedure and identifies the major facets. Thus we can define:

(a) *Total fabric isotropy*. Random or flocculated (non-oriented) array of particles in the fabric units and corresponding random or flocculated array of fabric units. The system is thus completely isotropic.

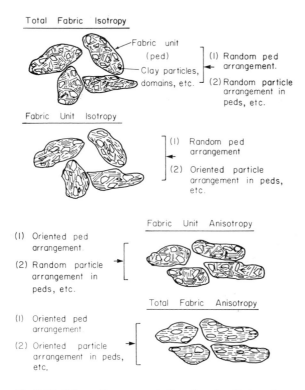

Fig.3.12. Schematic representation showing first- and second-order characterization of fabric.

(b) *Fabric unit isotropy.* Random or flocculated array of fabric units with dispersed (i.e., semi-oriented) or oriented arrangement of particles within the fabric units. The total system will behave as an isotropic system, i.e., the bulk behaviour pattern is isotropic.

(c) *Fabric unit anisotropy.* Dispersed or oriented arrangement of fabric units with random or flocculated arrangement of particles within the fabric units. The bulk behaviour of the total system is thus anisotropic.

(d) *Total fabric anisotropy.* Oriented arrangement of particles within fabric units with corresponding oriented arrangement of fabric units. The system is thus totally anisotropic.

3.5. PORE SPACES AND FABRIC

Two kinds of pores or pore spaces can readily be identified in many of the electron photomicrographs shown in the previous Figures. The pore spaces between fabric units are larger than the pore spaces between particles within the fabric units (Fig. 3.13).

We define the pores between fabric units as macropores, and the pore spaces between particles within the fabric units as micropores. Thus, interfabric unit pores are macropores and intrafabric unit pores are micropores. In identifying and characterizing the fabric of clays, it is necessary to take into account particle and fabric unit spacing in defining the geometry of the solid network of fabric units. A knowledge of the distribution of pore spaces provides an appreciation of "packing" of the fabric units and the gradation of the units. The use of pore-size distribution is not unlike that used in grain-size-distribution analyses of granular soils. On a larger scale, fabric units are considered as "granular" particles, i.e., clay aggregates. The fabric of clays composed of aggregates can be considered as the "packing" of clay aggregates and the models used to describe this situation are identical to those described in Section 3.3.

Soil behaviour in relation to water flow, pore-water extrusion, soil distortion and consolidation requires a knowledge of characteristics of water movement in the macro and micropores and also of the rearrangement of fabric units and distortion of individual units. The movement of water in macropores and micropores will be controlled by different sets of forces and conditions. Overall permeability will reflect the average value of permeability measured through physical testing which in fact is due to two separate physical phenomena of fluid flow through macro and micropores. The importance of pore-volume description and assessment will be examined in greater detail in succeeding chapters.

Micropores
(intra fabric unit pores)

Macropore
(inter fabric unit pore)

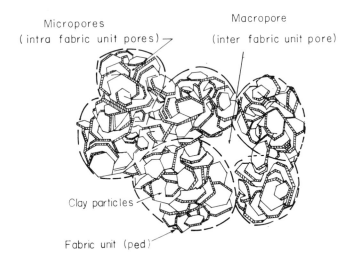

Clay particles

Fabric unit (ped)

Fig.3.13. Schematic diagram showing macro- and micropores.

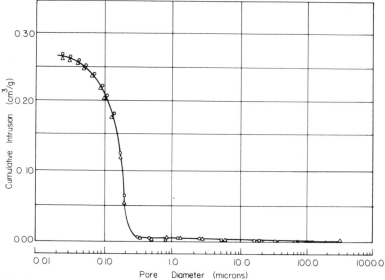

Fig.3.14. Average pore-size distribution from several samples of Georgia kaolin. (From Diamond, 1971.)

Measurements of pore-size distribution are obtained through techniques involving forced intrusion of a non-wetting fluid such as mercury, or through capillary condensation methods which involve interpretation of adsorption and desorption isotherms. Since capillary condensation methods are more difficult to implement experimentally, and in view of the limitation of a maximum pore-size measurement of the order of 1000 Å, which is smaller than the general macropore size, the mercury intrusion method is favoured (Diamond, 1970). The experimental procedure requires that the soil sample to be intruded by the non-wetting fluid be completely dry. By applying increments of pressure to force mercury into the sample, the volume intruded at that applied pressure can be measured. Successive increments of pressure and corresponding measurements will provide information leading to interpretation in terms of cumulative volume distribution and pore diameter. A typical curve from several tests is shown in Fig. 3.14. The limitations of the porosimetry technique using intrusive methods relate to accessibility of pore spaces, connectivity of spaces and pressure capacity of the dilatometer used.

3.6. TECHNIQUES FOR DIRECT FABRIC VIEWING

The three common methods available for a direct study of fabric are: (a) viewing of ultra-thin sections of clay under polarized light using the light microscope; (b) transmission electron microscopy of metal shadowed carbon replicas of sample surfaces; and (c) scanning electron microscopy of fractured surfaces suitably dry and coated with a conductive coating of silver or gold. The detailed techniques for these various methods can vary according to personal experience and preference of individuals (Pusch, 1966; Smart, 1969; Barden and Sides, 1971; McKyes and Yong, 1971). The general requirements in regard to preparation and viewing of samples, however, remain similar (see Fig. 3.15).

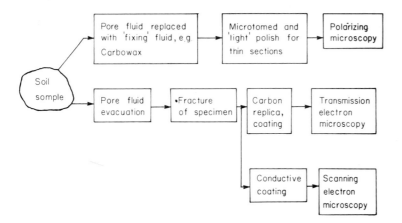

Fig.3.15. Flow diagram showing procedure in certain fabric viewing techniques.

   The preparation of samples for viewing, using any one of the three techniques, requires that extreme care be exercised to preserve the fabric. Since the samples are small, it is understood that a representative area or volume of the test sample must be selected.
   Preparation of thin sections for study in the polarizing-light microscope requires that the water in the clay sample be exchanged with sulfonated alcohol or polyethylene glycol. Volume changes in the test samples due to replacement technique (penetration and subsequent drying of the exchange fluid) must be minimized. The thin sections obtained by microtome and polishing can be suitably mounted on glass slides for viewing as in Fig. 3.15.
   Electron bombardment of the surface of the sample using scanning techniques in the electron microscope will cause local charging of the bombarded site; this would not only interfere with the viewing procedure but will also damage the delicate instrumentation. The necessity for supporting the sample on a conducting mount and for coating the surface of the sample with a conductive coat of silver or gold palladium becomes obvious, as shown in Fig. 3.15. Since a high vacuum is needed for this part of the procedure, the sample must be dry. As in other techniques, the problem of volume changes arising from preparation

Fig.3.16.

Fig.3.17.

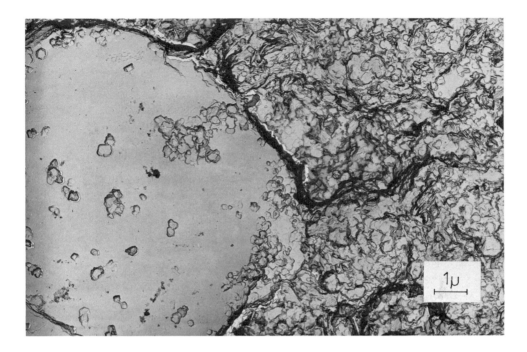

Fig.3.18. Transmission electron micrograph of carbon replica of sample of Edgar kaolin used to obtain picture shown in Fig.3.17. Note that the clearer resolution at higher magnification allows for identification of shapes of the smaller particles.

techniques becomes very important. The general procedures include either freeze-drying, slow high vacuum dessication, critical point drying, or polyethelene glycol impregnation. The general procedures for sample preparation and viewing are described in detail in specialized publications and manuals, e.g., Gillott (1968).

A comparison of the kinds of information obtained from the three different techniques is shown in Fig. 3.16—3.18. The ped units and arrangements are easily seen in the light micrograph picture of the kaolin thin section.

Fig.3.16. Microscopic view of fabric using polarizing microscopy. Sample shown is a thin section cut from a carbowax impregnated kaolin sample tested in triaxial shear. Diagonal lighter coloured band is the shear plane. Note light "specks" which represent fabric units oriented in different direction from other groups of fabric units which appear darker coloured (see Fig.3.19 for mechanism of retardation demonstrated in this picture). Sample viewed under crossed nicolls.

Fig.3.17. Scanning electron micrograph of Edgar kaolin showing large particles in the left part of the picture. Note indistinguishable shapes of smaller particles in the background and on the larger particles.

3.7 QUANTIFICATION OF FABRIC

The fabric pictures show that it is difficult to establish an absolute measure of fabric, even in terms of a numerical value on a scale representing fully oriented to fully random array for fabric units and particles. The very many different kinds of second-order characterization, when combined with the equally varied and complex first-order characterization, make for a very difficult and tedious procedure in quantifying fabric.

One way in which a numerical value for fabric can be obtained is to use the birefringent properties of most clay minerals in conjunction with polarizing-light microscopy. The birefringent nature of the clay mineral causes the incoming polarized light to be split into two components, both of which are polarized, but at right angles to each other (the ordinary and extraordinary rays, or o and e-rays—see Fig. 3.19). Since the light travels at different velocities along the different axes of the mineral, on exit these two rays will still be vibrating at right angles to each other, but one will have lagged behind so that they will produce interference with each other at the analyzer. In the analyzer both will be repolarized into the same plane, but they will retain the initial phase separation. This phase separation, known as the retardation (Δ) is a function of the indices of refraction of the mineral (n_1 and n_2), whose directions are normal to the path of the incident light and to each other. The functional relationship may be described by the equation:

$$\Delta = t\,(n_1 - n_2)$$

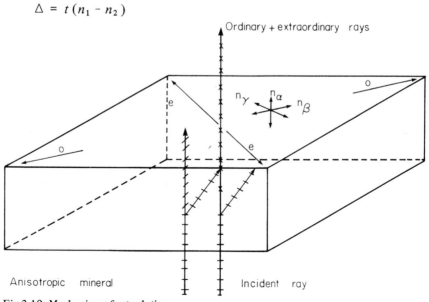

Fig.3.19. Mechanism of retardation.

where: Δ = retardation, t = thickness of the sample under examination, n_1 and n_2 = the pertinent indices of refraction of the mineral.

The actual value of the retardation may be varied through a wide range by: (a) varying the thickness of the mineral; (b) changing the operative values of n_1 and n_2. This may be accomplished by changing the orientation of the sample, a process which would allow the values of n_1 and n_2 to range between n_α and n_γ as an upper bound, and between n_α and n_β as a lower bound (see Fig. 3.19). It is this latter that may be used to determine the orientation of a known mineral within a soil. Since the thin sections prepared for polarizing microscopy will consist of many particles, the thickness of the thin section must be considered as equivalent to that of one mineral, i.e., the thickness of the thin section represents the thickness of an equivalent large representative mineral. The orientation of the representative mineral may be deduced from the measured value of the retardation by considering that the values of the pertinent indices of refraction (n_1 and n_2) may (a) vary between the limits of n_β and n_γ of the equivalent representative mineral, when the incident light follows a path coinciding with the axis of optical isotropy, (b) vary between the limits of n_γ and n_α or n_β, when the path of light coincides with an axis of maximum anisotropy. For intermediate paths, the critical values of the indices of refraction vary directly with the angle of inclination of the particle's axis of isotropy with the path of the light. Thus with the assumption of equivalent representative mineral, the angle of inclination may be found from:

$$\Delta = t\left[n_\gamma - \left(n_\alpha + \frac{\beta\theta}{90}\right)\right]$$

where β = birefringence of the mineral ($n_\gamma - n_\alpha$), and θ = angle of inclination of the mineral.

The use of retardation measurements, or birefringence of clay soils in establishing a measure of fabric orientation has been developed in greater detail and sophistication by Sheeran (1972) and Lafeber (1967).

Another technique for quantifying fabric is to use X-ray diffraction methods (Martin, 1966; Moore, 1968). This technique uses the intensity of X-ray diffraction to indicate particle orientation (Chapter 2). If the standard diffractometer is used, only one degree of specimen rotation freedom is obtained. However, if a pole figure device is used (Martin, 1966) the specimen examined can be rotated in two directions in the vertical plane in addition to horizontal rotation. This allows for three-dimensional sampling of particle orientation.

The X-ray diffraction patterns shown in Fig. 3.20 demonstrate the effect of ordered arrangement of the mineral particles on the peak heights of the diffraction trace.

To obtain the orientation index (Diamond, 1971), peak ratios as indicated in

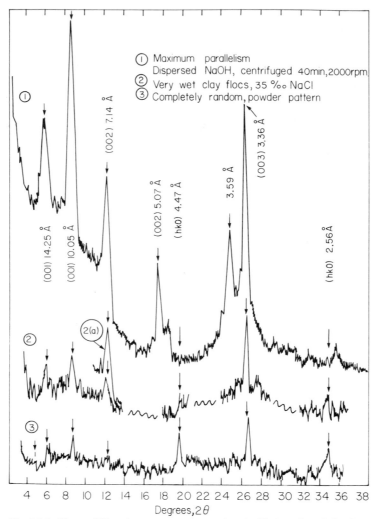

Fig.3.20. X-ray diffraction patterns for a glacial till, less than 0.4 microns. (From Quigley, 1961.)

the example shown in Table 3-2 are obtained. The degree of orientation which is defined as:

$$\text{Degree of orientation} = \frac{\begin{bmatrix} \text{orientation index} \\ \text{of} \\ \text{sample surface} \end{bmatrix} - \begin{bmatrix} \text{orientation index} \\ \text{of} \\ \text{random mount} \end{bmatrix}}{\begin{bmatrix} \text{orientation index} \\ \text{of full} \\ \text{oriented mount} \end{bmatrix} - \begin{bmatrix} \text{orientation index} \\ \text{of} \\ \text{random mount} \end{bmatrix}}$$

TABLE 3.2

X-ray orientation indices for impact-compacted kaolinite specimens
(From Diamond, 1971)

Orientation of surface with respect to axis of compaction	Moisture content (%)	Peak ratio[1] (001)/ (020)	S.E.[2]	Peak ratio[1] (002)/ (020)	S.E.[2]	Peak ratio[3] (001)/ (060)	S.E.[4]	Peak ratio[3] (002)/ (060)	S.E.[4]
Normal	30	5.91	0.11	2.99	0.09	10.73	0.22	5.59	0.14
	26	6.64	0.17	3.30	0.04	10.94	0.35	5.59	0.11
	22	6.40	0.37	3.05	0.14	10.35	0.35	5.14	0.12
Parallel	30	4.60	0.17	2.22	0.07	7.31	0.24	3.60	0.15
	26	4.39	0.22	2.04	0.09	7.97	0.30	3.95	0.17
	22	4.05	0.11	2.19	0.09	6.84	0.15	3.50	0.08

[1] Mean of eight replicate determinations, each on a separate surface.
[2] Standard error of the mean of eight determinations.
[3] Mean of sixteen replicate determinations, each on a separate surface.
[4] Standard error of the mean of sixteen determinations.

may be obtained by using the orientation indices for specially prepared fully
random and fully oriented kaolinite fabric arrangements shown in Table 3.3.

Assessment of fabric and soil structure roles in soil response behaviour should
rely on more than one identification and quantification technique. The procedures
given in Fig. 3.21 show the methods available for the study of soil fabric and
structure in assessment of soil behaviour. The need for a proper modelling of the

TABLE 3.3

X-ray orientation indices for "random" and for "fully-oriented" kaolinite fabric[1]
(From Diamond, 1971)

Specimen type	Peak ratio (001)/ (020)	S.E.	Peak ratio (002)/ (020)	S.E.	Peak ratio (001)/ (060)	S.E.	Peak ratio (002)/ (060)	S.E.
Random powder mount	2.28	0.04	1.27	0.03	4.41	0.07	2.45	0.03
"Fully-oriented" layer	118.3	5.2	76.9	3.9	170.0	6.9	110.0	4.8

[1] Data are means and standard errors of the means of eight replicate determinations, each on a
separate specimen.

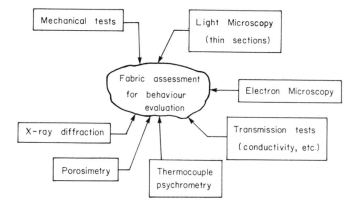

Fig.3.21. Schematic representation showing fabric identification input for assessment of soil fabric.

physics of soil behaviour establishes the need for a proper understanding of the components of interaction in the discrete system. At least two methods should be used. The application of many of the fabric and structure concepts will be shown in later chapters.

3.8 FABRIC CHARACTERISTICS FROM SEDIMENTATION

Fresh-water deposition

Deposition of sedimentary clays can occur in either fresh water (rivers and lakes) or in a marine environment. In fresh-water deposition, dissolved salts are assumed to be absent or have minimal effect. This situation lends itself to the optimum development of forces of repulsion between clay particles. For highly-active clays such as montmorillonite, vermiculite and chlorite (in decreasing order of activity), these forces of repulsion can be highly developed. The forces of attraction will be small by comparison, and particle arrangement will in general be dictated by the forces of repulsion. To show how this will affect particle arrangement, we consider the interaction of the diffuse ion-layers of two adjacent parallel particles or domains as shown in Fig. 3.22. The net repulsion is proportional to the midplane potential, i.e. the potential midway between the parallel particles or domains. The particles or domains in effect repel each other because of the interpenetration of their diffuse ion-layers. In schematic form, we might depict this action as arrows pointing away from the particles (Fig. 3.23), which portrays the effective action between particles and domains. If this action

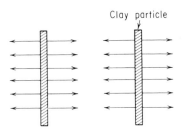

Fig.3.22. Interaction of diffuse ion-layers giving rise to swelling pressure.

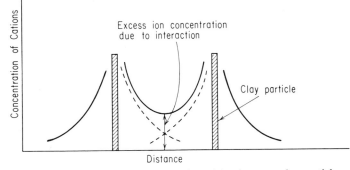

Fig.3.23. Schematic representation of repulsion between clay particles.

occurs between all the particles and domains sedimenting in fresh water, the resultant fabric obtained would be semi-oriented.

There are many other clay minerals where repulsive forces are not as dominant as in montmorillonite or vermiculite, e.g., kaolinite and illite. The fabric obtained in fresh-water deposition would still be semi-oriented but with a smaller void ratio as compared to the more active clays. Domains will tend to be more dominant and will be less oriented in view of their size. The degree of orientation of other particles will vary in accordance with presence of domains. The relative difference in fabrics for clays deposited in fresh water is shown in Fig. 3.24.

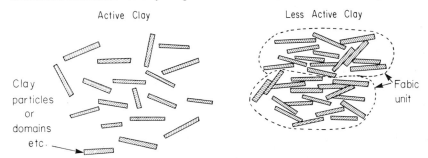

Fig.3.24. Fabric of clay deposited in fresh water.

The diagrams shown thus far have been conspicuous in that no silt particles are shown. This has been done specifically to present the theory underlying arrangement of clay particles. Silt particles incorporated into the systems shown will serve as focal points for orienting clay particles around them. The resultant effect is shells of clay particles encasing silt particles.

Deposition in salt water

The presence of dissolved salts in a marine environment serves to depress the diffuse ion-layer. This can be shown schematically for the active clays, where the diffuse ion-layers are well developed, using the model of two adjacent parallel particles (Fig. 3.25). .

With the depression of the diffuse ion-layer, interparticle repulsion is decreased. This allows the attractive forces to become more dominant. Deposition

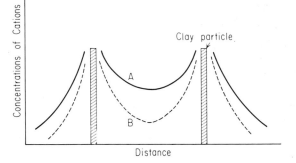

Fig.3.25. Interaction of diffuse ion-layers. *Curve A* = interaction in fresh water or solution with low salt concentration; *curve B* = interaction in solution with high salt concentration.

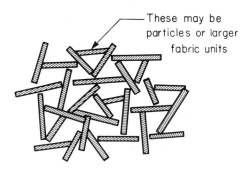

Fig.3.26. Typical edge-to-surface flocculated fabric characteristic of marine deposition.

in salt water (marine environment) tends to give a characteristic flocculated structure. The resultant fabric becomes more open with typical edge-to-surface arrangements (Fig. 3.26).

3.9 FABRIC ALTERATION BY COMPACTION AND COMPRESSION

In laboratory testing, the original structure and fabric of the test clays may be retained or reworked. In undisturbed samples, some change in fabric, and hence, in structure can occur because of sample disturbance either during sampling or during extrusion from the sampling tubes. In remoulded clays, compaction and altering water contents will induce preferred fabrics.

If the standard method of compaction (ASTM D698-58T), is used for a clay soil with water contents appreciably lower than the optimum water content, a randomly oriented fabric is obtained. Increasing the moulding water content will reduce the randomness in particle arrangement. If one compacts at water contents above optimum, the compacted clay soil would have a partially oriented fabric. The greater the moulding water content, the more oriented the fabric would be. Fig. 3.27 shows pore-size distribution curves for compacted soils influenced by moulding water content and compactive effort (Diamond, 1971).

Fabric changes due to application of external stresses will be discussed in later chapters to illustrate initial fabric control on final performance of the material. Fig.

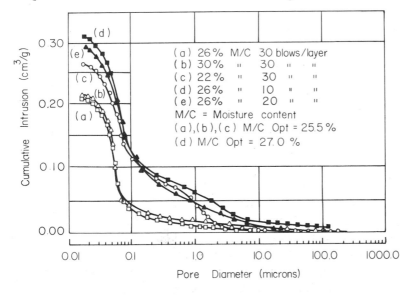

Fig.3.27. Pore-size distribution curves for impact-compacted kaolinites. (From Diamond, 1971.)

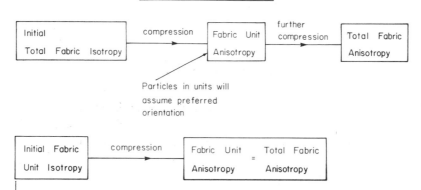

Fig.3.28. Flow diagram showing fabric change in one-dimensional compression.

3.28 shows the sequence achieved in one dimensional compression of isotropic and anisotropic fabrics. The electron photomicrographs for these will be seen in Chapter 7.

3.10 SUMMARY

One of the most important parameters in the study of soil behaviour is soil structure. The importance of an understanding of the basic soil fabric models cannot be overemphasized. While it may not be easy to obtain direct quantitative measures of soil fabric and soil structure, there are ways in which the basic fabric units can be evaluated and identified. The use of electron photomicrographs and other techniques such as X-ray diffraction and the study of thin slices, have provided us with an appreciation of formation and existence of fabric units which interact to provide for overall system stability. The mechanics of interaction of the units will describe the response of the soil system to probing stresses.

From our understanding of the nature of the interparticle forces of attraction and repulsion and of clay—water interaction, it is possible to postulate the fundamental fabric units. If interparticle repulsion is reduced, the role of the attractive forces becomes more important. In some clays, because of the existence of positive edge charges it is possible to create bonding between the edge and surface, of such a nature that this will be more stable than that which can exist between particles in the surface-to-surface arrangement. In natural clays cementation will occur, which produces bonds stronger than those from interparticle attraction.

SOIL WATER

4.1 INTRODUCTION

Soils have many characteristics which make them different from other materials familiar to the engineer and scientist. The most striking of these is the large change in soil properties with change in water content. Properties such as strength, compressibility, plasticity, and hydraulic conductivity change markedly with changing water content. The volume of many soils also changes with changing water content.

Water is retained in the voids of a soil; a changing water content results from a changing proportion of water and air in the voids or from a changing volume of voids. Water is held in soils against gravitational forces draining water out or against evaporation of water in drying. This water retention is due to capillary forces arising from curved air–water interfaces in the voids, or due to surface forces bonding water molecules. The capillary forces depend upon the size of voids, and the surface forces upon the amount and nature of the surfaces of soil grains.

In this chapter we will discuss the measurement of water content of soils, the nature of the forces holding water in soils, and the measurement of this energy of water retention. The movement of water in soils is discussed in the next chapter.

4.2 WATER CONTENT AND ITS MEASUREMENT

Definitions

The water content of a soil on a weight basis, ω, is defined as the grams of water contained in the sample per gram of oven-dry soil, multiplied by 100 to convert to per cent. The moist soil sample is weighed:

$$\omega = \frac{W_w}{W_s} \times 100$$

dried to constant weight at 105°C to 110°C, cooled in a dessicator and weighed again to get W_w, the water lost, and W_s. This is the standard method and is the measurement against which all indirect methods are calibrated. It has the disadvantage of being time consuming, since drying can take from eight to 36 hours. The calculations are routine and can be programmed for a computer if large numbers of samples are involved. A further disadvantage is the small volume of soil sampled, usually less than 50 cm^3. Since the soil is heterogeneous, and water content varies by several per cent from place to place even within a cubic yard, this method may lead to a serious error in estimating water content of a large volume of soil. This can be overcome only by taking a large number of samples. Repeated sampling in the same place is also impossible because samples are removed, i.e., the measurement is destructive.

Water content on a volume basis, θ, is defined as volume of water per volume of moist soil:

$$\theta = \frac{V_w}{V_s + V_v}$$

where: V_s (solid volume) + V_v (void volume) = V (total volume). This is a more useful expression for irrigation and drainage calculations, and for some theoretical considerations of water retention and flow in soils. It is the fraction of the soil volume occupied by water, and it can be converted directly to centimetres of water per metre of soil (or inches/foot).

The relationship between ω and θ is:

$$\theta = \frac{\omega}{100} \times \gamma_d \times \frac{1}{\gamma_w}$$

where γ_d is the bulk density or dry density, defined as the weight of soil divided by the total volume of the soil:

$$\gamma_d = \frac{W_s}{V_s + V_v}$$

and γ_w = density of water.

The volumetric water content is usually obtained by measuring ω and γ_d, and calculating θ. If a known volume of soil is sampled, θ can be measured directly since the density of water is known. The bulk density, γ_d, can be obtained by measuring the volume occupied by a known weight of soil, or by measuring the weight of soil in a known volume.

For non-shrinking soils, the bulk density does not change with water content, and the calculations are straightforward. For clay soils, however, where the volume decreases as water content decreases and hence bulk density increases, the bulk density must be measured at each water content. Fig. 4.1 illustrates the changes. The calculations are as follows: Assume that a soil has a water content $\omega = 38\%$ and

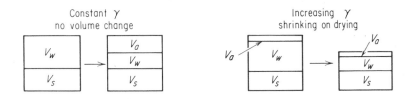

Fig.4.1. Changes in volume on drying wet soils.

a bulk density γ_d = 1.31 g/cm³. The volumetric water content θ = (38/100) × 1.31 × (1/1.00) = 0.50. The soil at this water content is nearly saturated, i.e., the voids contain little air. This is calculated as follows: If the density of the soil grains is 2.68 g/cm³, the 1.31 g of solid in 1 cm³ of soil occupies 1.31 g/(2.68 g/cm³) = 0.49 cm³. Since water occupies 0.50 cm³, only 0.01 cm³ is filled with air. At ω = 25%, θ would be 0.33 and 0.18 cm³ would be filled with air.

If, however, the soil shrinks on drying from ω = 38% to ω = 25%, the bulk density increases, and the change in θ is not as large. If the soil shrinks by an amount equal to the water loss, i.e., 0.13 cm³ water per gram of soil, then γ_d becomes 1.31/[1 - (1.31 × 0.13)] = 1.58 g/cm³. Then θ = 0.40.

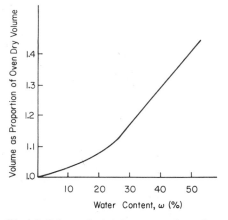

Fig.4.2. Schematic shrinkage curve for a clay soil.

Bulk density or dry density can be measured on samples taken at different water contents, or for some purposes the change in bulk density can be estimated from the shrinkage curve for a soil. An example of such a curve is shown in Fig. 4.2, from which the change in volume for a changing water content can be calculated.

Indirect measurement of water content

Drying and weighing to determine water content has the disadvantage that it is time consuming and destructive, i.e., the sample is changed by the measurement. Drying can be speeded up by using an infrared heat source, and commercial units are available which have a weighing balance built in with the drying unit.

The disadvantages of drying and weighing to determine water content have led to many studies and to many devices for indirect measurement of soil water content. Many methods are available. Any soil property which changes with water content can be the basis of an indirect method of measurement. These methods are best suited to measuring changes in water content of a soil with time. Being indirect methods, they require calibration with actual water content. These calibration curves can be influenced by soil properties other than water content, and hence one calibration curve may not offer sufficient precision for spot measurements of water content of different soils. While the slope of the calibration curve may remain the same, the intercept usually changes. When used for measuring changes in water content with time in one soil, these differences in soil properties are not involved, and the method is more satisfactory.

The best such method for measuring water content in undisturbed soil in the field is with the neutron moisture probe. While it has disadvantages, this method has been used increasingly since about 1955. The method is based on the fact that fast neutrons in moist soil are slowed down primarily by the small hydrogen atoms. Most of the hydrogen in soils exists as part of water molecules, although some is part of organic molecules. The neutron moisture probe consists of a source of fast neutrons, e.g., radium–beryllium, and a counter sensitive only to slow neutrons. The fast neutrons are slowed and scattered by collision with hydrogen atoms. The proportion of neutrons returning to the counter is related to the water content. The geometry of placing the source and counter determines the proportion of neutrons returning and the volume of soil sampled.

The energy lost by a neutron in a collision depends upon both size and capture cross section of the atom. Although H is the only small atom in soils, several atoms such as B (boron), K (potassium) and Fe (iron) have relatively high capture cross sections. They can be present in some soils in sufficient amounts to absorb energy from the neutrons and hence to change the calibration of the instrument.

The apparatus and a representative calibration curve are shown schematically in Fig. 4.3. An access tube of aluminum or thin stainless steel is installed in the soil. The neutron source and counter are lowered into the tube and readings taken at various depths. A fairly large volume of soil, about 30 cm in radius, is sampled, and readings can be repeated in the same place. Because of the uncertainty of the

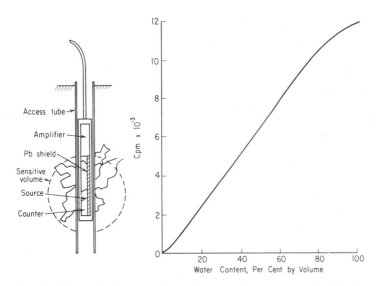

Fig.4.3. Schematic diagram of neutron moisture probe and calibration curve for Nuclear Chicago depth probe. (Nuclear Chicago Corp.)

calibration curve, which can vary with the soil, the neutron probe is best used to follow changes in water content during the season in a soil. It is not as useful for spot measurements of water content in different soils.

Another indirect method for measuring water content, used largely in the laboratory, is gamma-ray attenuation. The gamma rays from a source such as 137^{Cs} at one side of a soil column pass through the soil and are measured by a counter at the other side. Attenuation or absorption of gamma rays is by soil, water and container:

$$I = I_0 \exp\left(-\mu_s \gamma_d X_s - \mu_w \theta X_s - \mu_c X_c\right)$$

where: I = intensity counted; I_0 = Unattenuated intensity; μ = mass-absorption coefficient of soil (s), water (w) and container walls (c), respectively; γ_d = bulk density or dry density of soil; θ = volumetric water content; X = thickness of soil column (s) and container walls (c) respectively. The container term is constant, and if the soil mass does not change, the soil term is also constant. Then:

$$I = I_0 \, e^{-k_1 \theta - k_2}$$

where k_1 is a constant depending upon μ_w and X_s; k_2 is a constant depending upon μ_s, γ_d, X_s, μ_c and X_c. Intensity is then proportional to water content. This is a very useful method for measuring changing water contents in water movement studies in the laboratory. If the soil shrinks during drying, γ_d and possibly X_s are no longer constant. Measurements can then be made using two gamma-ray sources which have different mass absorption coefficients.

Several methods of measuring soil water content have been suggested which measure products of reaction of a chemical with water. When calcium carbide is mixed with wet soil, acetylene is formed:

$$CaC_2 + H_2O \rightarrow Ca(OH)_2 + C_2H_2$$

If this is done in a closed vessel, the resulting pressure from the acetylene gas produced can be measured with a diaphram and gauge. Pressure is directly related to water content. Such apparatus is available with a pressure gauge to measure the gas evolved. A weighed sample of wet soil and calcium carbide are mixed in the closed container. This method works best with granular soils.

Methods based upon measurement of such physical properties as dielectric constant or thermal diffusivity of soils are indirect measurements of water content. They have been described but are not widely used. Other indirect measurements are more closely related to suction than to water content and are discussed in Section 4.5.

The choice of the particular water content measurement to use depends upon factors such as accessibility of soil area, precision required, number of samples required and use to be made of the results. If the water content of a large area of soil is required, the precision will depend mainly upon the volume and number of samples taken. A small sample of an inherently inhomogenous material such as soil cannot provide a very reliable value for water content of a large soil mass.

4.3   THE CONCEPT OF SOIL-WATER POTENTIAL

Buckingham's capillary potential

The energy with which water is held in a soil at any water content can be specified as the soil-water potential. Buckingham (1907) explored the use of the energy concept, or the soil-water potential concept, in his study of movement of soil water. He defined the capillary potential of the soil as the work required per unit weight of water to pull water away from the mass of soil, with the idea that this potential is due to capillary forces holding water. This potential decreases as the water content increases, i.e., the water is held more strongly by dry soil than by wet soil. The movement of soil water from a point of low potential to one of higher

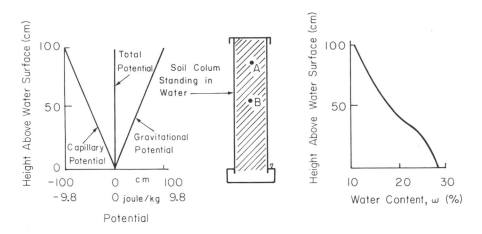

Fig.4.4. Illustration of potential concept of soil water.

potential could then be considered analogous to the flow of heat in a conductor. The subsequent study of the relationship between potential and water content by many workers is one of the significant contributions in soil physics.

Buckingham defined the potential in terms of the soil, but it has been found to be more convenient to define the potential in terms of the water phase, and speak of soil-water potential. This changes the algebraic sign; a drier soil has a lower soil-water potential, or the free energy (eq. 4.2) of the water is lower.

The energy concept can be visualized by analysing one of Buckingham's experiments. A column of dry soil with its lower end placed in a pan of water will be moistened by water moving up into the dry soil. If the top of the column is covered to prevent evaporation, equilibrium will be established. At equilibrium the water content will decrease with height as shown in Fig. 4.4.

A sample of soil at point B has a water content of about 17%. If this sample is transferred from point B to point A, it will lose water until it is at equilibrium with the water content of about 12% at height A. The capillary potential of the soil decreases, i.e., becomes a larger negative number. With the definition based on the water phase, the capillary potential is always a negative number, becoming more negative as the soil becomes drier.

Referring again to Fig. 4.4, when water was added at the bottom of the dry soil column, water moved upward into drier soil where it had a lower free energy. Water moved along a gradient of soil-water potential. Since the water is also subject to the gravitational force, it would move downward in response to a gravitational potential. At equilibrium, by definition, the total potential must be the same at all points, since water does not move. The point of zero potential for soil-water systems is defined as a free water surface. This is the reference potential. At equilibrium the difference in total potential between water at any point in the soil column and in the free water is zero. Therefore, the sum of the capillary $\Delta\psi_c$, and gravitational, $\Delta\psi_g$, potentials is zero:

$$\Delta\psi = 0 = \Delta\psi_c + \Delta\psi_g = \Delta\psi_c + g\Delta h \tag{4.1}$$

where g = gravitational constant, and Δh = height above free water surface.

Since the gravitational potential can be calculated at any point above the free-water surface (eq 4.1) the capillary potential may be found:

$$\Delta\psi_c = -g\Delta h$$

These potentials are plotted in Fig. 4.4. The units of $g\,\Delta h$ could be erg/g but the numbers using joule/kg, would be 10^4 smaller which is more convenient. These are both energy units. The potential can also be expressed as a head by dividing by g. Units of cm head are also shown in Fig. 4.4.

The algebraic sign conventions adopted for these calculations are:

(1) In calculation of the gravitational potential, height is taken as positive above the free water surface.

(2) Capillary potential is defined as the work required to move a unit mass of water form the free water surface of zero potential to the point in the soil. The capillary potential is always a negative number. This is opposite in sign to Buckingham's original usage.

Component potentials

The total soil-water potential was divided into capillary and gravitational components in Fig. 4.4. The capillary potential is now termed the matric potential. Salts in the soil also decrease the water potential, so an osmotic potential can be added. Other component potentials can also be identified. Care must be taken to ensure that the component potentials are independent. The component potentials can be written as: $\Delta\psi = \Delta\psi_m + \Delta\psi_g + \Delta\psi_\pi + \Delta\psi_p + \Delta\psi_a + \ldots$ The value of specifying component potentials is that it ensures that parts of the soil-water potential are not overlooked. Also, different methods of measuring potential do not all measure the same components.

Potentials are given as differences from an arbitrary point defined as having zero potential. A free water surface is used as zero potential, but this point is defined in various ways, depending upon need. Here we shall use zero potential as that of a pool of free water under a pressure of one atmosphere. It can be further specified that the pool be at the same elevation as the soil and that the temperature be the same.

The various potentials are defined as follows:

(1) $\Delta\psi$, the total potential, is the work required to transfer a unit quantity of water from the reference pool to the point in the soil. It is a negative number.

(2) $\Delta\psi_m$, the matric potential, is a soil matrix property. This is the equivalent of Buckingham's capillary potential. It is the work required to transfer a unit quantity of soil solution from a reference pool at the same elevation and temperature as the soil, to the point in the soil. ψ_m cannot be calculated except for uniform spheres where it is related to curvature of air—water interfaces and $\psi_m = S[(1/r_1) + (1/r_2)]$, or in freely-swelling clay plates where $\psi_m = RT$ cosh $(y_c - 1)$. S = surface tension of air—water interface; r_1, r_2 = radii of curvature of air—water interface; R = universal gas constant; T = absolute temperature; y_c = electrical potential midway between two clay plates (see Chapter 2).

ψ_m can be subdivided into a component due to swelling forces and a component from air—water interface forces, but these components cannot be separated experimentally.

(3) $\Delta\psi_g$, the gravitational potential, is the work required to transfer water from the reference elevation to the soil elevation.

$$\psi_g = -\gamma_w g h$$

where γ_w = density of water.

(4) $\Delta\psi_\pi$, the osmotic potential, is the work required to transfer water from a reference pool of pure water to a pool of soil solution at the same elevation, temperature, etc.

$$\Delta\psi_\pi = n\,RTc$$

where: n = number of molecules per mole of salt; R = universal gas constant; T = absolute temperature; c = concentration of salt.

(5) $\Delta\psi_p$, the piezometric or submergence potential, is the work required to transfer water to a point below the water table.

$$\Delta\psi_p = \gamma_w g d$$

where d = depth below free water level.

(6) $\Delta\psi_a$, the pneumatic or a pressure potential, refers to transfer of water from atmospheric pressure to the air pressure, P, on the soil.

$$\Delta\psi_a = P$$

The three component potentials, matric, piezometric and pneumatic are often taken together as the pressure potential, $\Delta\psi p$.

$$\Delta\psi_p = \Delta\psi_m + \Delta\psi_p + \Delta\psi_a$$

Formulation from reversible thermodynamics

Buckingham used mechanical potentials, but recognized that chemical potentials might be required later. There are many formulations based upon thermodynamic reasoning. At equilibrium, the partial molar free energy of water is everywhere the same.

The Gibbs free energy, G, defined as follows will be used here:

$$G = U + PV - TS$$

where: U = internal energy of the system; P = pressure; V = volume; T = temperature; S = entropy. This can be differentiated and rewritten using the first and second laws of thermodynamics as:

$$dG = VdP - SdT + \Sigma\mu_i dn_i + \Sigma\, Ydx$$

where: μ = chemical potential $= \overline{G} =$ partial molar free energy; n = molecules; Y = force fields. This can be rewritten in terms of the partial molar free energy:

$$d\overline{G}_i = \overline{V}_i dP - S_i dT + \underset{j}{\Sigma} \left(\frac{\partial \overline{G}}{\partial n_i}\right) dn_j + \Sigma\overline{Y}_i dx \qquad (4.2)$$

This is the "parent equation", which can be used in different ways depending upon the independent variables selected. This presents a difficulty because all variables must be accounted for, but no variable must be contained as part of another and hence counted twice.

Considering the soil as a three-phase system on a microscale, the equation can be written as:

$$d\overline{G}_w = \overline{V}_w dP_p - \overline{S}_w dT + dn + \overline{V}_w dP_s + Mgdz \qquad (4.3)$$

where P_p is air pressure; P_s is soil suction which is numerically equal to soil-water potential but opposite in algebraic sign. Usually this equation is used at constant temperature. This is equivalent to the component potential equation.

More complete descriptions of component potentials and formulations are available in specialized books and papers on soil water such as Bolt and Miller (1958), and Hillel (1971).

Formulation from irreversible thermodynamics

Systems in which one driving force can cause more than one flux, can be conveniently described by phenomenological relations derived in thermodynamics of irreversible systems. For example, a temperature gradient in a soil may cause flow of heat, water, and possibly electric current and salt. If a system is not too far from equilibrium, forces, F, and fluxes, J, are related by linear relationships.

$$J_1 = L_{11}F_1 + L_{12}F_2 + L_{13}F_3 \ldots$$

$$J_i = \sum_n L_{ik}F_k$$

where L are the phenomenological coefficients. These equations are derived on the basis of entropy production:

$$\frac{\partial \Delta S}{\partial t} = \sum_i J_i F_i$$

For example, for simultaneous flow of heat, electricity, solutes, and water, we have:

$$J_H = -L_{HH}\frac{\Delta T}{T^2} - L_{He}\Delta\left(\frac{E}{T}\right) - L_{Hs}\Delta\left(\frac{\mu_s}{T}\right) - L_{Hw}\Delta\left(\frac{\mu_w}{T}\right)$$

$$J_e = -L_{eH}\frac{\Delta T}{T^2} - L_{ee}\Delta\left(\frac{E}{T}\right) - \ldots$$

$$J_s = -L_{sH}\frac{\Delta T}{T^2} - \ldots$$

$$J_w = -L_{wH}\frac{\Delta T}{T^2} - L_{we}\Delta\left(\frac{E}{T}\right) - L_{ws}\Delta\left(\frac{\mu_s}{T}\right) - L_{ww}\Delta\left(\frac{\mu_w}{T}\right)$$

where "H" refers to heat; "e" to electricity; "s" to solute and "w" to water. Under isothermal conditions, in the absence of electrical and concentration gradients, the last equation reduces to:

$$J_w = -\frac{L_{ww}}{T}$$

which is Darcy's law (Chapter 5).

Applications of these equations to fluxes in soils can be found in Hillel (1971).

Geometric concepts and soil-water potential

Water retention in soils has so far been considered from the standpoint of potentials. A complete specification requires also a physical description of void geometry and its influence on water retention. In early geometric concepts developed in the late 1800's, soil voids were considered as bundles of capillary tubes. Water was held in them by surface tension forces, and the equation for capillary rise was applied to soils (Fig. 4.5).

$$h = \frac{2T\cos\alpha}{r\gamma_{w}} \tag{4.4}$$

where: h = height of· water rise in the capillary; T = surface tension of water; α = angle of contact of water and soil; γ_{w} = density of water; r = radius of capillary.

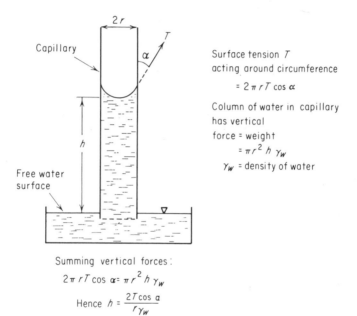

Fig.4.5. Height of rise in a capillary.

If, the values of r for average void size in soils are substituted into this equation, calculated values of h are from fifty to several hundred feet. For example, voids of 0.01 mm diameter are common in soils. The calculated capillary rise would be:

$$h = \frac{(2)\,(73\text{ dynes/cm})\,(1)}{(980\text{ cm/sec}^{2})\,(1\text{ g/cm}^{3})\,(0.0005\text{ cm})} \simeq \frac{0.15}{r} = 300\text{ cm}$$

Laboratory experiments on height of rise of water in soil columns give maximum values considerably less than this. Measurements of movements of water up from a water table confirm the laboratory experiments.

The contact angle between water and soil particles is generally taken to be equal to zero, but finite contact angles are found for some soils and some soil components, which are difficult to wet. Certain sandy soils with organic matter fall into this category.

Movement of water in soils according to the capillary tube concept should be described by the Poiseuille equation (see Chapter 5) for flow through a tube. This was applied to soils (see Keen, 1931, for a discussion of these studies) but it neither fitted experimental measurements very well nor predicted anything new about water flow.

The basic weakness of the capillary tube concept is that soil voids are not bundles of capillaries of different sizes, but are cellular voids interconnected through openings of various sizes. The description of this system requires a statistical approach which allows for the different sizes and different interconnections. The further development of geometric concepts was to approximate soil particles with uniform spheres. If a soil is assumed to consist of spheres of uniform size in a reproducible and constant system of packing, the geometry of the pore space can be specified. This can illustrate water retention in soils. The closest packing of spheres of a uniform size results in 26% pore space. A regular open packing in which spheres are arranged in rows, results in 48% pore space. With irregular random packing, voids can occupy over 80% of the volume.

From a consideration of the shape of pores between close-packed spheres Haines (in Keen, 1931) related water content to pressure calculated from the curvature of the air—water meniscus (Fig. 4.6). The pressure difference across a spherical curved air—water surface is given by the equation:

$$\Delta P = 2T/r$$

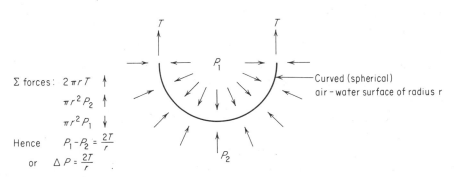

Σ forces: $2\pi r T$ ↑

$\pi r^2 P_2$ ↑

$\pi r^2 P_1$ ↓

Hence $P_1 - P_2 = \dfrac{2T}{r}$

or $\Delta P = \dfrac{2T}{r}$

Curved (spherical) air–water surface of radius r

Fig.4.6. Pressure difference across a spherical curved air—water surface.

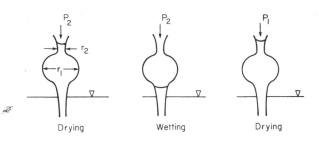

For $r_1 = 0.3$ mm and $r_2 = 0.05$ mm, P_2 must be > 30 cm

for drying and $P_1 < 5$ cm for complete wetting

Fig.4.7. Illustration of hysteresis in capillaries of varying sizes.

This is the pressure which must be applied to remove water from the sample, and hence it is the potential of the soil water. At low water contents, the water exists as rings at the points of contact between spheres. These rings increase in size until they touch at certain points forming connected wedges of water. With further increases in water content, water fills the voids and excludes the air. This happens when the pressure is low enough, i.e., when the potential is nearly zero. The values of pressure at which these changes occur can be calculated for spheres of known simple packing. On drying, the same water contents are reached at different pressures. Drying, or emptying of a series of interconnected pores does not occur until the pressure is high enough to overcome the surface tension forces in the pore with the smallest diameter (Fig. 4.7). Complete wetting does not occur until the pressure is so low that the largest pore will fill. Therefore, at the same pressure or at the same potential a sample has a higher water content on drying than on wetting. This is a manifestation of hysteresis, which is a general term indicating that a property of the system is influenced by its previous history or treatment.

While results for uniform spheres in regular packing do not apply to soils, the experiments did demonstrate that water retention in sands was explained by surface tension forces and that hysteresis was to be expected. Entrapped air would decrease the amount of water retained by a soil. It was also shown that an analysis of the retention curve would give the pore-size distribution of the soil. Values of pressure are equivalent of height to rise in the capillary rise eq. 4.4, and, therefore, values of pressure can be converted to values of radius of pore which would be emptied at that pressure. From the amount of water held between two values of pressure, the amount of pore space between two sizes can be calculated and hence the pore-size distribution of the soil calculated.

For example, if 6 cm^3 of water drained out of 100 g of soil with bulk density 1.2 g/cm^3 when the pressure on the soil was increased from 60 to 100 cm water head, then 6 cm^3 of voids had equivalent radii (using eq 4.4) between $r = 0.15/60 =$

0.0025 cm and $r = 0.15/100 = 0.0015$ cm. This is $6/(100×1.2)× 100 = 5\%$ of the total soil volume.

Water retention by spheres of uniform size has been studied occasionally since the early work. Recently Waldron et al. (1961) showed that at low water contents (below 5%), where small wedges of water would be present at points of contact, the experimentally measured water contents are much higher than those calculated from curvature of the air—water interface. This results from neglecting the adsorption forces which hold a thin film of water on all the surfaces and in effect provide anchorage for the air—water interface. At higher water contents this is a relatively small correction.

Miller and Miller (1955) have described a theory of capillary flow based on surface tension and classical viscous flow, which can be considered as a geometric approach to the specification of soil water. The shapes of the air—water interfaces are taken into account to explain hysteresis in terms of pores filling at a lower pressure than that at which they would be emptied. They express their equation in terms of reduced variables, which permits scaling to describe different systems. For example, the water retention curves of different sizes of uniform glass beads when expressed in terms of reduced variables could be superimposed.

Most of the advances in describing soil water have come through the potential concept. This can be applied without a knowledge of how water is retained in soils. Measurements of potential, are "black-box" measurements. A complete specification of soil water would require both geometric and potential descriptions.

Terminology and units

Many different terms have been used to describe the energy with which water is held in soils. This has come about because terms were required both in research work where thermodynamic terminology was usually used, and in practical soil-water work where descriptive terms were coined. Soil-water potential has been used so far in this chapter. The other terms will now be described briefly and will be used where suitable.

In a more descriptive usage, the terms soil-water tension and soil suction are used. They indicate respectively that the soil water is in equilibrium with a pressure less than atmospheric, and that the soil exerts a force to take in water. They can be expressed in units of head, pressure or energy. An attempt is being made to standardize usage, using potential (total and component) in theoretical treatments of soil water, and suction (total and component) in practical usage. Soil suction values are positive. Therefore soil-water potential and soil suction are numerically the same but opposite in sign. Component suctions can be defined analogous to component potentials. Matric suction is often used.

Various units have been used to measure soil-water potential. Potentials are usually given in units of erg/g, or more conveniently, because of the magnitude, in joule/kg. Soil suction is given in units of pressure such as atmospheres or bars, and in units of height of an equivalent water column in cm or equivalent mercury column in cm. Since these numbers may vary over many orders of magnitude, it is often convenient to use a logarithmic scale for suction. The log of soil suction expressed in cm of water is defined as the pF, in analogy to pH.

The conversion factors are:

1 bar (= 10^6 dynes/cm^2) is equivalent to a head of water of 1022 cm or about 0.99 atmospheres pressure.

1 atmosphere = $1.013 \cdot 10^6$ dynes/cm^2 = 1035 cm head.

1 millibar = 1.022 cm water head.

The relationship between energy and pressure units can be illustrated with the gravitational potential (Fig. 4.4

$$\psi_g = gh \, \frac{cm}{sec^2} \cdot cm \cdot \frac{g}{g} \quad \text{or erg/g}$$

$$\psi_g = \gamma_w \, gh \, \frac{g}{cm^3} \cdot \frac{cm}{sec^2} \cdot cm \quad \text{or dynes/cm}^2$$

$$\psi_g = h \text{ cm head}$$

Taking the density of water as 1 g/cm^3, 10^6 dynes/cm^2 = 1 bar = 10^6 erg/g = 10^2 joules/kg.

4.4 WATER RETENTION IN SOILS

Retention curves for different soils

Generalized water retention curves for soils with different grain size are shown in Fig. 4.8. Changes in water content occur over about seven orders of magnitude change in soil-water potential. Maximum rate of change of slope, which indicates the dominant void size holding water, occurs at lower potentials as grain size decreases. Some of the factors affecting the shape of the water-retention curve will be discussed in this section.

An undisturbed clay soil has natural planes of weakness, which as a result of shrinkage are often enlarged into voids between soil units. These natural soil units are called peds. Interparticle bonding holds the particles together in the peds. When such a soil is remoulded, the inter-ped voids and the interparticle bonds are destroyed. The soil then retains much more water at low potentials. Such a comparison is shown in Fig. 4.9.

Consolidation of a soil produces changes in the water-retention curve (Fig.

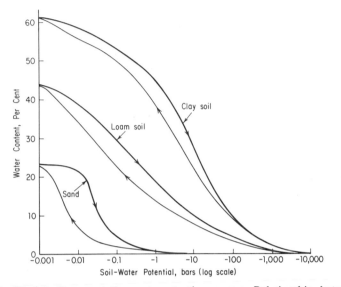

Fig.4.8. Representative water-retention curves. Relationship between potential and water content for different soils.

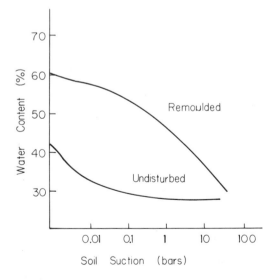

Fig.4.9. Effect of remoulding on water retention of a clay soil.

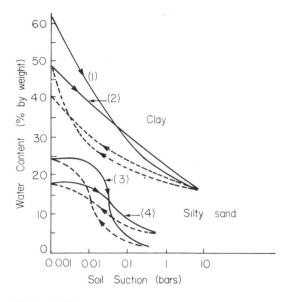

Fig.4.10. Effect of compaction on water-retention curves. Clay soil aggregates compacted at 50 p.s.i. (*1*), and 1,000 p.s.i. (*2*) (results from Chang and Warkentin, 1968). Silty sand at low (*3*) and high (*4*) dry density (results from Croney and Coleman, 1954).

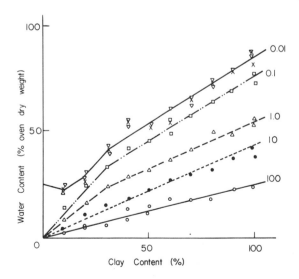

Fig.4.11. Water content, *w*, of Ca–kaolinite–glass bead mixtures at various suction values given in bars on drying run. (De Jong and Warkentin, 1965. *Soil Sci.,* © Williams and Wilkins Co., Baltimore.)

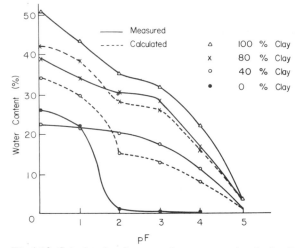

Fig.4.12. Calculated and measured water retention for kaolinite–glass bead mixtures.

4.10). The volume of large voids is decreased, and the small voids are increased. The compacted soil therefore retains more water at high suction but less at low suction.

In clay–sand mixtures, the water content at a given suction increases as clay content increases above a minimum proportion, which in Fig. 4.11 is 20%. However, the resulting water retention curves cannot be predicted from the curves for the components (Fig. 4.12). The mixing which occurs, with one grain size dispersed in the other, must be considered.

Forces of water retention in soils

The soil-water potential can be specified without regard to the matrix forces by which water is held in soils. However, it is also necessary to under stand these forces. They vary depending upon the soil, and this section will present a discussion of the dominant matrix forces in sands, clays and mixed soils.

Sands

A sand is defined by the size of its constituent particles or grains. The soil mass consists of sand grains of different sizes in different packing arrangements, leaving voids of different sizes and varying types of intervoid connections. When this sand is placed in contact with water, the surface attracts a few layers of water molecules by hydration forces. The energy of hydration of sand surfaces is high, and it is usually assumed that the contact angle between water and the particle surface is zero. These hydration layers meet where particles touch, to form a wedge of water with a curved air–water interface. Further water is taken up by the sand as a result of this pressure deficiency or surface-tension force. Water retention in sands is then identical with water retention in a capillary and for this reason the forces have been called capillary or surface tension forces.

The pressure is lower on the convex side of a curved air–water interface. This pressure difference (ΔP) is given by the relationship:

$$\Delta P = T\left(\frac{1}{r_1} + \frac{1}{r_2}\right)$$

Where T = surface tension; r_1, r_2, = radii of curvature of the interface. For a circular pore $r_1 = r_2$ and $\Delta P = 2T/r$.

The pressure difference can be calculated for a capillary tube, and equals the pressure which would have to be applied to the capillary to force water back to the level of the water outside, i.e., to empty the capillary.

Sands and sandy soils do not have voids which are capillaries, and the pressure difference can not be accurately calculated. However, if the water-retention curves are experimentally measured, they can be analyzed to get equivalent capillary pore sizes. Fig. 4.8 shows the water-retention curve for a sand with nearly uniform particle size. Most of the water is lost over a small pressure or potential increment and the corresponding capillary radius could be calculated. From the water-retention curve for a sand with varying particle sizes, the volume of pores having equivalent radii between certain values can be calculated as in Section 4.3.

Clays

Clays are characterized by the volume change which accompanies water content changes. High-swelling clays which do not have a structure with stable large pores remain water saturated to high values of suction; the decrease in water content is accompanied by an equal decrease in volume. No air–water interfaces are present except at the boundaries of the sample. Air enters the sample when the volume has been reduced to where interparticle interference prevents shrinkage. On rewetting, the soil swells or increases in volume due to swelling forces. The forces of water retention are the same as the swelling forces, and soil suction is equal to swelling pressure at any water content. Swelling and surface forces are discussed in detail in Chapters 2 and 6.

Subsoils with a high content of clay, especially if it is montmorillonite, have water-retention characteristics which approach the above description.

Water retention in soils

Most soils have a range of grain sizes and water retention is partly due to swelling forces and partly due to capillary forces. Fig. 4.13 shows drawings of thin sections of soils, which will be used to illustrate forces of water retention. Such soils

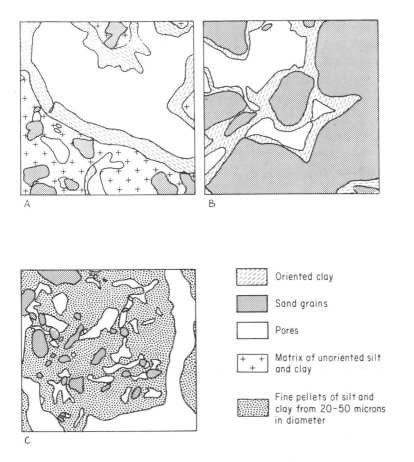

Oriented clay

Sand grains

Pores

Matrix of unoriented silt
and clay

Fine pellets of silt and
clay from 20-50 microns
in diameter

Fig.4.13. Thin-section drawings of soils showing particle arrangement. A. Coatings of oriented clay lining large pore and defining ped surface. Small pores within matrix of unoriented particles. B. Sand grains held together by oriented clay skins and clay bridges. C. Crumb-structure unit with large pore defining it, and small pores within crumb. (From *Soil Classification, A Comprehensive System*, U.S. Department of Agriculture, 1960.)

contain some large pores which are filled with water only at saturation; at suctions of a few centimeters of water they are emptied and contain air. Water retention here is by capillary forces. These may be the pores between crumbs in a clay soil with a stable crumb structure, or pores defined by sand particles in loam soils. When further water is lost, some shrinkage occurs and these pores increase in size as the crumbs shrink together.

   Within the crumbs or peds are pores which may be stable but have irregular and narrow connections with the inter-ped pores. The water is not held by swelling forces but may be considered to be trapped by the narrow connections in which water is held by capillary forces. Clay soils have a marked bimodal void-size distribution curve due to intra and inter-ped voids.

Water is retained also in the matrix of silt, fine sand and randomly oriented clay. Water will be held to the clay by swelling forces, and trapped in the "dead volume" into which the surface forces do not extend.

In the oriented layers of clay, or clay skins, which occur in some soils surrounding the voids or sand grains, water retention approaches that of high swelling clays. With parallel particle orientation the volume change is greatest and the "dead volume", the least. Most of the water is held by swelling forces and little is trapped.

Even though it is impossible to calculate water retention theoretically, except from diffuse ion-layer theory for some high-swelling clays in parallel orientation (Chapter 2) and from capillary theory in samples of spheres of uniform sizes (Section 4.3), it is helpful to consider, even if qualitatively, water retention due to different mechanisms. The measurement of water retention can be made without regard to the forces holding water but the interpretation of these measurements can not.

Hysteresis

The studies on uniform spheres showed that hysteresis in the water content vs. potential curves could be predicted. These water-retention curves may be determined either by decreasing the soil-water potential for a wet soil, or increasing it for a dry soil. The manifestation of hysteresis is that the wetting and drying curves for water content as a function of potential do not coincide.

Hysteresis in sands and silts can be satisfactorily explained as resulting from interconnection of voids of different sizes as illustrated in Fig. 4.7 and discussed for glass beads in Section 4.3. This is known as the "ink-bottle" effect. During drying, a void remains filled with water until the suction is large enough to empty the smallest interconnecting void. Then all the water empties because the equivalent suction for larger voids has already been exceeded. During wetting, the voids will not fill until the suction is low enough to fill the largest voids. Then all the voids fill rapidly. Changes in water content of uniform silts and sands occur by steps. These jumps in water content, called "Haines jumps", can actually be observed when sand of fairly uniform grain size, or glass beads, are subjected to changing soil-water potential.

Hysteresis in clays is more difficult to explain. "Ink-bottle" effects are present, especially in inter-ped voids. Part of the hysteresis is due to movement of peds and particles into different arrangements which accompanies the volume changes. This fabric distortion is "plastic readjustment". Interparticle contacts and

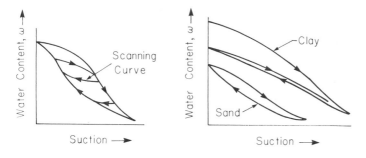

Fig.4.14. Soil-water-energy characteristic surface. (Yong et al., 1971)

the forces at the points of contact differ on wetting and drying. When volume change occurs, the fabric changes associated with the volume change will produce changes in soil structure with resultant changes in fabric, pore-size distribution, and clay—water interaction arising from physico-chemical forces. Fig. 4.14 shows the suction-water content surface for a clay soil obtained by associating the hysteresis performance with measured changes in the soil volume.

The first drying curve for a clay soil can be greatly different from the first rewetting (Fig. 4.15). The original water content is not regained at zero suction. These are not reversible changes, and hence are not hysteresis. The second drying and wetting can result in a closed loop and hence show hysteresis. The large, irreversible change on first drying is due to fabric changes.

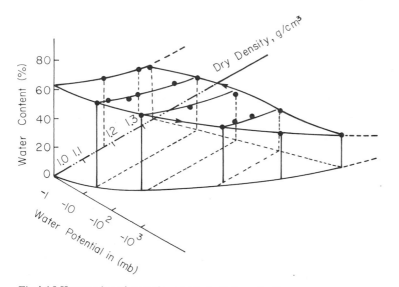

Fig.4.15 Hysteresis and scanning curves in water retention.

If the soil is rewet or dried from some intermediate position on the retention curve, a new path is followed which is called a scanning curve (Fig. 4.15). These scanning curves for wetting or drying can occur anywhere within the envelope defined by the outer curves. These scanning curves define the water-content-potential changes as the soil dries due to evaporation and wets due to irregularly spaced rain. Any calculation of water redistribution in the soil, or of water stored, must use the scanning curves.

4.5 MEASUREMENT OF SOIL-WATER POTENTIAL

Soil-water potential measurements are made for two purposes: in the laboratory to define the equilibrium relationship between potential and water content for a soil, and in situ to measure soil suction for irrigation, water movement, water availability, or other purposes. The potential, rather than water content, determines the availability of water to plants and also such engineering properties as strength. Therefore, a measurement of potential is often required. This can be obtained in situ by direct measurement, or by measuring water content and converting to potential from the relationship determined in the laboratory. Measurements of water content vs. potential will be discussed first.

Measurement of water content at applied potentials

Soil-water potential is the work required to transfer a unit mass of water to the soil. Fig. 4.4 shows a plot of water content against potential expressed in cm of water head. Representative water content vs. potential curves for several soils (Fig. 4.8) show the wide range of potential required to be measured. Here potential is expressed in pressure units of bars (1 bar = 10^6 dyne/cm^2) and is a negative number. Basically the measurement of potential can be made by any method in which a measured force is applied to the soil water and the resultant change in water content is measured. Because of the range of forces required, certain methods are applicable in parts of the range. Three basic methods, which cover the full range, will be described here. These are methods for applying a specific potential and measuring the resulting equilibrium water content. Methods for measuring the potential which exists in a moist soil sample will be discussed later.

The apparatus in Fig. 4.16A applies a negative pressure directly to the soil water, and can be used from 0 to −1 bar potential. The sample is placed firmly on the ceramic plate; water is continuous from the sample through the pores of the ceramic and into the measuring tube. Two methods of applying negative pressure are shown. The one on the left is the classical Haines' method. The change in water

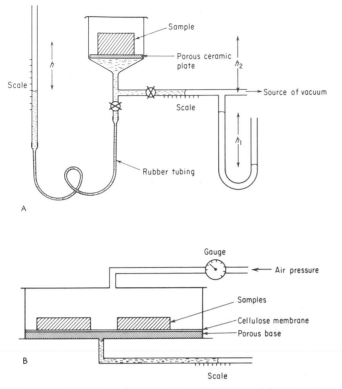

Fig.4.16. Apparatus for measuring soil-water potential vs. water content. A. Unit for use to − 1 bar. B. Unit for use to − 25 bar.

content of the sample can be directly observed by measuring the changing position of the meniscus in the tube. It is not possible to apply a negative pressure greater than − 1 atmosphere (− 14.7 psi) gauge pressure, or zero absolute pressure, so this determines the limit of applicability of this method. A requirement of this method is that no air moves through the porous ceramic to enter the measuring tubes. This necessitates use of a ceramic plate with small pores, which leads to slow water movement through the ceramic and consequently a longer time to establish equilibrium for any pressure change. These conflicting requirements must be balanced and in practice this method is used to about − 0.8 bars.

Instead of applying a negative pressure to the bottom of the sample a postive pressure can be applied to the top as shown in Fig. 4.16B. The lower surface is at atmospheric pressure, and the water in the soil comes to equilibrium with the excess pressure. The upper airpressure limit for this apparatus is determined by the strength of the chamber and the air-entry value of the porous base. Ceramic bases are usually used up to several bars although they are available with air-entry values of 15 bar. Cellulose membranes with air-entry values in excess of 100 bar are used for the higher range. The strength of the chamber is usually such that the apparatus with membrane is used to 20 bar air pressure.

Air pressure is applied to wet soil samples placed in the chamber. When water has ceased to move from the sample through the membrane, the soil water is in equilibrium with the applied pressure. When the pressure is released, the soil water has a potential of the same magnitude but opposite sign to the applied air pressure. The sample may then be removed, weighed and dried to determine water content. Alternatively, the volume of water moved out may be measured. This apparatus is often called "pressure plate" and "pressure membrane", and is available commercially. It is widely used in soil-water studies, since its development by Richards and others in the 1940's (see Richards, 1949, for a review of these methods).

The soil samples used in this apparatus are usually 1–2 cm high and 5–7 cm diameter. Medium or coarse-grained soils reach equilibrium within a day or two, but clays may take 5–10 days. Vapour losses may be high with these extended periods. It is also difficult to get representative, undisturbed samples in these small sizes. For these reasons, the measurements are less satisfactory for clays than for loams or sands.

Air bubbles almost always accumulate in the water below the ceramic or membrane. This may be due to leaks, but is probably the movement of air through the water phase due to a pressure gradient. This air interferes with the measurement of water moved out, and also carries water vapour away from the sample.

In the third part of the range, below −25 bar potential, indirect methods are most easily used. A controlled vapour pressure of water is admitted to a chamber containing the sample. The soil either loses or gains water until the potential of the soil water is the same as that of the air around it. The sample is then weighed to determine water content.

A simple method consists in placing soil samples in a dessicator containing a solution, e.g., sulphuric acid, which maintains a definite vapour pressure which can be regulated by regulating the concentration of the acid (Table 4.1). The air is evacuated from the dessicator to allow water molecules to move more rapidly. The sample is weighed periodically to determine when it has come to equilibrium.

The vapour pressure is related to soil-water potential by the hypsometric formula:

$$h = - \frac{RT}{Mg} \ln \frac{P}{P_o}$$

where: h = soil-water potential expressed as cm of water column; R = gas constant, $8.3 \cdot 10^7$ ergs degree^{-1} mole^{-1}; M = molecular weight of gas, taken as 16 g/mole;

TABLE 4.1

Relationship between density of sulphuric acid and vapour pressure
(From Croney et al., 1952)

Density of acid (g/cm^3)	Relative vapour pressure, P/P_0	Soil-water potential (bars)
1.050	0.977	−32
1.110	0.930	−100
1.205	0.795	−320
1.344	0.484	−1,000
1.578	0.101	−3,200
1.840	0.001	−10,000

g = gravitational constant; T = absolute temperature; P = vapour pressure of water in the soil; P_0 = vapour pressure of pure water.

This relationship, plotted in Fig. 4.17, shows that the useful range of the method is below −20 bars where large relative humidity changes occur with changes in potential. Over the range of water contents for soils between field capacity and wilting percentage at −15 bars, the relative humidity is nearly 100%.

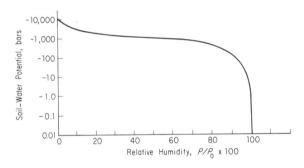

Fig.4.17. Soil-water potentials at different vapour pressures with which soil is in equilibrium. (Croney et al., 1952.)

These three methods allow one to obtain, in the laboratory, the entire range of soil-water potentials shown in Fig. 4.8. Several methods will now be discussed which allow a measurement of potential in the field or in the laboratory.

Measurements of potential in situ

Soil-water potential can be measured directly in the field with "tensiometers". Water in the tensiometer is in contact with soil water through the

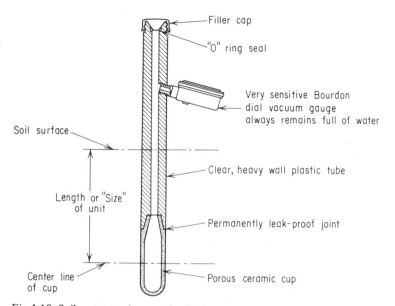

Fig.4.18. Soil-water tensiometer for field use. (Soilmoisture Equipment Co.)

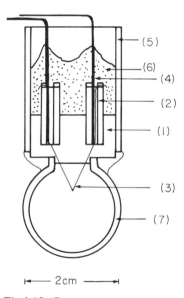

Fig.4.19. Cross-section of a thermocouple psychrometer for in-situ measurement of soil-water potential showing teflon insert (*1*), copper heat sink (*2*), chromel-constantan thermocouple (*3*), copper lead wires (*4*), acrylic tubing (*5*), epoxy resin encapsulating heat sinks (*6*), and ceramic bulb (*7*). Water is condensed on the thermocouple junction by Peltier cooling immediately preceding each measurement. (Rawlins, 1967. *Soil Sci.*, 31, pp.297–303, © Williams and Wilkins Co., Baltimore.)

ceramic tip (Fig. 4.18). As the soil dries, a tension is exerted on the water in the tensiometer, and this is measured with a manometer or Bourdon gauge. The tensiometer could also be used to measure positive pressures or heads, in which case it would be an ordinary piezometer, and the porous cap which prevents air movement is not required.

The tensiometer will function until a negative pressure is reached at which air bubbles through the ceramic. This is the air-entry value, which is about -0.85 bar for most tensiometers. Ceramic tips with higher air-entry values have smaller pores, and a consequent slower exchange of water, which causes a lag in readings.

Tensiometers are widely used to indicate when irrigation is required. For maximum yield, coarse-grained soils must be irrigated before the potential has decreased to -0.85 bar. Up to three quarters of the water available to plants has been used at this potential.

One of the newest important instruments for soil-water potential measurement is the thermocouple psychrometer which measures vapour pressure. Determination of soil-water potential by measurement of vapour pressure has many theoretical advantages. The vapour pressure is a well-defined thermodynamic quantity which depends upon both the matric and osmotic components of total potential. Tensiometers and membranes allow movement of salt through the porous membrane or ceramic and hence they do not measure the osmotic potential. The difficulty with vapour pressure has been the extreme precision of measurement required in the range from 99 to 100% relative humidity, which is the range of vapour pressures corresponding to field water contents (Fig. 4.17). Recently, sensitive psychrometric procedures have been developed for measurements of relative humidity in this range.

The thermocouple end is fixed inside a small ceramic bulb about 1 cm. diameter (Fig. 4.19). A cooling current is applied which condenses a minute amount of water on the junction. As this water evaporates, it cools the junction. The rate of evaporation is inversely related to vapour pressure in the bulb. The cooling, or temperature drop, is measured as the voltage output of the thermocouple. A very sensitive voltmeter, measuring to 0.01 μV is required. Since the vapour pressure is temperature dependent, a second thermocouple is added to measure the temperature. The vapour in the ceramic bulb is in equilibrium through its pores with the vapour in the soil. Hence the measurement of vapour pressure can be converted to soil-water potential.

The thermocouple psychrometer is calibrated with salt solutions for which the vapour pressure is known. The relationship between osmotic pressure and vapour pressure is:

$$1 - \frac{P}{P_o} = \frac{\overline{V}}{RT}\pi = 7.3 \cdot 10^{-4}\pi$$

where: $\bar{V}$ = partial molar volume of water; π = osmotic pressure in bars.

The precision of the psychrometer used in the laboratory is usually ±0.1 bar, but precision in the field is more usually ±0.5 bar. It is, therefore, most useful at low potentials, that is, where the tensiometer will not work. At potentials around −15 bar, vapour equilibrium is established quickly and the readings take only a few minutes, but at higher potentials such as −1 bar the readings may take hours. This method is described in a number of publications, e.g., Rawlins (1971).

Indirect measurements of potential

There are several indirect methods of measuring potential which are useful in certain field applications.

The most widely used indirect method is measurement of the resistance between the two electrodes embedded in a porous gypsum block which is buried in the soil. These units are commonly called gypsum blocks or Bouyoucos moisture blocks after the man who developed them. As the soil dries, the pores of the gypsum block, in contact with the soil, lose water and the resistance between the electrodes increases.

Gypsum blocks are easily made. Plaster of Paris and water are mixed in a fixed proportion, about 1:1 by weight. A higher proportion of water results in a lower porosity and finer pores in the dried block. This slurry is poured into a mold around the electrodes, which may simply be bared ends of multistrand copper wire. When the plaster of Paris sets, the block is ready for use. The resistance is measured with a Wheatstone bridge arrangement using an a.c. power source to prevent electrode polarization. This meter, when battery powered with transistors, is a small, easily portable instrument.

The blocks may be calibrated by determining resistance vs, potential or water content for the soils to be used. Calibration can be carried out by placing the blocks in soil in a pressure membrane apparatus. However, the low precision of the gypsum blocks does not usually warrant such precise calibration. Within the limits of precision of the blocks, an adequate calibration can be obtained by taking samples near the block when it is buried in the field, or by using the generalized curve in Fig. 4.20.

When the resistance of the blocks is plotted against available water, i.e., water content at field capacity is 100% and at permanent wilting percentage is 0%, the calibration for many soils will fall on the same curve. This, along with occasional checks by sampling in the field, provides sufficient precision for those soil-water measurements for which the blocks are suited. The saturated blocks have a resistance of about 500 ohms. This increases to about 75,000 ohms at the water content at which plants wilt.

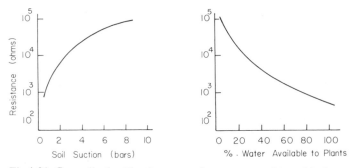

Fig.4.20. Generalized calibration curves for gypsum moisture blocks.

The greatest source of error is the hysteresis of both the soil and the block. The water content on wetting a dry soil to a certain potential is not the same as that on drying a wet soil to the same potential. The same is true for the blocks. This makes the blocks unfit to measure water content during rapid wetting and drying regimes. The blocks do indicate when the soil has dried to a certain water content, and when it has been rewet to field capacity. Consequently, they are used in irrigation studies.

The blocks are more sensitive in the dry range, as seen from the slope of the resistance vs. available water curve. This sensitivity can be altered by using nylon or fibre glass rather than gypsum as coating for the electrodes. Blocks with nylon are available and are recommended where measurements in the wet range are required. Blocks are also available in which the electrodes are wrapped in nylon before being embedded in the plaster of Paris.

The current between electrodes in the gypsum blocks is carried by ions in a saturated gypsum solution. This makes the blocks less sensitive than nylon units to changes in salt content of the soil solution. However, it has been experimentally found that when the salt concentration of irrigation water exceeds 0.2%, the measured resistance is influenced. This is due mostly to the fact that part of the current path between electrodes is in the soil outside the blocks.

Another indirect method for measuring soil suction in situ is to weigh some porous material placed in contact with soil and for which the relation between potential and water content is known. Ceramic plugs and filter paper have been used. This method requires no expensive equipment and may be useful under certain conditions.

4.6. SECONDARY EFFECTS ON WATER RETENTION

Entrapped air

Entrapped air can influence the measurement of equilibrium water content at a given suction. If suction is applied in the Haines apparatus, the air expands, forcing out some water. If air pressure is applied, the entrapped air contracts. The soil can then have a higher water content under applied air pressure than under applied tension. This effect is most noticeable in the fine silt fraction (Fig. 4.21).

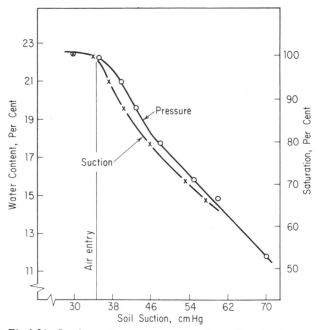

Fig.4.21. Suction-water content curves for a fine silt showing difference due to method of removing water. (Chahal and Yong, 1965. *Soil Sci.,* 99, ©Williams and Wilkins Co., Baltimore.)

Temperature

Temperature has a large influence on the rate of water movement, but only a small influence on the amount of water retained at equilibrium. Surface tension decreases as temperature increases; hence, water retained by sands decreases. Water retained by swelling forces increases with temperature, and hence water retention in clay soils increases. Careful measurements have confirmed these effects.

Water retention in soils is by both surface tension and swelling forces. Since the temperature dependence is opposite, the effects are cancelled out, and there is little temperature dependence of water retention in soils. Many of the large effects reported were effects on the measuring instruments. Most instruments have a large temperature coefficient which must be taken into consideration.

A sudden temperature change can act as a "temperature shock" and result in decreased water retention. This is not reversible.

Effect of drying

Drying has a small and unpredictable effect on the water-retention curves of most soils. Drying can change the fabric and hence the water retained at any suction. Slow swelling of a dried sample may decrease the water held at any suction.

Allophane soils are the only group which show a large, irreversible change in the water-retention curve on drying (Fig. 4.22). The marked change in surface properties of allophane soils on drying was discussed in Chapter 2. Both the volume of small voids and the swelling forces are decreased on drying. The large decrease in water content at a particular suction from a moist to a dry sample can be used to rate the allophane content of a soil.

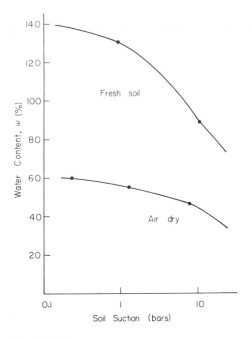

Fig.4.22. Water-retention curves for allophane soil. (Redrawn from Yamazaki and Takenaka, 1965.)

Rate of potential change

The size of pressure or suction steps influences the equilibrium water content; the larger the steps, the lower the water content. In clays this can be explained from the greater readjustment of particles, i.e. fabric change, from one big load. In sands it may be due to non-equilibrium of the air—water interface.

Equilibrium water-retention curves may not give the correct suction-water content values to use in the calculation of water flow. Water flow is a non-steady state process. Unsteady state water retention is higher than equilibrium retention (Fig. 4.23).

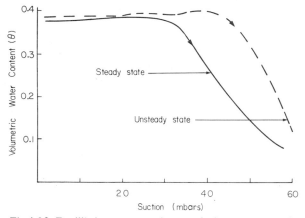

Fig.4.23. Equilibrium vs unsteady state drying-water-retention curves for fine sand. (Topp et al., 1967. *Soil Sci. Soc. Am., Proc.,* 31.)

Overburden load effects

A subsoil sample at 100 cm depth is under an overburden load from the weight of soil above it. The soil has intergranular contacts which take part of the load; hence, the load on the water is only a fraction of the overburden mass. The "effective stress" concept is discussed in Chapter 10. In a plastic, compressible clay where intergranular contacts are absent, the load could be a maximum of 100 cm × 1.2 g/cm^3 = 120 g/cm^2 × 980 cm/sec^2 ≃ 1.2 · 10^5 dynes/cm^2 ≃ 0.12 bar. Such a sample at zero suction in situ would be at a suction of 0.12 bar when it was removed. The potential water-content curve measured on the sample is, therefore, not the curve which applies to the soil in situ. The water content is lower in situ than predicted from the water-retention curve. This can be expressed as:

$$S_{\text{in situ}} = S_{\text{unloaded}} - \alpha P$$

where: S = matric suction (osmotic suction is not changed); α = compressibility factor $(0 \leqslant \alpha \leqslant 1)$; P = stress due to overburden.

Values of α of 0.2 are common in soils, and α increases with increasing water content. It is, therefore, only at low suction where this correction is important.

Fabric effects

Different fabric leads to different void-size distribution, and also to a different proportion of water held by swelling and by surface tension forces. Compaction decreases total void volume and large voids, but increases small voids. These differences determine the shape of the water retention curve of a soil.

A problem arises when the sampling or measuring procedure changes the fabric of the sample and hence its water-retention characteristics. Disturbing a soil usually breaks up particle arrangements, exposes more surface area to water retention and breaks interparticle bonds which are then only slowly reformed. All this leads to a higher water content at a given suction. Some clay soils, however, in which the particles exist in loose random arrangement, show the opposite effect. Disturbance brings the particles closer together and lowers the water content at constant suction.

Water-retention curves are often measured on samples which are dried, sieved, and then repacked. It is difficult to achieve the bulk density which existed in the field. For coarse-grained soils the error will be large at low suction but negligible beyond 1 or 2 bar. A disturbed clay soil sample retains more water than an undisturbed sample, even at 15 bars. Interpretation of tensiometer readings in the field and of field capacity suctions (see Chapter 5) are affected by these differences.

Fabric changes occur during drying and wetting. This can result in delayed equilibrium for clay soils under an applied potential. Water loss results in a fabric change which has a lower equilibrium water content, which then results in further water loss, more fabric change, etc.

4.7 USE OF THE POTENTIAL CONCEPT

General

The potential concept has been very useful in visualizing soil-water phenomena. Work must be done in order to extract water from a soil. The amount of work required per unit mass of water extracted increases as the amount of water remaining in the soil decreases. The soil-water potential is defined as the work done on a unit mass of water, required to move it from a free water surface to a point in the soil. The free water is defined as having zero potential. This total potential includes the work required to remove water against the suction forces of the soil (matric suction) and also the osmotic forces (osmotic suction) of the soil solution. The matric suction may result from surface-tension forces, from forces binding water to clay mineral surfaces and from forces holding water to exchangeable ions. These components cannot be separately measured but the total water potential can be measured.

The potential is a property by which quantitative comparisons of water relations can be made between soils. The difference in potential between two points in a soil represents the work required to move the water between these two points. The difference in potential between water in the soil and in the plant and the atmosphere can be used to indicate the work done by the plant in removing water from the soil, and the work required in evaporating water from the leaves. If the potential is thought of as a measure of the security with which water is held in soils, many applications become obvious. A sandy soil and a clay soil each at 10% water by weight will contain very different amounts of water available to plants, but these two soils at the same soil-water potential would contain water equally available to plants. The same statement applies to mechanical behaviour. This is a matter of quality and not quantity. The analogy between temperature and heat capacity can be drawn here. Temperature is a qualitative measure, and heat content is a quantitative measure. Consequently, a specification of the potential of water in a test sample is much more meaningful than a figure for water content.

Shear strength of cohesive soils is more closely related to soil-water potential than to water content, as discussed in Chapter 10. Water movement to the freezing zone in soils determined by the potentials created at the ice—water interface is discussed in Chapter 11.

The concept of potential will be used in Chapter 5 to explain the movement of soil water. This has been the greatest contribution of the potential concept, making possible the application of the extensive mathematics of diffusion to the water flow problem in unsaturated soils. Water will flow from a point of higher potential to one of lower potential, or from lower suction to higher suction. If a clay at 10% water and a sand at 10% water were placed in contact, the water would flow from the sand into the clay because the sand would have the higher water potential.

The relation between soil-water potential and soil water content is a fundamental water property of a soil. This is a smooth curve, and hence any classification of soil water, for example into gravitational, capillary and adsorbed water is arbitrary. The soil-water potential decreases as the water content decreases and there are no sharp breaks in the curve.

Predicting water under covered areas or in swelling soils

As an illustration of the use of component potentials, the potentials above and below the water table in a soil will be calculated. The total potential is equal to the sum of the matric, gravitational and osmotic potentials:

$$\Delta\Psi = \psi_p + \psi_g + \psi_\pi$$

If evaporation from the soil surface is absent and if sufficient time has been allowed for water to move up from the water table, the soil water is static, and hence the total potential must be the same at all points in the soil. Since the potential at the free water surface is zero by definition, the total potential is everywhere zero. The following calculations show the magnitude of the component potentials for the conditions in Fig. 4.24.

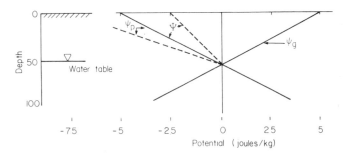

Fig.4.24. Illustration of component potentials.

Assume $\psi_\pi = 0$. Therefore: $\Delta\Psi = \psi_p + \psi_g = 0$. The value of ψ_g is taken as zero at the water table (at 50 cm) and positive above it. At the soil surface:

$$\psi_g = gh = 980 \times 50 = 4.9 \cdot 10^4 \ \frac{ergs}{g} \cdot 10^{-4} \ \frac{joules}{kg} \ \frac{ergs}{g}$$

$$= 4.9 \ joules/kg$$

$$\psi_p = -\psi_g = -4.9 \ joules/kg$$

At a depth of 50 cm below the water table:

$$\psi_g = gh = 980\,(-50) = -4.9 \ joules/kg$$

$$\psi_p = -\psi_g = 4.9 \ joules/kg$$

The pressure potential is negative above the water table and positive below. Below the water table this is the hydrostatic pressure. The potentials are plotted in Fig. 4.24.

If an additional potential gradient is now established due to evaporation at the surface, and this additional gradient has the value of 0.05 joules kg^{-1} cm^{-1}, the component potentials can again be calculated as follows. The total potential is now not equal to zero at the surface but has a value of:

$$\Delta\Psi = -0.05 \text{ joules kg}^{-1} \text{ cm}^{-1} \times 50 \text{ cm} = -2.5 \text{ joules/kg}$$

The gravitational potential remains the same since it depends only upon position of the water table. Then:

$$\Psi_p = \Delta\Psi - \Psi_g = -2.5 - 4.9 = -7.4 \text{ joules/kg}$$

at the soil surface. These values are shown in dashed lines in Fig. 4.24. The potentials below the water table are not changed.

The osmotic potential can be readily included where values for salt concentrations are known.

The application of the water-potential theory and measurements to predict water content under roads is described in publications by Croney and Coleman (e.g., Croney et al. 1958).

Use of soil-water potential for clay soils

Measurement and theoretical development of soil-water potentials have been based largely on coarse-grained soils. Several problems arise in measuring and using potentials for clay soils. For example, application to water movement usually assumes a rigid medium with no interaction between water and the medium. This assumption cannot be used for clay soils.

The long time required to reach equilibrium in a pressure plate apparatus is a serious disadvantage in measuring water-retention curves for clay soils. A sample 1 to 2 cm high can still be losing water after two weeks. Such thin samples may not be representative of the soil, but thicker samples require even longer time to equilibrium. Drainage time increases approximately with the square of the thickness. The water content—time curve at early times can be extrapolated to find the equilibrium-water content, similar to the technique used in analyzing consolidation. This can overcome part of the time problem, however this must be used with caution since the load increment affects the final water content. The slow drainage is partly due to the low hydraulic conductivity of clay soils, but partly also to fabric changes accompanying water-content changes. As water is lost, shrinkage causes a change in fabric; the new fabric has a lower equilibrium water contnet and hence more water is free to move out. Direct measurement of soil-water potential with a psychrometer does not have this time limitation because the water content does not have to be changed during measurement.

Fabric changes and volume changes which accompany water-content changes in clay soils make measurement of water-retention curves difficult, but even more important they make the resulting values less reliable measurements of what happens in the soil. Volume change in a sample will differ from that in situ because the boundary conditions differ. Slight remoulding at the surface of a sample where it contacts the porous plate in the pressure plate apparatus can greatly reduce rate of water movement. Shrinkage during water removal often breaks the contact between sample and measuring device, either pressure plate or tensiometer in the field. Also, rewetting is different for a laboratory sample than for the soil in situ.

For all these reasons the measured water-retention curves of clay soils are less useful than those for coarse-grained soils. Since measurements are time consuming, it is often not economical to use these methods. Direct measurements on samples in situ, as for example with psychrometers, do not have these disadvantages. The improved reliability of these measurements more than compensates for the extra difficulty of working directly with field samples.

4.8 SUMMARY

The relationship of soil-water content to the energy with which water is held in the soil is one of the fundamental physical properties of a soil. The physical description of how water is held is given for different soils. The effect of changes of soil structure on water retention is illustrated in this chapter.

The water-retention curve is different for drying than for wetting. The manifestation of hysteresis in water retention is related to the forces holding water in soils.

The free energy of water in the soil is lower than that of a pool of water. This lower free energy is expressed as a negative soil-water potential or as a positive soil suction. The total soil-water potential is divided into components which describe the sources of water retention due to the soil matrix (matric), to the salt concentration (osmotic), and others.

Direct and indirect measurements of water content and of soil-water potential are discussed, which can be used in the laboratory or in the soil in situ.

Chapter 5

WATER MOVEMENT IN SOILS

5.1 INTRODUCTION

The statics of soil water has been treated in Chapter 4 where the interactions of soil with water, the forces of water retention and the free energy of water in soils were discussed. The principles governing the movement of water in soil will be discussed in this chapter.

The water content of a soil is rarely static; water additions from snow-melt, rainfall, irrigation or condensation and water losses from evaporation, transpiration or drainage occur most of the time. Thus the distribution and migration of water in soils are dependent on many fluxes such as those arising from the internal energy of the water itself, and from external and surficial mechanisms and driving forces due to thermal, ionic, osmotic, gravitational, hydraulic and other gradients. These were discussed in Chapter 4. The rate of movement or migration of water will depend on the magnitude of the forces and gradients and also on the factors determining the transmission coefficient or hydraulic conductivity of the soil.

Water movement or moisture transfer in soils may be divided into two particular systems for general consideration: (a) the saturated system where all the voids are filled with water; (b) the partly-saturated system where both air and water are present. For partly-saturated soils, the mechanism for moisture transfer will depend upon whether the system is relatively dry or relatively wet. In the former, vapour transfer is greater than liquid transfer. This is especially important if large temperature gradients exist. In the latter case, where the soil-water system is relatively wet, liquid transfer will outweigh vapour transfer. The movement of water in liquid transfer is generally considered as viscous flow. Diffusive flow due to physical-chemical gradients is generally taken to be minor and insignificant in computations of liquid transfer. The absence of a vapour phase in saturated systems indicates that the movement of water in such systems is effected solely by liquid transfer.

Water entering a dry soil at the surface does not saturate the soil below and moves through unsaturated soil. Unsaturated flow is thus the most common kind of flow in soils, but saturated flow is generally easier to describe. Even "saturated" soil rarely has all its voids filled with water. It is not unusual to expect from 2 to 12% of air remaining in the voids.

Most of the water movement in soils in the liquid phase is due to gradients of matric or capillary potential (see Chapter 4), which arise from differences in water content. The term capillary flow is hence often used. This is not restricted to capillary rise of water above a water table. Capillary flow can occur in any direction where there is a gradient for water flow, if the soil is not saturated. The definition of saturation must be re-examined here. The saturated condition of a soil can be defined as the entire void volume filled with water, or as the water content at zero potential or suction. For sands there is little practical difference between the two definitions. However, clay soils can exist at fairly high suction values, one or more bars, and still have no air in the voids. The water lost on increasing the suction is compensated by a decrease in volume. Water movement in clay soils decreases with increasing suction, even when no air is present. The zero suction definition is, therefore, more suitable for clay soils, and capillary flow can occur due to suction differences when no air is present.

Soil water will also move under the influence of physico-chemical forces associated with the interaction between clay and water. When the concentration of a solution differs from that at another point, there is a tendency for the more dilute liquid to diffuse into the region of higher concentration. The potentials existing in clay soils as shown in Chapter 4 produce gradients which will induce moisture transfer. For example, there is a tendency for water to diffuse into regions of higher ionic concentration to attain a more uniform ionic distribution. The electrostatic forces near the particles maintain a local condition of high ionic concentration. This condition creates a condition for osmotic flow and is of interest in the swelling of clays due to the intake of water. This will be considered further in the latter part of this chapter.

Movement of moisture in soils is not restricted to the liquid phase but can also occur in the vapour phase. Vapour movement may occur under a thermal gradient and also under isothermal conditions. Movement in the vapour phase is by convective (bulk) flow of the soil air, or by diffusion of water molecules in the direction of decreasing vapour pressure. A vapour-pressure gradient may be established by such factors as temperature, salt concentration, or differential suction within the soil. The temperature gradient is by far the most important. Under isothermal conditions a vapour-pressure gradient can exist only if there is a moisture-content gradient. Differences in salt concentration give rise to a reduction in the vapour pressure of the soil solution in proportion to the salt concentration. This causes a vapour-pressure gradient which results in vapour transfer.

The movement of water under thermal gradients is important in the study of soil freezing and frost heave. Thermal gradients caused by exposed pavement and snow covered shoulders may create conditions that will favour vapour diffusion, thus causing moisture to collect beneath the pavement which will enhance ice-lens formation.

5.2 SATURATED FLOW

In groundwater and seepage problems the soil mass is considered as a homogeneous medium of interconnected pores. A material is said to be porous if it contains void spaces in which the material is absent. Pore spaces can be discontinuous or entirely continuous. Sands as an example of porous material are composed of macroscopic particles that may be rounded, subangular or angular in shape. Numerous voids of varying shapes and sizes exist between the individual particles. Each pore is connected by constricted channels to other pores.

In hydraulic flow problems such as flow through pipes or open channels, flow is confined within impermeable boundaries. In ground water flow and seepage, however, the channels consist of multiply connected pores of varying shapes and sizes and limited in extent by the geometry of the flow system. The flow channel cannot be considered as a pipe equivalent to a single series of connected pores because of the absence of impermeable boundaries in the connected pores of the soil. Assuming that the velocity distribution for viscous fluid flow across a pipe is parabolic, the velocity distribution across each pore space may also be assumed to be parabolic. Macroscopically, however, it is considered to be uniform. Thus the significant property of a porous medium with respect to its fluid-carrying capacity is its porosity.

Darcy's equation

Darcy, studying the rate of flow of water through sand filter beds, found that the macroscopic flow velocity was proportional to the hydraulic head or hydraulic gradient:

$$V = \frac{Q}{At} = k\frac{\Delta h}{\Delta x} = ki \tag{5.1}$$

where: Q = volume of water flow; A = cross sectional area of bed; t = time of flow; V = flow velocity; k = constant, defined as "hydraulic conductivity"; $i = \Delta h/\Delta x$ = hydraulic gradient. This has become known as Darcy's equation. The hydraulic conductivity, k, is a measure of the resistance of the soil to flow of water. If the properties of water affecting flow are included specifically in the equation, it can be written as:

$$V = k' \frac{\gamma g}{\eta} i \qquad\qquad\qquad\qquad (5.2)$$

where: γ = density of fluid; η = viscosity of fluid; k' = intrinsic permeability, or permeability; g = gravity.

To avoid confusion, the value of k in Darcy's equation will be referred to as hydraulic conductivity. Permeability will be used either for the value of k' or as a general term denoting the rate at which water moves through a soil.

Derivation of the fluid-flow equation shows that Darcy's equation should be valid at the low flow velocities of water in soil (Childs, 1969). The limitations of Darcy's law fall into the same general category as those for the Kozeny-Carman relationship. For granular particles, which are relatively spherical in shape, if the particles are uniform in size and if the surfaces are smooth, the results obtained by applying Darcy's law are quite reasonable. However, if we deviate from these and further impose other constraints, the law becomes less and less valid. Only gravity forces are considered in the soil-water system. Further, both laminar flow and average gradients are presumed, with the latter being a constant. Swartzendruber (1962) has discussed the modification required in Darcy's equation for small gradients $\Delta h/\Delta x$, especially in compacted clay soils where flow is not linearly related to gradient. In general, Darcy's equation applies with sufficient precision for use in problems of water movement in soil, for example, drainage.

The value of "k" can be determined experimentally in the field or on laboratory samples. The values for saturated hydraulic conductivity are used in the design of drainage systems, e.g., depth and spacing of tiles, and are best obtained in the field because the sample will be undisturbed, it will be much larger than a laboratory sample, and it will have natural boundary conditions. Laboratory samples are used to study the influence of soil treatments on permeability, e.g., the influence of water with different salt concentrations. Both field and laboratory measurements are discussed in a number of books on land drainage and groundwater.

Kozeny-Carman relationship

If both the surface area of soil particles and porosity of the soil are considered in the determination of hydraulic conductivity, the form of analysis will require the Poiseuille equation for viscous flow of fluids through narrow tubes as a basis for development, i.e.:

$$v' = \frac{\gamma r^2 \Delta\psi}{8\eta \Delta l}$$

where: v' = mean effective flow velocity through a tube; r = radius of tube; η = viscosity of fluid or permeant; $\Delta\psi$ = potential difference between the ends of the tube; Δl = length of the tube; γ = density of fluid.

For purposes of extending the equation for use in porous media, we can introduce a shape factor C_s and rewrite the equation as:

$$v' = C_s \frac{\gamma r^2}{\eta} \frac{\Delta\psi}{\Delta l} \qquad (5.3)$$

There are several points to be considered if we wish to adapt eq.5.3 for flow through porous media. Recognizing that the effective path length in porous media is governed by the total volume of pores, and that these pores are not circular, the modifications required involve (a) path length, (b) pore volume or porosity n, and (c) surface area of particles per unit volume, S. We may, with the use of eq.5.2 formulate a flow equation as follows.

The apparent velocity V is in fact not the flow velocity in the soil pores. If porosity n defines the pore volume per unit volume of soil mass, then:

$$v' = V/n \qquad (5.4)$$

Eq.5.4 must be modified if the path length Δl is not direct, i.e., if an effective path length Δl_e exists because of the distribution and connection of pore spaces. Thus $\Delta l_e > \Delta l$. Hence,

$$v' = \frac{V}{n} \frac{\Delta l_e}{\Delta l} = \frac{V}{n} T \qquad (5.5)$$

where $T = (\Delta l_e / \Delta l)$ = tortuosity.

If S_w is the wetted surface area per unit volume of the porous medium, then n/S_w will be equivalent to the volume divided by the wetted surface area. This will serve to replace the radius r in the Poiseuille equation. Since the pores are not circular, the shape factor C_s previously introduced may be used to account for this. Thus: $r = n/S_w$. Distinguishing the wetted surface area per unit volume S_w from the total surface area S per unit volume, i.e., the surface area wetted by the flowing fluid, the porosity n may be introduced to evaluate S, since it defines the pore volume, i.e., $S_w = S(1 - n)$. Hence:

$$r = \frac{n}{S(1-n)} \qquad (5.6)$$

With the substitutions provided for by eq.5.6 and 5.5, eq.5.3 may now be written as:

$$V = \frac{nv'}{T} = \frac{n^3}{S^2(1-n)^2} \cdot \frac{C_s}{T} \frac{\gamma}{\eta} \frac{\Delta\psi}{\Delta l}$$

Using the effective path length Δl_e instead of Δl to describe the potential gradient:

$$V = \frac{C_s\gamma}{\eta TS^2} \frac{n^3}{(1-n)^2} \frac{\Delta\psi}{\Delta l_e}$$

Thus:

$$V = \frac{C_s\gamma}{\eta T^2 S^2} \frac{n^3}{(1-n)^2} \frac{\Delta\psi}{\Delta l}$$

Since $V = ki$ from eq.5.1 and:

$$\frac{\Delta\psi}{\Delta l} = i = \frac{\Delta h}{\Delta x}$$

then:

$$k = \frac{C_s\gamma}{\eta T^2 S^2} \frac{n^3}{(1-n)^2} \qquad\qquad (5.7)$$

Eq.5.7 which is a form of the Kozeny-Carman equation expresses the hydraulic conductivity k in terms of surface area, a flow factor, and porosity. The units and factors in eq.5.7 are:

C_s = shape factor = 0.5 for a circle, 0.33·for a strip, 0.56 for a square. C_s may be taken as 0.4 for a standard value with a possible error of less than 25% in computation of k;

T = tortuosity = $\dfrac{\Delta l_e}{\Delta l} = \dfrac{\text{effective flow path}}{\text{thickness of test sample}} = \sqrt{2}$;

n = porosity;

S = surface area per soil volume of particles, cm^2/cm^3.

The primary considerations needed for the valid application of eq.5.7 are (a) relatively uniform particle size, (b) laminar flow of liquid through the pores, (c) validity of Darcy's law, and (d) absence of long and short-range forces of interaction.

Ideally, the quantitative prediction of k will be closest for granular particles. The constant T which must be introduced to account for tortuosity is best evaluated experimentally with controlled tests. The Kozeny-Carman relationship represents a refinement of Darcy's law accounting for certain soil properties and characteristics. However, if we introduce the interaction characteristics known to exist in clay-water systems, and add to that the fact that clay particles are platelike in shape and form fabric units (Chapter 2) such that overall permeability between and within fabric units are different, it is apparent that eq.5.7 becomes less valid.

Layered soils

Water flow in layered soils is not determined by the permeability of the least permeable layer in so far as quantity is concerned. Calculations using the value of k for the least permeable layer will produce flows that are too low. Fig.5.1 illustrates the mechanisms causing flow changes in layered soils. The volume of flow in the layered soil shown in Fig.5.1 is higher because the pressure gradient is not uniform; a larger head drop occurs through the least permeable layer. The volume of water flow, which is the product of the gradient and the conductivity, is therefore higher

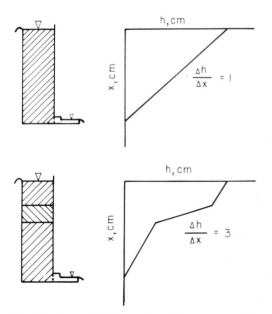

Fig.5.1. Schematic illustration of pressure gradient change due to layer of lower hydraulic conductivity.

than calculated if a uniform gradient is assumed. In the example shown in Fig.5.1, it is assumed that the gradient in the less permeable layer is three times as large as it would be if the gradient were uniform. Thus, the flow through this layer would be three times as large.

For a soil profile of depth L with layers of different thickness L_i, the analysis by Swartzendruber (1960) shows that an equivalent conductivity K_e for the entire profile can be defined from Darcy's equation as:

$$K_e = L \sum_{i=1}^{n} L_j / K_j \tag{5.8}$$

The result is independent of the order of summation, so it is a general relationship valid for any pattern of layering. The amount of flow Q through a layered profile of length L_o is:

$$Q = -K_e A \, \Delta h / L_o$$

Taking the ratio of this flow Q to the flow Q_o through a uniform profile of the same length and conductivity K_0 gives:

$$\frac{Q}{Q_o} = \frac{-K_e A \, \Delta h / L_o}{-K_o A \, \Delta h / L_o} = \frac{K_e}{K_o} \tag{5.9}$$

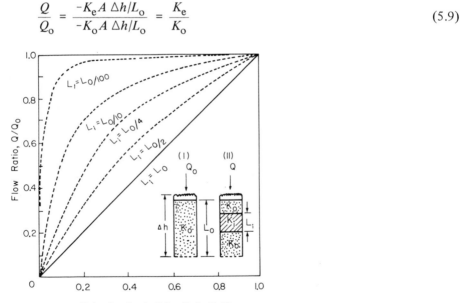

Fig.5.2. Curves of flow ratio vs. hydraulic conductivity ratio. Flow and soil-profile conditions are shown diagrammatically in *I* and *II*. (Swartzendruber, 1960. *J. Geophys, Res.*, 65: 4038, © Am. Geophys. Union.)

Assuming the layered profile contains one section of length L_i the remainder having a conductivity K_o, and substituting eq.5.8 for K_e into eq.5.9 gives:

$$\frac{Q}{Q_o} = \frac{K_i/K_o}{L_i/L_o \ K_i/K_o \ (1 - L_i/L_o)} \tag{5.10}$$

This equation is plotted in Fig.5.2 which shows by how much Q/Q_o is larger than K_i/K_o. If the permeability of the least permeable layer determines flow, Q/Q_o would equal K_i/K_o.

Factors affecting saturated flow

An examination of eq.5.1 and 5.3, shows that the factors and parameters determining permeability fall into two groups: (a) those factors and parameters associated with the permeant, and (b) factors and parameters associated with the physical properties of the soil. In addition, a more precise evaluation of permeability and hydraulic conductivity, requires the addition of a further group of factors – (c) factors associated with the forces holding water to soils, and clay–water interaction. These factors can be listed as:

(a) Factors associated with permeant: viscosity, pressure, density.

(b) Soil properties: tortuosity, void ratio (porosity), soil-water potential, pore-size distribution, fabric.

(c) Soil-water interaction: heat of wetting, ionic concentration, thickness of layers of water held to soil particles.

It is difficult to assess the relative importance of each of the factors, since many of them are interdependent. One may with controlled experiments vary many of these factors directly or indirectly and make the necessary measurements. However, a complete separation of individual effects may not be attained because the effect of a change in one factor on the performance of other contributing factors cannot be assessed precisely. This demonstrates the interdependence between many of the variables in the system.

In eq.5.7, if the viscosity of the permeant is kept constant, k will increase if γ increases. On the other hand, if γ is constant and η increases, then k will decrease. Pore-size distribution and particle-size influence the porosity n. The role of porosity n may be seen in the form that n takes in eq.5.7. In the case of soil-water interaction, the conductivity k will depend upon the degree of development of the clay-water forces. This is discussed in greater detail when flow through clay soils is considered.

The effect of temperature on liquid transfer in saturated flow is not fully known, since temperature affects both the permeant and clay-water interaction.

Water moves under the influence of a thermal gradient. The resultant changes in the soil-water system arising from a change in temperature will contribute to the total movement of water.

Saturated flow in clays

The limitations of both the Darcy and Kozeny-Carman relationships may be shown experimentally in controlled laboratory experiments. The two implied conditions in the Darcy relationship are (a) flow rate is directly proportional to hydraulic gradient, and (b) the relationship between flow rate and hydraulic gradient is linear through the origin. The condition of linearity between flow rate and hydraulic gradient is valid for porous media and for many clay soils. However, the condition specifying linearity through the origin has been shown not to be valid for all clays (Hansbo, 1960). The saturated flow tests performed by the Swedish Geotechnical Institute on four undisturbed natural clays show the existence of both a critical gradient and an apparent threshold gradient (Fig.5.3).

The existence of a critical gradient in saturated flow in clays suggests that the factors associated with the soil properties become increasingly important at gradients below the critical value.

Plotting the results of measured flow rate and relating this as a ratio of measured to predicted flow ratio to porosity (Fig.5.4) for several soils, we can see graphically the divergence between reality and simplified theory as represented by the Kozeny-Carman relationship.

From the previous discussions on factors affecting flow and from Fig.5.3 and

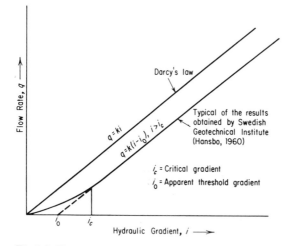

Fig.5.3. Hydraulic flow rates vs. hydraulic gradient. (Olsen, 1961; from Hansbo, 1960.)

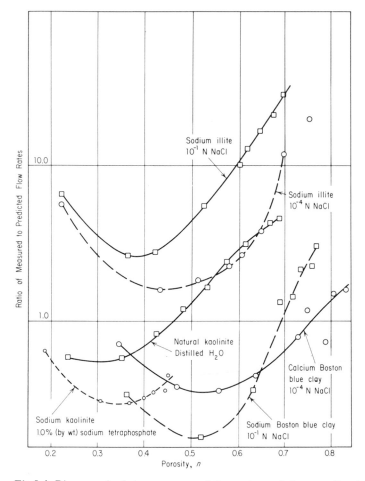

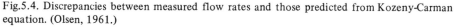

Fig.5.4. Discrepancies between measured flow rates and those predicted from Kozeny-Carman equation. (Olsen, 1961.)

5.4, we must conclude that for a given permeant (water) and under standard conditions, soil and clay-water forces affect moisture migration. The presence of these forces demonstrated in terms of double layer and diffuse ion-layer water (see Chapter 6) in effect create immobilized hydrodynamic layers of water surrounding each particle. The thickness of these immobilized hydrodynamic layers is dependent on the interaction characteristics of the soil-water system. By altering the concentration, for example, of sodium chloride in illite samples, the thickness of the immobilized layers is correspondingly affected as shown by the change in flow rate for the same porosity.

Consider the problem of tortuosity as influenced by fabric and clay-water

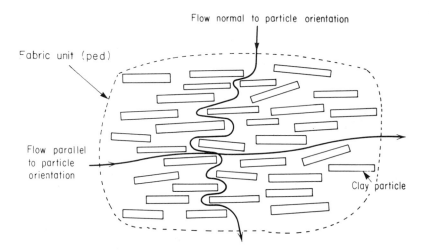

Fig.5.5. Effect of particle orientation on flow path.

forces. In a well oriented clay (Fig.5.5) tortuosity will be increased if water movement is normal to the orientation of particles.

The ratio of l_e/l, i.e., ratio of effective flow path to sample thickness, is larger for flow in the normal direction.

In a random structure, l_e/l, is insensitive to direction of flow. For flocculated structures, the possibility of clusters or packets forming will influence flow. In general, however, l_e/l is still insensitive to direction of flow (Fig.5.6).

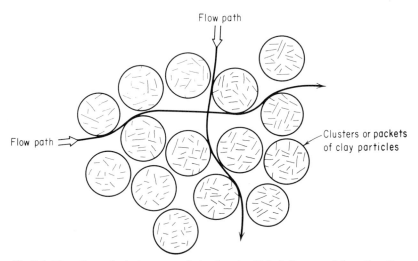

Fig.5.6. Flow through clusters or packets, showing little influence of flow direction.

The role of soil structure and of soil fabric is an important aspect in studying hydraulic conductivity. Since fabric units of various sizes constitute integral components of most clay soils, it is clear that liquid transfer or fluid flow between fabric units will be different from fluid flow through fabric units. This is shown schematically in Fig.5.7.

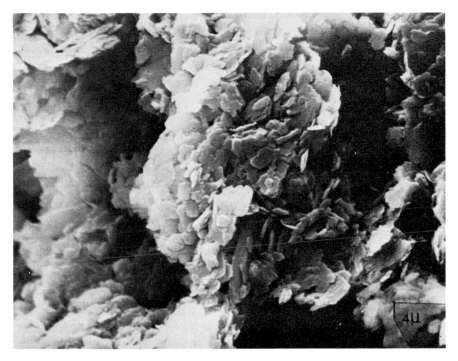

Fig.5.7. Kaolin soil ped arrangement showing macro and micropores. Picture shows that fluid flow through macro and micro pores will be subject to different boundary constraints and forces. Micropores within ped units (e.g., in middle ped) by and large show higher tortuosity and resistance to flow.

It is apparent that fluid flows dominantly between fabric units. The arrangement of these units would obviously determine the degree of tortuosity. Fluid flow through the individual fabric units would be very slow under low pressure heads. This aspect of flow which involves a separate mechanism controlling flow becomes important in highly precompressed soils, in consolidation studies, and also in chemical treatment problems where treatment through the soil-water regime is desired.

The clay-water forces will influence flow in Fig.5.5 and 5.6 to the extent that driving forces for flow must exceed the forces tending to hold water to the soil particles. At the midplane between two particles, the midplane potential must be

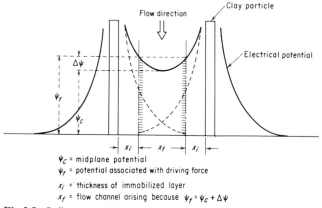

Fig.5.8. Influence of diffuse ion-layers of clay particles on thickness of immobilized water layers.

exceeded before liquid transfer can occur. The restraint imposed by the hydrodynamic immobilized layers is dependent on the driving forces inducing flow. If the driving forces are large (as, for example, a high pressure gradient) the midplane potential will be greatly exceeded thus allowing for a larger flow channel (Fig.5.8). In other words, the hydrodynamic immobilized layer is not a unique layer. Rather its thickness is dependent on not only the clay-water forces but also on the driving forces inducing flow. A change in the intensity of the clay-water forces will cause a corresponding change in the thickness of the immobilized layer.

The primary physical factors considered responsible for the discrepancies between predicted and measured flow rates have been discussed. The results from studies available are not in total agreement as to the degree of influence of the factors, but there is common agreement as to the need for more study in this area.

Steady-state flow

For considerations of steady-state flow, it is assumed that the soil is as near to being saturated as possible to allow for the flow to occur.

Consider a soil cube of dimensions dx, dy, and dz as shown in Fig.5.9. If M_i = mass input, then $\partial M_i/\partial t$ = mass input flux.

For simplicity in the analysis, we will consider that the fluid is incompressible, with a density of δ_w. Assuming that Darcy's equation is valid and that $k_x = k_y = k_z$; then in the x direction:

$$\frac{\partial M_{ix}}{\partial t} = \delta_w V_x\, dy\, dz$$

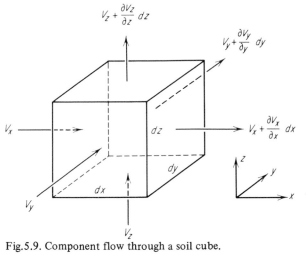

Fig.5.9. Component flow through a soil cube.

the mass outflow flux $\partial M_o / \partial t$ is given by:

$$\frac{\partial M_{ox}}{\partial t} = \delta_w \left(V_x + \frac{\partial V_x}{\partial x} \, dx \right) dy \, dz$$

The rate of storage in the x-direction is the difference between the mass input and output flux. Thus:

$$\frac{\partial M_{ix}}{\partial t} - \frac{\partial M_{ox}}{\partial t} = \text{rate of storage}$$

$$= - \delta_w \frac{\partial V_x}{\partial x} \, dx \, dy \, dz$$

By symmetry, the rate of storage arising from flow in the x, y, and z directions will be given as:

$$\text{rate of storage} = - \delta_w \left(\frac{\partial V_x}{\partial x} + \frac{\partial V_y}{\partial y} + \frac{\partial V_z}{\partial z} \right) dx \, dy \, dz$$

By assuming steady-state flow, the rate of storage becomes zero. Hence:

$$\delta_w \left(\frac{\partial V_x}{\partial x} + \frac{\partial V_y}{\partial y} + \frac{\partial V_z}{\partial z} \right) dx \, dy \, dz = 0$$

Since γ_w and dxdydz cannot be zero, it follows that:

$$\frac{\partial V_x}{\partial x} + \frac{\partial V_y}{\partial y} + \frac{\partial V_z}{\partial z} = 0 \qquad (5.11)$$

Applying Darcy's equation (5.1), eq.5.11 may be written as:

$$\frac{\partial^2 h}{\partial x^2} + \frac{\partial^2 h}{\partial y^2} + \frac{\partial^2 h}{\partial z^2} = 0 \qquad (5.12)$$

if $k_x = k_y = k_z = k$.

For a semi-infinite mass it is not inappropriate to consider a two-dimensional flow problem. Thus eq.5.12 will be reduced to the following form:

$$\frac{\partial^2 h}{\partial x^2} + \frac{\partial^2 h}{\partial z^2} = 0 \qquad (5.13)$$

This derivation for saturated steady-state flow pays no attention to driving forces other than the pressure potential or hydraulic head term given as $\partial h/\partial x$ or $\partial h/\partial z$. The assumption of steady-state flow relegates to other potentials such as osmotic and temperature a minor role.

The hydraulic conductivity term k may be accounted for in eq.5.13 to allow for differences in hydraulic conductivity in the x-, y-, or z-directions. The step from eq.5.11 to eq.5.12 may include k, in which instance:

$$k_x \frac{\partial^2 h}{\partial x^2} + k_y \frac{\partial^2 h}{\partial y^2} + k_z \frac{\partial^2 h}{\partial z^2} = 0 \qquad (5.14)$$

or for a two-dimensional case:

$$k_x \frac{\partial^2 h}{\partial x} + k_z \frac{\partial^2 h}{\partial z^2} = 0$$

Eq.5.12 is generally known as the Laplace equation. The solution to a physical problem on groundwater flow involves finding a function $f(x,y,z)$, or $f(x,z)$ for a two-dimensional case, which satisfies both the Laplace equation and the boundary conditions associated with the physical problem. Of the many methods available for computing the quantity of groundwater flow or seepage, e.g., the use of complex variables and numerical methods, the approximate sketching technique is generally the simplest to use. Methods of solution of eq.5.14 for various kinds of problems (i.e., initial and boundary conditions) may be found in specialized books on seepage and groundwater flow.

5.3   UNSATURATED FLOW

General considerations

Most of water movement in soil occurs when both water and air are present in the voids — unsaturated flow. This is true even where water moves into the soil from ponded water at the surface. The value of k is no longer constant but decreases as the water content decreases or as the soil-water potential (negative) decreases. This relationship is shown in Fig.5.10. The rapid decrease of hydraulic conductivity with decreasing water content occurs because the large pores empty first when the soil becomes unsaturated. Since permeability varies directly as a power of the pore radius as seen from the development of the Kozeny-Carman relationship (eq.5.7), the larger pores make the largest contribution to water flow. This explains why a soil will reach an approximately constant water content called

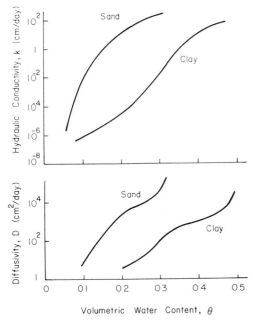

Fig.5.10. Variation of hydraulic conductivity and diffusivity with water content for two soils. (Staple, 1969. *Soil Sci. Am. Proc.,* 33, pp.842 and 843.)

the field capacity. The rate of movement of water becomes too slow to further decrease the water content appreciably. For the same reason it is impossible to wet a soil uniformly to a water content below the field capacity by adding water at the surface. With a limited amount of water, the top layer of soil will be at field capacity and the lower layer will remain dry. At this point the large pores are emptied, the water exists as films around particles or in small voids, and the connections between films are limited.

Two parameters, k and D, are used in unsaturated flow studies. They are defined for one-dimensional flow by eq.5.15:

$$v = -k(\theta)\,\frac{\partial \psi}{\partial x}$$

$$\text{(5.15)}$$

$$v = D(\theta)\,\frac{\partial \theta}{\partial x}$$

The diffusivity, $D(\theta)$, is used because water-content gradients are sometimes easier to measure, and also because some water-flow equations are more easily solved with diffusivity rather than conductivity. The term diffusivity does not indicate moisture transfer by diffusion.

Unsaturated fluid flow in soils can be classified into three general types according to the different phenomena involving soil fabric change occurring during the flow process and development of associated potential gradients. The three systems are:

CASE 1. No change in soil fabric, during flow, i.e. no change in pore geometry or porosity.

CASE 2. A change in soil fabric during flow but resulting in no change in overall porosity for: (a) non-swelling soils, (b) swelling soils under confined conditions.

CASE 3. A complete change in fabric, i.e. a change in pore geometry as well as in porosity and volume change in the soil.

Case 1 describes flow in a non-swelling soil with unchanging pore geometry and with the physical boundaries perfectly confined. This case fits the situation encountered in cemented or highly compact material. Case 2 relates to (a) unsaturated flow in non-swelling or low-swelling soils with changes in pore geometry due to advance of the fluid front, but with no significant development of swelling pressures; and (b) unsaturated flow in swelling clays where the physical boundaries are perfectly confined, thus allowing for the development of swelling pressures. In Case 3, unsaturated fluid flow in a swelling clay with free swelling allowed is described. Changes in pore volume and overall soil volume would obviously alter fluid transmission characteristics.

Unsaturated flow equations for no volume change

Unsaturated fluid flow in soils satisfying the physical condition of little or no volume change during flow is generally described by an equation analogous to the heat-flow equation. The formulation for analysis of flow presumes that Darcy's equation holds for unsaturated flow. Hence, generalizing the hydraulic gradient, grad h, and replacing it with the soil-water-potential gradient, grad ψ, the Darcy equation is written as:

$$v = -k \text{ grad } \psi \qquad (5.16)$$

where:

v = vector flow velocity;

$$\text{grad } \psi = \left(\frac{\partial}{\partial x} i + \frac{\partial}{\partial y} j + \frac{\partial}{\partial z} k \right) \psi$$

The negative sign in eq.5.16 accounts for the fact that water flow is in the direction of decreasing potential. The equation of continuity, which states that the flow of water into or out of a unit of soil equals the rate of change in water content implicitly indicates that the system is preserved, i.e., no volume change. Thus:

$$\text{div } v = - \frac{\partial \theta}{\partial t} \qquad (5.17)$$

where:

$$\text{div } v = \left(\frac{\partial}{\partial x} i + \frac{\partial}{\partial y} j + \frac{\partial}{\partial z} k \right) \cdot v$$

θ = volumetric water content of the soil.

Substitution of eq.5.17 into eq.5.16 will yield the general three-dimensional diffusion equation. Thus:

$$\frac{\partial \theta}{\partial t} = \text{div} (k \text{ grad } \psi) \qquad (5.18)$$

For horizontal water flow, the gravity potential ψ_g representing gravity forces is zero, i.e. $\psi_g = 0$, and assuming only matric potentials (no osmotic or other potential gradients) then $\psi = \psi_m$. If ψ is a unique function of θ, eq.5.18 can be reduced to:

$$\frac{\partial \theta}{\partial t} = \frac{\partial}{\partial x}\left(k\,\frac{\partial \psi}{\partial \theta}\,\frac{\partial \theta}{\partial x}\right) \tag{5.19}$$

where x is the horizontal coordinate axis.

Solution of this equation is facilitated by introducing the diffusivity or diffusion coefficient of water $D(\theta) = k(\theta)\,(\partial\psi/\partial\theta)$, which makes the equation analogous to those describing thermal diffusion of heat.

$$\frac{\partial \theta}{\partial t} = \frac{\partial}{\partial x}\left(k\,\frac{\partial \psi}{\partial \theta}\,\frac{\partial \theta}{\partial x}\right) \tag{5.20}$$

$$= \frac{\partial}{\partial x}\left(D(\theta)\,\frac{\partial \theta}{\partial x}\right)$$

For vertical water flow, e.g., during infiltration, $\partial\psi_g/\partial z = 1$, since the gravitational potential is just equal to the height, z, above the reference level. Thus eq.5.20 becomes:

$$\frac{\partial \theta}{\partial t} = \frac{\partial}{\partial z}\left(k\,\frac{\partial\psi_m + \partial\psi_g}{\partial z}\right) \tag{5.21}$$

$$= \frac{\partial}{\partial z}\left(k\,\frac{\partial \psi}{\partial z} + \frac{\partial k}{\partial z}\right)$$

or, introducing the diffusivity, D:

$$\frac{\partial \theta}{\partial t} = \frac{\partial}{\partial z}\left(D\,\frac{\partial \theta}{\partial z}\right) + \frac{\partial k}{\partial z} \tag{5.22}$$

Graphically (experimental) aided solutions to eq.5.20 and 5.21 can be obtained in the form of values of θ vs x or z at different times, t (Fig.5.12). Values of k as a function of θ are determined experimentally, or may in some cases be calculated. Values of $\partial\psi/\partial\theta$ are measured from the slope of the soil-water potential versus water-content curve for the soil such as those shown in Chapter 4. The diffusion equation is solved analytically subject to the boundary conditions imposed by the problem under consideration.

It is pertinent and useful to point out that the diffusion equation depends upon the conditions leading to its formulation, i.e., (1) Darcy's law is valid, and (2) $D = D(\theta)$, which results from the definition:

$$D(\theta) = k(\theta) \frac{\partial \psi}{\partial \theta} \qquad (5.23)$$

If swelling of the soil sample occurs during and as a result of infiltration, it becomes obvious that the rate of movement of water relative to a fixed coordinate system does not remain the same as that with respect to the soil particles. Then the specialized form of the continuity equation given in eq.5.17 which requires fluid incompressibility and no volume change, is no longer valid. Thus the form of diffusion equations utilizing this continuity condition (eq.5.18, 5.20, 5.22) not admissible.

For the case where conditions leading to the development of eq.5.18, 5.20, 5.22 remain valid, i.e. no volume change, infiltration test results such as those shown in Fig.5.11 indicate that advantage can be taken, for horizontal infiltration, of the fact that linearity is obtained between x, the distance of the wet front from the source, and $\sqrt{t}$, the square root of time taken for the wet front to reach x. By resorting to a similarity solution technique, invoking the Boltzmann transform:

$$\lambda = \frac{x}{\sqrt{t}} = \lambda(\theta) \qquad (5.24)$$

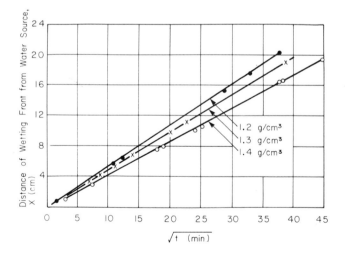

Fig.5.11. Wetting front advance. x vs. square root time, t, for a kaolin clay at three densities.

it becomes possible to reduce eq.5.20 to an ordinary differential equation:

$$\frac{\lambda}{2}\frac{d\theta}{d\lambda} = \frac{d}{d\lambda}\left(D\frac{d\theta}{d\lambda}\right) \tag{5.25}$$

We note that the physical requirements for the similarity solution are adequately met, i.e. actual physical linearity is obtained between x and $\sqrt{t}$ as demanded by the Boltzmann transform.

Subject to the usual boundary conditions:

$$t = 0, \quad 0 < x < \infty \text{ (thus } \lambda = \infty), \quad \theta = \theta_i$$

$$t > 0, \quad x = 0 \text{ (thus } \lambda = 0) \quad \theta = \theta_0, \text{ i.e. } \theta \text{ at } x = 0$$

the total amount of water entering per unit area $(q_t)_0$ at the plane $x = 0$ can be obtained from a solution of the combination of eq.5.25 and 5.20 as:

$$(q_t)_0 = t^{\frac{1}{2}}\int_{\theta_i}^{\theta_0} \lambda\, d\,\theta = \int_{\theta_i}^{\theta} x\,dx \tag{5.26}$$

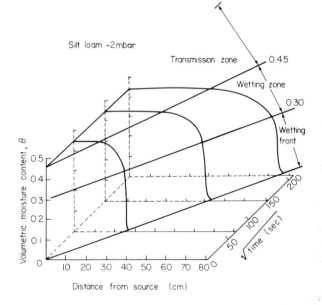

Fig.5.12. Time-wetting front profiles for a silt loam with negative intake at -2 mbar.

The physical performance demonstrated by eq.5.26 may be seen in the three-dimensional representation of flow into an unsaturated soil (Fig.5.12) depicting time, distance from source of water and volumetric moisture content. The total quantity of water at any time (and thus at any position) will be indicated by the volume under the $\theta - x - \sqrt{t}$ surface shown in Fig.5.12. We note that projection onto the horizontal plane provides a linear relationship between x and $\sqrt{t}$ thus satisfying the assumption used in eq.5.25 and allows for admissibility of eq.5.26.

The wetting front profiles shown on the vertical plane projection in Fig.5.12 are characteristic for the situation of unsaturated flow with little or no volume change.

Unsaturated flow equations for volume-change cases

Linearity between x and $\sqrt{t}$ as required in the application of the Boltzmann transform used in arriving at eq.5.25, allows use of the ordinary differential

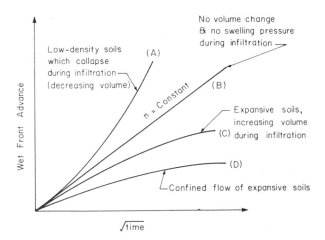

Fig.5.13. Wetting front advance vs. square root of time showing influence of volume change and swelling pressure on time rate of advance profile.

equations leading to the solution shown in eq.5.26. It is clear from Fig.5.13 that if curves *A*, *C* and *D* are encountered, application of the Boltzmann transform as a similarity solution technique is no longer possible. It becomes necessary therefore to work with the generalized diffusion eq.5.18 or the other specialized forms shown as eq.5.20 and 5.22.

Such being the case, it is necessary to examine why curves *A*, *C* and *D* deviate from *B*. For *A* and *C*, the primary reason for departure from linearity is seen to be the result of volume change occurring during flow. This may be (1) a decrease in volume as in the case of low-density soils which will collapse in the presence of increased water availability (curve A), or (2) an increase in volume as in the case of swelling soils allowed to swell in the presence of available water. This is represented by curve C.

The situation for curve D represents the condition of unsaturated flow in swelling soils where no volume change is allowed to occur. Under such circumstances, swelling pressures will result and flow characteristics will be altered accordingly. This will be discussed separately in a later section.

Thus, the primary factors giving rise to curves A and C would require that the continuity condition shown in eq.5.17 must be altered to suitably account for volume change. Working with the horizontal infiltration study, the no volume change continuity condition is represented as:

$$\frac{\partial V_x}{\partial x} = - \frac{\partial \theta}{\partial t}$$

where V_x = velocity of flow in the x direction.

If S = degree of saturation and n = porosity, θ will be given as nS. Thus:

$$\frac{\partial V_x}{\partial x} = - \frac{\partial}{\partial t} (nS)$$

$$= - \left[S \frac{\partial n}{\partial t} + n \frac{\partial S}{\partial t} \right]$$

and the diffusion equation for horizontal unsaturated flow now becomes:

$$S \frac{\partial n}{\partial t} + n \frac{\partial S}{\partial t} = \frac{\partial}{\partial x} \left(D \frac{\partial \theta}{\partial x} \right) \tag{5.27}$$

Eq.5.27 is approximate in that it does not regard the actual physics of flow which would require that D be suitably modified to account for the influence of volume change on D and also on k. However, where small changes in porosity n are involved the above approximate relationship may be used.

To correct for D and $\partial \theta / \partial x$ in view of volume change, we may introduce a change in the value of x which recognizes, in effect, a changing porosity. Thus, defining x' as:

$$x' = x + (\Delta n / n) \tag{5.28}$$

eq.5.27 may be written as:

$$S \frac{\partial n}{\partial t} + n \frac{\partial S}{\partial t} = \frac{\partial}{\partial x} \left(D' \frac{\partial \theta'}{\partial x'} \right) \tag{5.29}$$

where θ' relates to the volumetric water content in view of x' and $D' = D(\theta')$, i.e., the diffusivity D' is a function of θ'.

Instead of using the simple correction for x, as in Eq.5.28 for a changing porosity (i.e., changing volume), we may use instead, a material coordinate system similar to that used by Philip and Smiles (1969) which accounts for volume change. The merit in this kind of treatment is its simplicity since the analytical formulation is similar, in a sense, to the no volume change analysis. Specifying the relation between material coordinate m and x as:

$$m = \int_{-\infty}^{x} \frac{dx}{1+e}$$

where e = void ratio.

The corresponding continuity condition will then be:

$$\frac{\partial}{\partial t} [\theta (1+e)] = -\frac{\partial V_{ws}}{\partial m} \tag{5.30}$$

where V_{ws} = velocity of water relative to moving soil particles.

By using Darcy's equation in terms of water movement relative to the moving soil particles:

$$V_{ws} = -k \frac{\partial \psi}{\partial x} \tag{5.31}$$

Eq.5.30 now becomes:

$$\frac{\partial}{\partial t} [\theta (1+e)] = \frac{\partial}{\partial m} \left(k \frac{\partial \psi}{\partial x} \right) \tag{5.32}$$

and thus:

$$C \frac{\partial \theta}{\partial t} = \frac{\partial}{\partial m} \left(D_m \frac{\partial \theta}{\partial m} \right) \tag{5.33}$$

where e and ψ are considered functions of θ only, and:

$$C = 1 + e + \theta \frac{de}{d\theta}$$

$$D_m = (1 + e)^{-1} k \frac{d\psi}{d\theta}$$

We note the similarity in form of eq.5.33 with eq.5.20. Thus, the techniques for solution of eq.5.33 can be considered to be similar to those used to solve eq.5.20. The similarity transformation technique used previously will now be used to relate m with $\sqrt{t}$ to provide:

$$\lambda' = m/\sqrt{t}$$

Eq.5.33 can now be reduced to an ordinary differential equation and solved subject to appropriate specification of boundary conditions.

A generalized unsaturated flow equation

A way in which changing and non-changing soil volume can be incorporated into unsaturated flow analyses, without utilizing modifications in coordinate systems, is to account for the movement of soil particles in view of pressures developed in the soil due to the presence and flow of water. Thus if we consider the status of an elemental soil volume of dimensions dx, dy and dz, and if one dimensional unsaturated flow in the x-direction is examined, soil particles flowing into and out of the elemental cube must satisfy conservation principles. Thus:

$$\frac{\partial V_{sx}}{\partial x} = \frac{1}{1 - n} \frac{\partial n}{\partial t}$$

where V_{sx} = velocity of soil particles in the x direction, i.e., soil particle flux in the x direction. However, since:

$$\frac{1}{1 - n} \frac{\partial n}{\partial t} = \frac{\partial \nu}{\partial t}$$

where ν = volumetric strain, we will thus obtain:

$$\frac{\partial V_{sx}}{\partial x} = \frac{\partial \nu}{\partial t} \qquad\qquad (5.34)$$

Using the flow condition which states that the total fluid velocity is composed of (a) fluid flow relative to the moving soil particles and (b) soil particle flow contained in the fluid phase which will be in a position to move:

$$V_x = V_{wsx} + \theta V_{sx} \qquad\qquad (5.35)$$

where V_{wsx} = water movement relative to the moving soil particles.

Differentiating eq.5.35 and using eq.5.31 and 5.35 we obtain:

$$\frac{\partial V_x}{\partial x} = - \frac{\partial}{\partial x}\left(k\,\frac{\partial \psi}{\partial x}\right) + \theta\,\frac{\partial v}{\partial t}$$

Making the same assumptions for the diffusion coefficient D as in eq.5.23:

$$\frac{\partial V_x}{\partial x} = - \frac{\partial}{\partial x}\left(D\,\frac{\partial \theta}{\partial x}\right) + \theta\,\frac{\partial v}{\partial t} \qquad\qquad (5.36)$$

The continuity condition now specifies that the rate of change in water content, i.e. $\partial\theta/\partial t$, must be equal to $\partial V_x/\partial x$ in eq.5.36 since the attendant volume changes have been suitably accounted for. Thus, we will obtain from:

$$\frac{\partial V_x}{\partial x} = - \frac{\partial \theta}{\partial t}$$

the resultant diffusion equation for one dimensional horizontal unsaturated flow which states that:

$$\frac{\partial \theta}{\partial t} = \frac{\partial}{\partial x}\left(D\,\frac{\partial \theta}{\partial x}\right) - \theta\,\frac{\partial v}{\partial t} \qquad\qquad (5.37)$$

Eq.5.37 represents a generalized diffusion equation which allows for volume change. If the soil-water system remains rigid, i.e. if no volume change occurs during flow, $\partial v/\partial t = 0$. Thus eq.5.37 is reduced to the form shown in eq.5.20. Eq.5.37 has also been obtained (with certain simplifying assumptions) from a more rigorous base, (in Appendix 3), using the requirement that the flux of soil particles must satisfy (a) continuity, (b) Newton's law of motion, and (c) a rheological equation of state.

The usefulness of the various analytical expressions derived to study unsaturated flow depends on the ability of the investigator to measure or determine the appropriate soil properties, e.g., D, k, ν, etc. Application of the various equations together with the generalized form shown as eq.5.37 will be discussed in the next section.

5.4 MOISTURE PROFILES AND WETTING FRONT ADVANCE

Prediction of the rate of advance of the wetting front in unsaturated soils requires that the equations developed above be properly used. Fundamental to the requirements of any of the equations shown previously is the need for a knowledge of the variation of the diffusion coefficient D with water content θ. As stated in the previous section, for unsaturated flow into low or non-swelling soils where little or no volume change occurs, the characteristic moisture profile development shown in Fig.5.12 can be expected. For cases where volume change becomes significant, or where other mechanisms exist which are not accounted for in the simple formulations provided previously, the moisture profiles can demonstrate many

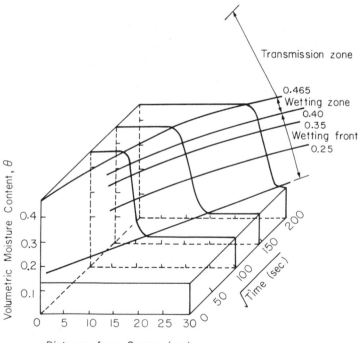

Fig.5.14. Time-wetting profiles for unsaturated flow into a low density kaolinite. Volume decrease occurs during infiltration due to collapse of soil on wetting.

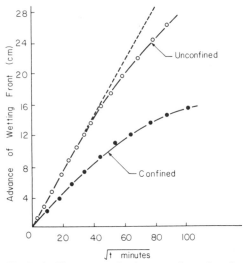

Fig.5.15. Time rate of advance of wetting front for unsaturated flow into a swelling soil showing difference in profile if volume change is allowed. In confined flow into the swelling soil, no volume change is allowed.

different shapes for the profile. Fig.5.14 and 5.15 show some profiles for soils with volume change. In Fig.5.14, because of the initial low density of the soil, collapse occurs during flow. In Fig.5.15, the swelling characteristics of the soil will produce volumetric expansion if no constraints to swelling are applied. The shapes of the

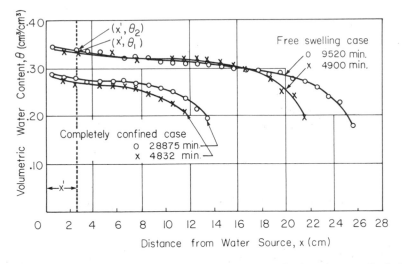

Fig.5.16. Comparison of moisture profiles for confined and unconfined flow into an unsaturated swelling soil. (Wong, 1971.)

profiles shown in Fig.5.14 and 5.15 can be compared with those seen in Fig.5.12 to highlight the "non classical" flow characteristics. If samples are confined to prevent swelling, the shapes of the wetting fronts do not appear to be affected. The rate of advance of the wetting front, however, is significantly changed if an unsaturated swelling soil is not permitted to expand during fluid flow into the soil (Fig.5.16).

Diffusivity functions

The influence of various kinds of relationships between D and θ can be seen by writing the right-hand side of eq.5.20 in finite difference form and solving for certain specific cases governing θ and D.
Thus writing eq.5.20* as:

$$\frac{\partial \theta}{\partial t} = \frac{1}{(\Delta x)^2} [D_{n-\frac{1}{2}} (\theta_{n+1} + \theta_{n-1} - 2\theta_n) +$$

$$(D_{n+\frac{1}{2}} - D_{n-\frac{1}{2}}) (\theta_{n+1} - \theta_n)] \ldots$$

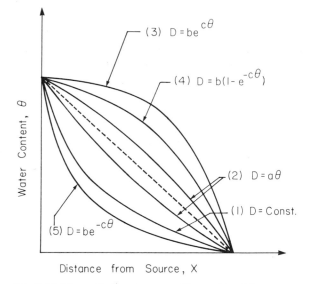

Fig.5.17. Theoretical moisture profiles predicted from evaluation of diffusion equation.

* Note that eq.5.20 can be used since $(\theta(\partial v/\partial t))$ does not really enter into or need participate in the characterization of D. The volume change term in eq.5.37 will affect the numerical value of D but has an insignificant effect on characterizing (shape) the $\theta - D$ relationship.

where the subscript n denotes space iteration such that $x_n = n(\Delta x)$. Considering only the absorption process, Yong and Wong (1973) have examined five relationships between θ and D, the results of which are shown in Fig.5.17. They indicate the very strong participation of the $\theta-D$ relationship in characterizing the shape of the moisture profile.

The cases considered in Fig.5.17 are:

CASE 1. D = constant.
CASE 2. D increases linearly with θ in the fashion shown as:

$$D = a\theta$$

where a = positive constant.

CASE 3. D increases more rapidly than the corresponding increase in θ as flow progresses, i.e.:

$$D = be^{c\theta}$$

where b and c are positive constants.

CASE 4. D increases less rapidly than the corresponding increase in θ as flow progresses, i.e.:

$$D = b[1 - e^{-c\theta}]$$

CASE 5. D decreases as θ increases. There are many functions that will satisfy this requirement. A typical function could be:

$$D = be^{-c\theta}$$

In the solution given in eq.5.26, the simplifying assumption corresponding to case 1 has been made to allow for a simple analytic solution. By and large, solution of the diffusion equations with cases 2 to 5 will require numerical and machine computational techniques. It is clear from Fig.5.14–5.17 that the simple situation of D = constant is not correct.

The diffusion coefficient can be calculated from an experimentally measured moisture profile. The profile is divided into n equal parts (Wong and Yong, 1973) as shown in Fig.5.18.

$$(\theta_o - \theta_i)/n = \Delta\theta \quad \text{and} \quad \theta_r = (n-r)\Delta\theta + \theta_i$$

Since the part of the curve between A and B in Fig.5.18 is continuous and differentiable, therefore, by the Mean Value Theorem, there will be a point C in between A and B such that the tangent (i.e. slope) at C is equal to the slope of the straight line joinint A and B, i.e.:

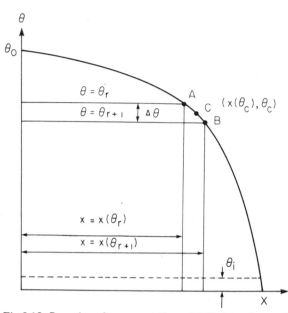

Fig.5.18. Procedure for computation of diffusivity from soil moisture profile. (Wong and Yong, 1973.)

$$\left(\frac{d\theta}{dx}\right)_{\theta_c} = \frac{-\Delta\theta}{x(\theta_{r+1}) - x(\theta_r)} \tag{5.38}$$

By taking $\Delta\theta$ sufficiently small, C can be taken approximately as the point $[x(\theta_{r+\frac{1}{2}}), \theta_{r+\frac{1}{2}}]$ on the part of the curve between A and B. Integrating eq.5.25 with respect to λ, with θ varying from θ_i to θ_c will give the diffusivity at θ_c as:

$$D(\theta_c) = \frac{1}{2}\left(\frac{d\lambda}{d\theta}\right)_{\theta_c} \int_{\theta_i}^{\theta_c} \lambda \, d\theta$$

$$= -\frac{1}{2t}\left(\frac{dx}{d\theta}\right)_{\theta_c} \int_{\theta_i}^{\theta_c} x \, d\theta \tag{5.39}$$

Since θ_c is approximately equal to $\theta_{r+\frac{1}{2}}$, it follows that $D(\theta_c) = D(\theta_{r+\frac{1}{2}})$; substituting eq.5.38 into eq.5.39 gives:

$$D(\theta_c) = D(\theta_{r+\frac{1}{2}}) = \frac{x(\theta_{r+1}) - x(\theta_r)}{2t \, \Delta \theta} \cdot \int_{\theta_i}^{\theta_{r+\frac{1}{2}}} x \, d\theta$$

$$= \frac{x(\theta_{r+1}) - x(\theta_r)}{2t \, \Delta \theta} \cdot \sum_{j=n-1}^{j=r} x(\theta_j) \, \Delta \theta + R \qquad (5.40)$$

where:

$$R = \int_{\theta_i}^{\theta_{n-\frac{1}{2}}} x \, d\theta = x(\theta_i) \, (\theta_{n-\frac{1}{2}} - \theta_i)$$

The maximum error (E_{max}) introduced in the calculation of $D(\theta_c)$ by taking C as the point $[x(\theta_{r+\frac{1}{2}}, \theta_{r+\frac{1}{2}}]$ will be given by:

$$E_{max} = \pm \left[\frac{x(\theta_{r+1}) - x(\theta_r)}{2 \, \Delta \theta} \right] \left[\frac{x(\theta_r) \, \Delta \theta}{2} \right] / t$$

$$= \pm \frac{1}{4} \, [x(\theta_{r+1}) - x(\theta_r)] \, x(\theta_r) / t$$

The error is small except for very small values of θ. Eq.5.40 can be used directly to calculate values of D using machine computation.

5.5 MEASUREMENT OF UNSATURATED HYDRAULIC CONDUCTIVITY, k OR D

The methods used for measuring conductivity or diffusivity can be divided into steady state flow and transient flow methods. Some steady state methods make use of a long soil column, generally saturated at one end with controlled water loss at the other. The amount of water flowing through the soil and the potentials within the soil are measured and the conductivity calculated. If only water contents are measured, then D can be calculated. In another steady state method a soil sample is clamped between two porous plates which are maintained at different soil-water potentials. The amount of flow and the potentials within the soil are again measured and the soil conductivity calculated.

In transient flow or outflow methods, the outflow of water is measured as a function of time after change in equilibrating pressure. Usually these measurements are made in a pressure membrane apparatus. The average hydraulic conductivity between initial and final water contents is calculated. The measurements are repeated for successive increments of pressure, i.e., at successively lower water contents. Results can be obtained in a shorter time with the outflow methods, and different water contents are easily obtained. However, there are more uncertainties in this method than in the steady state methods. Diffusivity can also be calculated from inflow or infiltration measurements.

Steady state, k

A potential gradient is maintained across a sample by applying a different suction to the end plates (Fig.5.19). When the flow rate is constant:

$$v = \frac{Q}{At} = -k\,\frac{\Delta\psi}{\Delta x}$$

The potentials can be taken from the tensions at the end plates or from tensiometers inserted in the soil. This apparatus can be used for cores of un-disturbed soil. It is limited to suctions below 1 bar, and in practice to below 0.5 bar.

Steady state, D

A greater range in potentials can be maintained by supplying water under small suctions at one end of a soil column and allowing evaporation at the other (Fig.5.20). At steady state flow, for a horizontal soil column:

$$v = \frac{Q}{At} = D\,\frac{\Delta\theta}{\Delta x}$$

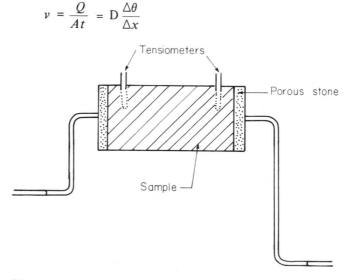

Fig.5.19. Steady state measurement of *k* showing tensiometers and flow measurement.

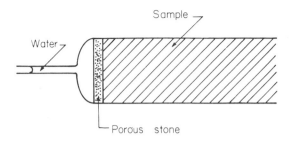

Fig.5.20. Steady state measurement of D.

The soil column can be sliced for θ determination or θ can be measured non-destructively by gamma ray attentuation. A range of water contents is obtained from wet at the inlet to nearly dry at the evaporation end. For any two adjacent slices of soil:

$$D = \frac{Q}{At} \frac{L}{\theta_2 - \theta_1}$$

where L = distance between slices; θ_1, θ_2 = volumetric water content of adjacent slices.

From these values, a curve of D vs θ can be drawn. Values of k can be obtained with the aditional measurement of the water retention curve.

If potentials could be measured in the soil column, k could be calculated directly. Newer methods such as thermocouple psychrometers which can measure potentials between -1 and -20 bar may be applied.

Outflow method for k

The rate, $\partial\theta/\partial t$, at which water moves out of a soil sample under a potential gradient can be used to measure k. Air pressure is applied to the sample (Fig.5.21) and θ vs. t measured. After equilibrium is reached, a higher pressure is applied, and the measurements repeated. For small pressure steps, i.e., small $\Delta\psi$ values, the conductivity, k, can be assumed to be constant over the interval and the flow equation becomes:

$$\frac{\partial\theta}{\partial t} = \frac{\partial\left(k \frac{\partial\psi}{\partial x}\right)}{\partial x} = k \frac{\partial^2\psi}{\partial x^2}$$

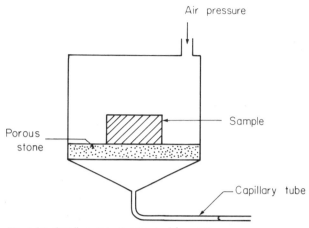

Fig.5.21. Outflow measurement of k and D.

The solution is an infinite series, which converges rapidly for this problem, and the first term alone can be used:

$$Q_t = Q_\infty \left(1 - \frac{8}{\eta^2} e - \frac{\pi^2}{4L^2} Dt \right)$$

where: Q_∞ = total outflow for pressure step; L = length of sample.

A plot of $\ln Qt/Q_\infty$ vs t is used to get the slope $(\pi^2/4L^2)D$. Since $\Delta\psi/\Delta\theta$ is also being measured simultaneously, k is also calculated directly.

The method requires careful measurements and control over evaporation losses. The solution of the equation requires a matching technique which is difficult for some measurements.

Infiltration method for D

One of the simplest experimental methods is to measure the water-content profile at a measured time after infiltration has produced a wet front in a horizontal column of soil (Fig.5.12). The solution of the flow equation for this case involves a change of variables known as the Boltzmann transformation. This is described in detail in the previous section.

Field measurement of k

Values of conductivity measured on small and usually disturbed samples of soil in the laboratory are often poor estimates of conductivity of heterogeneous soil

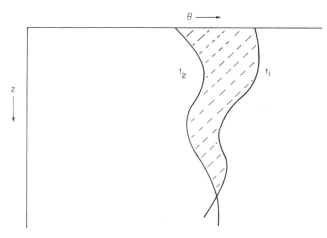

Fig.5.22. Water content profiles at two times after wetting at the surface. Layering in the soil accounts for the shape of the $z - \theta$ curves.

in the field. A procedure has been developed for measuring k in the field which can be applied even to layered soils (Rose et al., 1965). It involves wetting a soil and measuring the change in water content profiles as redistribution takes place (Fig.5.22).

The conservation equation for water content changes in the soil can be written as:

$$\int_{t_1}^{t_2} (P + I - E - v_z)\, dt = \int_{t_1}^{t_2} \int_{z_1}^{z_2} \frac{\partial \theta}{\partial t}\, dz dt$$

where: P = precipitation rate; I = irrigation rate; E = evaporation rate; v_z = vertical flux of water $cm^3\ cm^{-2}\ sec^{-1}$. Also:

$$v_z = -k \frac{\partial \psi}{\partial z} = k \frac{\partial(\psi_m + z)}{\partial z} = k \left(1 + \frac{\partial \psi_m}{\partial z}\right)$$

$$k_z = \frac{\int_{t_1}^{t_2}\left(P + I - E - \int_0^z \frac{\partial \theta}{\partial t}\, dz\right) dt}{\left(\frac{\partial \overline{\psi}_m}{\partial z} + 1\right)_z (t_2 - t_1)}$$

where a mean value of the potential gradient at depth z is used.

The measurement is usually made after rainfall or irrigation, when $P = I = 0$. If a bare surface is covered to prevent evaporation, $E = 0$. Otherwise E can be estimated from energy measurements. The values of $\theta(t_2, z)$ are most easily obtained from neutron probe measurements. The suction is measured directly with tensio-meters for values below 0.8 bar, or taken from a water-retention curve at the measured θ for higher suction.

Calculation of k from void-size distribution

The difficulties in measuring k have lead soil physicists to look for methods of calculating k from other soil properties. Conductivity obviously depends upon size of void filled with water. An average neck size obtained from statistical models can be combined with the Poiseuille equation to give conductivity in terms of void radius. Void radius is usually estimated from water retention values and the equation for height of rise in a capillary. The equation for conductivity can then be more conveniently written in terms of suction. One such equation which has been used successfully is:

$$k_{cm/min} = 1.88 \cdot 10^4 \, F\epsilon^{4/3} \, n^{-2} \, [h_1^{-2} + 3h_2^{-2} + 5h_3^{-2} + \ldots$$

$$(2n-1)h_n^{-2}]$$

where ϵ = water filled void space cm^3/cm^3; n = number of intervals used to divide water retention curve; h = suction in millibars; F = matching factor. Usually the ratio of measured to calculated saturated hydraulic conductivity.

This equation works best for sandy soils, where the void geometry can be measured most meaningfully. The matching factor is a useful empirical addition when the equation is used for a variety of soils. The use of these equations is described in detail by Marshall (1959).

5.6   UNSATURATED FLOW MECHANISMS AND BEHAVIOUR

Interaction of salt and clay

The movement of water and salt in soils, and the associated changes in matric and solute potentials, are influenced by the initial distribution of salt and by the interaction between salt ions and clay. Fig.5.23 shows the moisture profile developed during unsaturated flow of distilled water into an initially unsaturated coarse kaolin soil with initial salt concentration of 0.6 meq/100 g.

The change in soil-water potential due to advance of the wet front shown in Fig.5.23 is given in Fig.5.24. The numbered positions on the diagonal $x-\sqrt{t}$ line

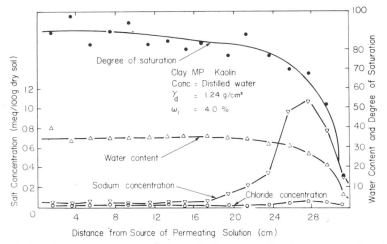

Fig.5.23. Water content, saturation and salt distribution for unsaturated distilled water flow into a coarse kaolin soil. (Yong and Sheeran, 1972.)

indicate the locations of the thermocouple psychrometers used to measure the water potential. At position 1 for example, prior to advance of wet front, the potential value is approximately -33 bars. Following passage of the wetting front, the equilibrium potential value of -2 bars is attained. The corresponding electrical conductivity measurements are shown in Fig.5.25. The numbered positions on the x-$\sqrt{t}$ line indicate the positions of the conductivity measuring cells.

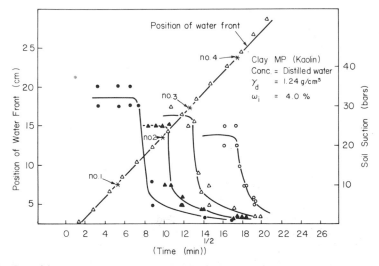

Fig.5.24. Variation of soil suction with advance of wet front. Numbers on wet front line refer to location of psychrometers. (Yong and Sheeran, 1972.)

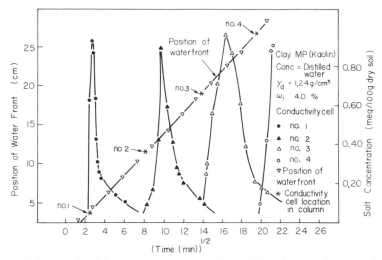

Fig.5.25. Conductivity measurements as influenced by advance of wetting front. Numbers on wet front line refer to location of conductivity cells. (Yong and Sheeran, 1972.)

Swelling in unsaturated flow

If free swelling is permitted during unsaturated flow in swelling soils, soil-water potential will reach a maximum when the wetting front has passed. If the sample is confined, the potential will remain lower. Fig.5.26 shows the volumetric expansion developed during horizontal unsaturated flow into a swelling soil. The information obtained from Fig.5.26 can be used in the general diffusion equation (i.e. eq.5.37) to predict the moisture profile for the swelling soil. The comparison between predicted and measured values may be seen in Fig.5.27.

The analysis or prediction of volumetric strains can also be performed by extending eq.5.34. We recall that:

$$\frac{\partial v_{sx}}{\partial x} = \frac{\partial v}{\partial t}$$

If Φ is the internal pressure responsible for movement of the soil particles in view of the developing fluid flow, we may write:

$$v_{sx} = +k_s \frac{\partial \Phi}{\partial x} = +k_s \frac{\partial \Phi}{\partial v} \frac{\partial v}{\partial x}$$

$$= +D_s \frac{\partial v}{\partial x} \qquad\qquad (5.41)$$

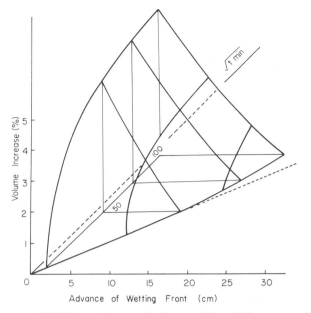

Fig.5.26. Volumetric expansion in a swelling soil due to unsaturated flow.

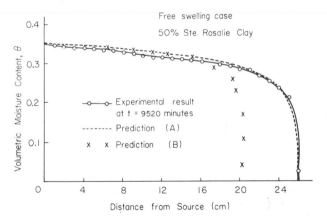

Fig.5.27. Comparison of actual and predicted moisture profiles for unsaturated flow in a swelling soil. Prediction (A) is by Wong (1972) and prediction (B) is obtained using uncorrected classical theory.

where: k_s = soil-particle conductivity coefficient analogous to k for fluid flow; D_s = soil-particle diffusivity coefficient.

Since Φ is responsible for particle movement in unsaturated flow of a swelling soil, the relationship between volumetric strain ν and Φ can be established. Thus:

$$D_s = k_s \frac{\partial \Phi}{\partial v}$$

Eq.5.41 can now be used in eq.5.34 to yield:

$$D_s \frac{\partial^2 v}{\partial x^2} = \frac{\partial v}{\partial t}$$

Subject to the boundary condition:

$$v = v_o \qquad x = 0 \qquad t > 0$$
$$v = 0 \qquad x > 0 \qquad t = 0$$

the analytical solution will be obtained as:

$$v(x, t) = v_o \left[1 - \mathrm{erf}\left(\frac{x}{2\sqrt{D_s t}} \right) \right] \tag{5.42}$$

Fig.5.28 shows the typical case where the theoretical predictions, assuming either a constant or variable D_s, are compared with the measured values of

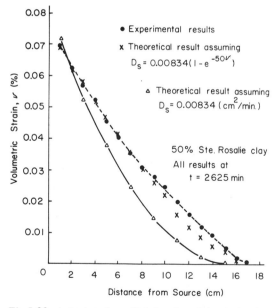

Fig.5.28. Actual and predicted volumetric strains in a swelling soil during unsaturated flow. (Wong, 1972.)

volumetric strain. The value for the constant soil diffusivity D_s can be obtained from a matching method (Wong, 1972) with reliance on available experimental information.

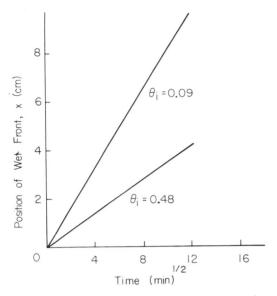

Fig.5.29. Rate of advance of wetting front for moist and air-dry samples of allophane soil from Dominica, W.I. (Maeda, 1970.)

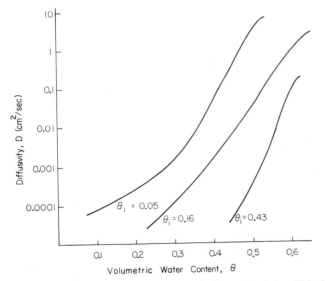

Fig.5.30. Diffusivity of allophane soil from Dominica, W.I. dried to different initial water content θ_i. (Maeda, 1970.)

Flow in allophane clays

Another group of clay soils with distinguishably different unsaturated flow behaviour is the group of allophane containing soils. Allophane, an amorphous clay mineral, differs from other clays in its large and irreversible changes on drying. A wet allophane soil behaves much as other swelling, plastic clay soils. After drying, the allophane behaves as an inactive clay. This accounts for the more rapid advance of the wet front in the dry soil (Fig.5.29), and the higher diffusivity values (Fig.5.30).

Flow due to thermal gradients

Water moves in the soil along a thermal gradient in both the vapour and liquid phase. Vapour movement is the larger component where thermal transport is high. The driving forces are easily visualized. Water evaporates at the hot end, moves by molecular diffusion and condenses at the cold end. Water will, therefore, move up and down the soil profile in response to diurnal and seasonal changes in soil temperature. Buildings or roads placed on the soil alter the temperature regime, and can lead to accumulation of water due to a temperature gradient. The relative effectiveness of thermal and suction gradients at low suctions is illustrated in Table 5.1. Thermal gradients are most important in relatively dry soils, where $1°C/cm$ can be equivalent to 10^3 cm water head/cm.

The temperature dependence of vapour pressure of water in saturated air is shown in Fig.5.31. Since this is a non-linear function, the magnitude of the gradient will depend upon the average temperature as well as upon the difference in temperature. Fig.5.32 is an illustration of the distribution of water in columns of soil in the laboratory which results from a temperature gradient. It is not yet possible to predict this distribution theoretically with an accuracy of better than an

TABLE 5.1

Relative importance of thermal and suction gradients for loam soil at bulk density of 1.2 g/cm^3 (From Cary, 1966)

Suction	mbar suction equivalent to 1°C	Average temp. (°C)	Ratio of vapour to liquid flow
0.07	3	19	0.25
0.24	13	8	0.5
0.24	14	33	1.2
0.45	250	25	—

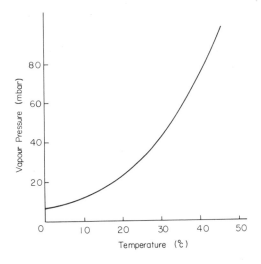

Fig.5.31. Saturated vapour pressure of water in contact with air.

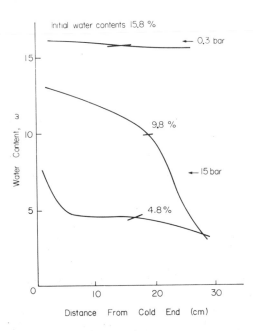

Fig.5.32. Water-content distribution in 30 cm soil column of very fine sandy loam at a bulk density of 1.3 g/cm³ subjected to a temperature of 10°C at the cold end and 25°C at the hot end. (Hutcheon, 1958.)

order of magnitude. It depends upon soil factors and the complicated mechanism of water movement in response to thermal potentials, which will be discussed below.

Diffusion of vapour due to a vapour pressure or concentration gradient can be described by equations based upon the classical laws of diffusion:

$$J = -D(dc/dx)$$

or for soil:

$$J = -\beta D(dc/dx) \tag{5.43}$$

where: J = vapour flux; D = diffusion coefficient in air; β = dimensionless constant accounting for lower porosity and increased path length in soil; c = concentration of water vapour; x = distance.

This equation adequately describes isothermal vapour transport, i.e., where vapour-pressure differences arise from suction differences. This can occur only in dry soils; the relative humidity at 15 bar suction is above 98%, so water contents must be lower to get significant vapour pressure differences.

If the vapour pressure gradient is due to a temperature gradient, eq.5.43 may be combined with the Clausius-Clapyron equation:

$$\frac{dp}{dt} = \frac{H}{TV} = \frac{Hp}{RT^2}$$

where: H = heat of vaporization of water; V = volume of vapour.

These equations predict vapour transport which is as much as 10 times lower than measured values. There are two reasons for this discrepancy. First the actual temperature gradients over interparticle distances in the soil may be from twice up to twenty times as large as the overall ΔT used in the equation. This will cause greater movement (Fig.5.31). Second, and most important, transport is not entirely by molecular diffusion but some condensation, movement in liquid form and re-evaporation occurs (Fig.5.33) to produce series liquid vapour flow. Liquid wedges and water films around soil particles, which would block diffusion, become regions where rapid liquid flow may occur. Several equations for water transfer under thermal gradients have been derived from different models, e.g., Philip and DeVries (1957).

Water flow in the liquid phase due to a thermal gradient also occurs, usually from hot to cold. The most obvious explanation for unsaturated soils is that increasing temperature decreases surface tension, and movement is due to a surface tension gradient. However, transport occurs also in saturated soils, where air—water

Condensation, decreasing curvature

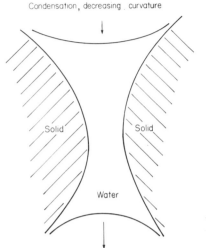

Evaporation, increasing curvature

Fig.5.33. Flow of water in liquid films due to change in curvature of films at air—water interfaces.

interfaces do not exist and surface tension differences could not be involved. Winterkorn (1958) has suggested flow along a gradient caused by changes in energy of adsorption of water with changes in temperature. The structure and other properties of adsorbed water would change with temperature. It is also observed that application of a thermal gradient to a soil gives rise to electrical potentials, so it may be electro-osmotic flow. It is also possible to have thermally induced osmotic gradients. Another effect in some unsaturated soils is that entrapped air expanding and contracting with increase and decrease of temperature can cause water movement. The cases where liquid flow from cold to hot has been observed can be explained for clay soils by the increase in the swelling component of matric suction with increasing temperature.

While it has been known for over fifty years that water moves in response to thermal gradients, the magnitude of transfer in the soil profile is hard to predict. Very high values are sometimes quoted. It has been shown that salt accumulation can occur from flow due to temperature differences. Water content increases over a period of months and years are also well documented where the surface is covered with a building, road, etc. Diurnal temperature changes occur relatively rapidly and reverse the direction of water transport. Also, evaporation at the hot end and condensation at the cold tend to decrease the temperature gradient because of the high latent heat of vaporization of water of 580 cal./g. In moist soils the amount of water transport is probably small, but above 10 bar suction, laboratory measurements indicate that there can be significant movement due to thermal gradients. A specific example of large water transport due to a thermal gradient is where part of the soil is frozen and water moves to the zone of freezing, causing ice lenses and frost heaving.

Thermal effects on water movement are complex because thermal, electrical and pressure potentials may all be generated. Heat, electricity, water and solutes may all be transported. Winterkorn (1955) recognized this problem and made some of the first measurements. A convenient way of summarizing such a system of coupled phenomena is through flux equations developed in the thermodynamics of irreversible processes. The flux of, e.g., water, has components due to each of the driving forces. The rate of internal entropy production is used to find equations for the driving forces. The following phenomenological equations can then be written down:

$$J_h = -L_{hh}\left(\frac{\Delta T}{T^2}\right) - L_{hs}\Delta\left(\frac{\mu_s}{T}\right) - L_{hw}\Delta\left(\frac{\mu_w}{T}\right)$$

$$J_w = -L_{wh}\left(\frac{\Delta T}{T^2}\right) - L_{ws}\Delta\left(\frac{\mu_s}{T}\right) - L_{ww}\Delta\left(\frac{\mu_w}{T}\right)$$

$$J_s = -L_{sh}\left(\frac{\Delta T}{T^2}\right) - L_{ss}\Delta\left(\frac{\mu_s}{T}\right) - L_{sw}\Delta\left(\frac{\mu_w}{T}\right)$$

where: J are fluxes and L the phenomenological coefficients. The subscripts "h", "w" and "s" refer to heat, water and solutes. T is temperature, μ_s is chemical potential of the solutes, and μ_w is electro-chemical potential of water.

An additional equation could be written for flux of electricity, which would add a fourth term, with the coefficient L_{ej}, to each equation.

The coefficient L_{ww} is related to the hydraulic conductivity of the soil, while the term L_{hh} is related to heat conductivity. It is usually assumed that the coefficients for coupling are equal, e.g., $L_{hw} = L_{wh}$.

While the same phenomena can be described by equations based upon mechanical considerations, e.g., Darcy's Equation, there is an advantage in the above approach because it clearly shows the interactions which can take place. The simultaneous flow of heat and water is a good example. Some of these interactions may be small or non-existent, but this needs to be determined and if the factors are neglected they are neglected knowingly. This is useful in planning experiments.

Effect of solute gradients

Differences in concentration of solutes set up osmotic gradients which cause water movement. These gradients can be very large near a salt granule or at a salt crust formed from evaporation of water. A gradient can also be set up by water movement through a part of the soil where salt sieving occurs i.e., where salt is held back and water moves through the pores.

As with thermal gradients, osmotic gradients found in field soils account for little water movement in wet soils. At suctions above 5 bar there can be significant movement, as there will be for unusually large gradients, e.g., near a salt granule.

Some comparisons are available on the relative efficiency in water transport of osmotic and suction gradients. They would be equal if water and salt moved entirely independently, i.e., flow of water across a membrane which entirely excluded solutes would be the same with a pressure gradient or an equivalent free energy difference due to an osmotic gradient across the membrane. The osmotic efficiency coefficient would be higher at high suction where thin water films could restrict solute, or for compacted clay layers where small voids would have the same effect.

Using the form arising from irreversible thermodynamics, the water flow equation can be written:

$$J = L_p \frac{\Delta S}{\Delta X} + L_{pd} \frac{\Delta \pi}{\Delta x}$$

where L_p and L_{pd} are the coefficients relating water flux to suction and osmotic gradients respectively; π is the osmotic pressure, and S the suction. The osmotic efficiency coefficient is:

$$L_{ps} / L_p = \sigma$$

Written in the usual form, the flow due to osmotic gradient would be given as:

$$q = \sigma k \, (\Delta \pi / \Delta x) \tag{5.44}$$

where k is the hydraulic conductivity.

Some results, showing low values of σ, but increasing with increasing suction are shown in Table 5.2. Water flow due to osmotic gradients does not increase as suction decreases, but it decreases less than hydraulic conductivity.

TABLE 5.2

Coefficients relating water flow to osmotic and suction gradients in sodium-saturated loam soil
(Letey et al., 1969. *Soil Sci. Am, Proc.*, 33)

Suction (bar)	Bulk density (g/cm³)	NaCl concentrations, N	L_p (cm/hr $\cdot$ 10^{-6})	L_{pd} (cm/hr $\cdot$ 10^{-6})	σ*
0.26	1.26	0.01 and 0.03	–	0	0
0.39	1.26	0.01 and 0.03	17.4	0.11	0.006
0.65	1.26	0.01 and 0.03	3.3	0.076	0.023
0.66	1.03	0.08 and 0.10	0.75	0.12	0.16

* Kemper and Rollins (1966) quote values of σ close to one for compacted clay plugs.

5.7 INFILTRATION INTO SOILS IN THE FIELD

Infiltration is the downward entry of water into the soil. The infiltration rate
is the rate at which a soil, in a given condition, can absorb water. It is defined as the
volume of water passing into a unit area of soil per unit time, with dimensions of
velocity (LT^{-1}). The infiltration velocity is the actual volume of water moving
downward into the soil per unit area per unit time. Its maximum is the infiltration
rate. The infiltration capacity of a soil is the infiltration rate which it will allow.

The infiltration rate obviously depends upon both surface and subsurface
conditions. The most important are stability of pores in the surface layers of soil
and water transmission rate of the soil body. Either the surface or the subsurface
may limit infiltration rate, and their relative importance depends upon the
particular soil. The initial water content of the soil has a large influence on initial
infiltration.

Distribution of water during infiltration

Water entering a dry soil from a constant supply on the surface distributes
itself as shown in Fig.5.34. This generalized distribution is independent of grain
size, even though the time required to wet a given depth of soil increases with
decreasing grain size. Bodman and Coleman (1943) have distinguished four parts of
the wetted zone. The surface 1 cm. layer is saturated, and swelling occurs with clay
soils which leads to even higher water contents. Below this layer, the water content
decreases rapidly to 70–80% of saturation, a water content between field capacity
and saturation. This is the transmitting zone and the water content remains
constant or decreases slightly with depth as the wetting depth increases. The length
of the transmitting zone increases with depth of wetting. Below the transmitting

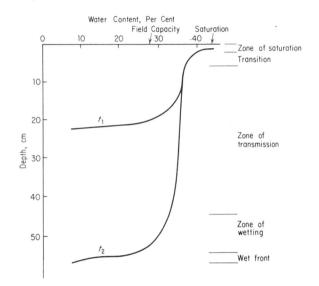

Fig.5.34. Water content vs. depth for two times during infiltration. (From Bodman and Coleman, 1943. *Soil Sci. Soc. Am. Proc.,* 8.)

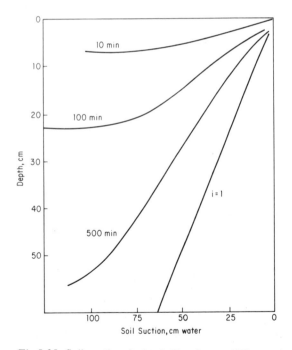

Fig.5.35. Soil suction during infiltration at different times for a sandy loam. (From Miller and Richard, 1952. *Soil Sci. Soc. Am. Proc.,* 16.)

zone is a wetting zone where the water content of the soil is increasing rapidly as infiltration continues. The wetting zone ends at the wetting front, which is sharply distinguished as long as infiltration proceeds. Philip (1957) has solved the diffusion equation for water flow for the case of infiltration into soils, and has shown that it predicts the zones observed and described by Bodman and Coleman.

From Fig.5.35 it is seen that water movement is by unsaturated flow, except in the surface layer. The soil suction in the transmitting zone increases with depth, but the gradient of the suction remains equal to one or increases only slightly. However, the gradient is initially high and decreases with time, as shown in Fig.5.35. This decrease explains the decrease in infiltration rate with time where soil properties remain constant. The gradient cannot become less than unity; at this stage water moves downward through the transmitting zone due only to gravity or gravitational potential. A further conclusion from these studies is that since the water is always under suction greater than zero, it will not move into larger pores where the suction would be zero. Therefore, root channels and worm holes will not increase the infiltration rate unless they extend to the surface to allow water entry.

The infiltration pattern is the same if the water is supplied at a rate below the infiltration rate, although the rate of advance to the wetting front will be slower.

The principles of the infiltration process are the same if the soil is initially moist. The suction gradient at the wetting front is lower and reduces the rate of entry of water. The rate of penetration of the wet front is faster, which could result from downward displacement of water already in the soil (Coleman and Bodman, 1944). These observed results may also be predicted from the water-flow equation (Philip, 1957).

In most natural soils the surface layer is more permeable than the subsoil. When the wetting front reaches a less permeable layer, the infiltration rate decreases and water moves laterally in the upper layer. If lateral movement is restricted, the water content of the transmitting zone increases until the upper layer is saturated. Water moves downward into the lower, less permeable layer, at a rate controlled by its permeability.

Where the difference in texture of the layers is large, e.g., movement of water from clay into sand or into drains, soil suction controls movement. This has been called the Outflow Principle (Richards, 1950), and is illustrated in Fig.5.36. Water must be under zero suction before it can move into very large pores which do not exert suction on the water. The wetting front moving through a clay would stop at a coarse sand or air boundary. The water content would increase by additions through the transmitting zone, until saturation occurred. Then the water, at zero suction, would move into the sand. If there is no restriction, water will move laterally in the clay because of the suction forces in dry clay, and delay accumulation of water to saturation at the boundary.

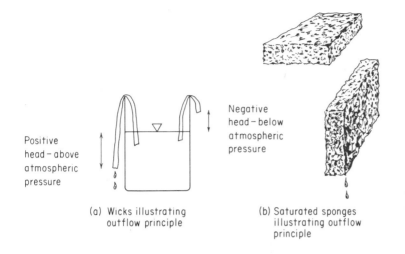

Positive
head – above
atmospheric
pressure

Negative
head – below
atmospheric
pressure

(a) Wicks illustrating
outflow principle

(b) Saturated sponges
illustrating outflow
principle

Fig.5.36. Illustrations of the Outflow Principle. (Richards, 1950. *Trans. Am. Geophys. Union*, 31, p.751, © Am. Geophys. Union.)

Infiltration equations

Infiltration measurements are usually made to determine the infiltration rate, which is significant in irrigation and hydrologic studies. This rate decreases with time and tends to approach a constant value. All the information can be shown on a graph of cumulative infiltration vs. time. Several empirical equations adequately fit many measurements:

$$I = a t^b$$

or:

$$I = c t^{\frac{1}{2}} + d$$

where: I = cumulative infiltration in cm^3/cm^2; t = time; a,c,d = constants. The constant b is found to be about one-half in many measurements of water movement into dry soil where suction forces predominate and gravitational forces are negligible. In wet soils, b approaches unity.

Factors affecting infiltration into soils

Most infiltration studies have been concerned with an evaluation of those soil properties which determine infiltration rate, and with obtaining values for use in irrigation and watershed studies. These papers have been reviewed by Parr and

Bertrand (1960). Factors influencing infiltration include such obvious soil properties as grain size, type and degree of development of structure, organic matter content, water content, porosity, volume weight, etc. Some properties in this list are interdependent. There are also considerations of stability of these properties, stability of porosity, swelling or volume change, dispersion of aggregates, etc. Often a thin layer of surface soil is altered by applied water, becoming compact and impermeable. Surface soils may swell and decrease the volume of large voids. Entrapped air will decrease permeability and may have the same effect as a layer of lower permeability. If the air cannot escape, pressure will build up due to slow advance of the wetting front until at a critical pressure air displaces some of the water and is released upward. Air is often entrapped in isolated pores, and reduces infiltration by decreasing effective porosity. Presence of salt in the water influences infiltration through its influence on volume change.

Infiltration rate is measured by ponding water on the surface and measuring the amount entering the soil, or by spraying water onto the soil surface at a rate high enough to cause surface runoff which is measured. Various devices have been used based on these two methods (Parr and Bertrand, 1960). Estimation of infiltration on soil cores brought into the laboratory is not recommended because of the inevitable sample disturbance during sampling and transportation, and because of the small volume of soil tested in this way. Because of the natural variability of soil properties, a large number of measurements on small samples are required to get a reliable average infiltration rate for a field.

Field capacity

Field capacity is defined as the amount of water remaining in a well-drained soil when the velocity of downward flow into unsaturated soil has become small (usually after one or two days).

After water is no longer added to the soil surface following infiltration, the wet front continues to move downward as water drains out of the transmitting zone. The movement becomes very slow after a time due to decreasing permeability in the transmitting zone, which is now drying, and to the decreasing suction gradient. The water content at the time when downward movement has virtually ceased is the water content at field capacity, or the soil is at its field capacity. It is not a constant equilibrium value, but a point on the drainage curve where drainage has become very slow. Its practical significance is that it is the upper limit of content of water which remains in the root zone long enough to be taken up in significant amounts by plants.

From this discussion it is obvious that some soils show a more pronounced field capacity than others. In sands the permeability decreases very sharply with decreasing water content, and sandy soils have a well-defined and reproducible field capacity. Clays have a higher unsaturated hydraulic conductivity than sands, and do not have a point where drainage rate changes sharply. Water moves down at a continually decreasing rate but movement continues for weeks. Layered soils also show unusual field-capacity values because the layering influences water movement as discussed above. A soil underlain by sand will retain more water than a soil with a uniform profile. These changes can be estimated from the hydraulic conductivity of the materials in the layers. The depth of soil into which water moves influences the drainage curves, and they will be different when the soil is wet than when it is dry. Water moving into dry soil will result in a lower field capacity than where a water table is present a few feet below the surface.

Field capacity is measured by thoroughly wetting a soil area, covering it to prevent evaporation, and measuring the water content after 24 or 48 hours. A neutron moisture meter is especially convenient for this purpose because the water content changes during drainage can be followed in one spot.

5.8 SUMMARY

The movement of water in soils is a complicated phenomenon to describe because soil surfaces interact with the water being transferred, and the soil changes during the water movement process. Water can move in either the liquid or vapour phase, with the latter being more important for relatively dry soil. Various driving forces in water flow exist in the soil, including differences in water content, in salt content, in temperature, and in void-size distribution. An overall description is required to encompass this flow system; for a particular soil some of the components can be neglected. Water entry into the soil is discussed, as well as the movement of the wetting front in the soil. Redistribution of water in the soil in the absence of entering water is also described.

Saturated flow takes place when no air is involved and when the water is under zero or positive potentials. Void-size distribution and soil structure is examined in relation to saturated flow.

The most common flow condition is unsaturated flow, where the water is at negative potentials or when air is present in the larger voids. The flow is described for different conditions of volume and fabric change in soils. Measurement of the conductance parameters, hydraulic conductivity and diffusivity, is discussed. Some flow equations are given, with emphasis on describing the physical process of flow under different conditions so that the correct flow equations can be chosen.

Chapter 6

VOLUME CHANGES IN CLAY SOILS

6.1 INTRODUCTION

Clay soils typically undergo changes in volume when the water content is changed. When clay soils are dried, shrinkage and cracking occur. If rewetting occurs subsequent to drying, swelling occurs. Large volume changes will occur in a climate with alternate wet and dry seasons, while smaller changes will result from alternating precipitation and drying periods. If the volume change is uniform it is not easily observed, and considerable vertical soil movements may occur without being recognized.

Soil movement leads to many serious problems. Engineering problems associated with volume changes arise not only from overall volume change, but also from uneven shrinkage or swelling of soils supporting loads. Cracking associated with shrinkage is of concern in embankment and earth dam stability. In agricultural soils, cracking may be beneficial in promoting aeration of plant roots or harmful in breaking roots and allowing extreme drying. Numerous small cracks promote formation of soil aggregates or crumbs.

A large amount of shrinkage occurs when water-deposited clays are first dried. The amount of reswelling in the presence of free water depends primarily upon the clay minerals present in the soil. Clay soils containing montmorillonite show an almost reversible swelling and shrinking on rewetting and redrying, whereas clays containing kaolinite or illite show an initial large volume decrease on drying with only a limited swelling on rewetting. The terms "high-swelling" and "low-swelling" are applied to the former and latter types of clay soils. The net decrease in volume after each drying and wetting cycle for low-swelling clays becomes smaller until an equilibrium is reached where swelling and shrinking occur between constant limits. This is illustrated in Fig. 6.1 for the change in water content of a low-swelling marine-deposited clay consisting largely of finely-ground primary minerals. This reversible volume change between fixed limits is also found for surface horizons of agricultural soils which have undergone numerous wetting and drying cycles.

In the following sections the mechanisms of volume change and the soil properties determining its magnitude will be discussed.

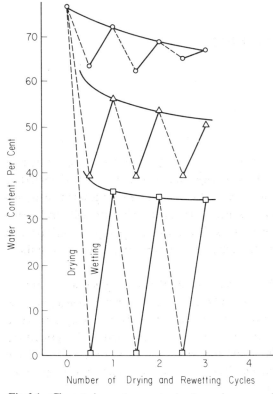

Fig.6.1. Change in water content at maximum swelling on repeated drying and wetting.
(Warkentin and Bozozuk, 1961.)

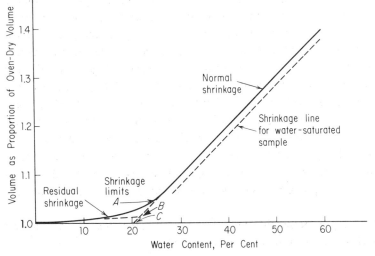

Fig.6.2. Shrinkage curve for a typical clay soil.

6.2 SHRINKING

If the volume of a saturated clay sample is measured as it is slowly dried, a relationship of the type shown in Fig. 6.2 is obtained between volume and water content. At high water content the slope of the line is $45°$; for each unit of water lost, the volume decreases by one unit and the soil remains saturated. This is defined as normal shrinkage (Haines, 1923). A soil sample need not be fully water-saturated to show normal shrinkage. Undisturbed samples taken from below the water table usually have a few per cent of air, but the volume of the gas phase remains constant on drying, resulting in normal shrinkage. The theoretical line for no air is shown below the shrinkage line for the sample in Fig. 6.2.

On further drying the slope of the shrinkage curve changes and air enters the voids at the shrinkage limit. A small amount of shrinkage, termed residual shrinkage, takes place on further drying. The shrinkage limit has been defined in several ways as shown by points *A, B,* and *C* in Fig. 6.2. The water content at the shrinkage limit which has a physical significance is at the point of unsaturation, point *A*. Point *C* gives the total shrinkage.

The force causing shrinkage arises from the pressure difference across the curved air—water interfaces of the voids at the boundaries of the sample. This type of force has been discussed in Chapter 4. As water evaporates from the surface, a curved interface is formed in the voids at the surface with a lower pressure on the convex inner side. Water is drawn from inside the sample due to this pressure difference. As long as this force exceeds the resistance of the clay particles to closer approach to each other, i.e., to shrinking, the sample remains saturated. Thus volume decrease is equal to the water-content loss. This is the range of normal shrinkage. Eventually a condition is reached where particle interaction restricts shrinking and further increments of water removed are partly replaced by air. Some additional shrinkage (residual shrinkage) occurs due to further fabric arrangement, and in some cases to bending of particles. Particle interaction resisting closer approach of particles may consist of (a) direct contact of particles or of hydration layers, and (b) interparticle repulsion due to forces causing swelling. This interaction has been discussed in Chapters 2 and 4.

Soil characteristics affecting shrinkage

Characteristics of the shrinkage curve vary with the nature of the soil. Total shrinkage increases with increasing initial water content. This is a function of the per cent clay in the soil, the kind of clay minerals, the mode of geological deposition, the depositional environment which determines both particle arrangement and overburden pressure, and the degree of weathering. Sand and

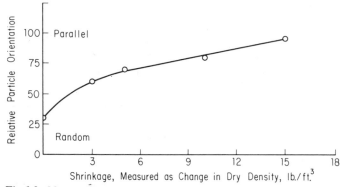

Fig.6.3. Measured shrinkage of compacted Boston clay as a function of particle orientation: parallel arrangement is 100% orientation and random arrangement is 0%. (Lambe. 1960. *Trans. Am. Soc. Civil Engrs., Pap.* 3041.)

silt-size particles reduce total shrinkage because they dilute the clay and decrease the volume of water held by the soil. High-swelling clays containing the mineral montmorillonite have high initial water content. Sediments deposited in salt water have a random or edge-to-face association of particles with large volumes of trapped water (see Chapter 3), resulting in high initial water contents even in the absence of high-swelling clays. Overburden pressure consolidates sediments and decreases the water content. Repeated cycles of wetting and drying will overcome the influence of depositional environment for surface soils and in consequence total shrinkage will depend primarily upon the clay minerals in the soil.

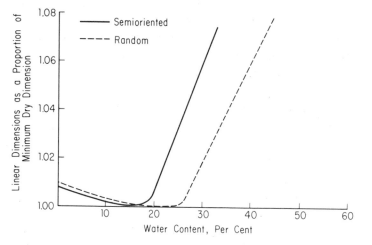

Fig.6.4. Measured shrinkage curves for kaolinite with flocculated and semioriented particle arrangement.

If we examine the influence of soil structure on shrinkage, we see that total shrinkage is less for random particle arrangement — as shown by shrinkage measurements for two clays in Fig. 6.3 and Fig. 6.4. Making comparisons at the same water content, the semioriented sample of kaolinite with more parallel particle arrangement shows a much larger volume change than the sample with random particle arrangement. Consequently, the final dry densities must also be different. The lower densities for random particle arrangement are shown in Table 6.1 for the two kaolinite samples of Fig. 6.4 and for a marine-deposited clay which contains chlorite and mica (Fig. 6.1).

TABLE 6.1

Shrinkage limit and oven-dry density as influenced by particle orientation

Clay	Qualitative orientation	Water content at shrinkage limit (%)	Oven-dry density (g/cm³)
Dispersed kaolinite (pH 10)	semioriented	19	1.70
Flocculated kaolinite (pH 4)	random	26	1.52
Remoulded marine clay	partly oriented	20	1.77
Undisturbed marine clay	random	30	1.64

The water content at which unsaturation of a soil occurs, the shrinkage limit, also depends upon the fabric and upon the clay minerals (Table 6.1 and Fig. 6.4). A more random arrangement increases the shrinkage limit; a more parallel arrangement decreases it. Particle interaction leads to unsaturation at a higher water content for the random edge-to-face arrangement of particles than for the more parallel arrangement. The additional volume of water is trapped between particles in the random arrangement and is not affected by the forces holding water at soil surfaces. The shrinkage limit for most clay soils occurs at soil suction values in excess of 10 bars. A low shrinkage limit is usually associated with large volume change, e.g. montmorillonite will have a value of 10—15% while kaolinite will be 20—25%. The exception to this is allophane, which shows high shrinkage limit of 70—90%, and also high volume change.

Some clays exhibit residual swelling. Kaolinite, for example, may show a residual swelling rather than shrinking (Fig. 6.4). This is probably due to elastic rebound of particles after the water films which connected the particles are broken.

Shrinkage will be anisotropic if the tabular clay particles have a preferred orientation in the sample. This is most likely to occur in sediments deposited under lacustrine conditions, or where overburden pressure has oriented the particles during consolidation. Orientation of particles in a soil sample can be achieved in the laboratory by one-dimensional consolidation or by drying a sample from suspension. In oriented samples shrinkage normal to particle orientation or bedding exceeds shrinkage in the plane of orientation or bedding, and the dimensional shrinkage is not linear with decreasing water content. This is illustrated in Fig. 6.5 with measurements on an undisturbed sample of a glacial lacustrine clay containing montmorillonite and illite. For a clay plate, which is longer in the horizontal than in the vertical dimension, shrinkage will be greater in the vertical direction even if a water film of uniform average thickness surrounds the particle. This occurs because the proportion of water to solid is greater along a line in the vertical direction. In addition, interparticle repulsion is greater between flat surfaces, resulting in a larger average distance in the wet sample between flat surfaces than between edges. As seen from the slopes of the shrinkage curves, the rate of vertical shrinkage gradually decreases while the rate of horizontal shrinkage increases until the shrinkage limit is reached. If the volume is calculated from the dimensional shrinkage, it can be shown that normal shrinkage occurs above 20% water.

Shrinkage of weathered surface soils, soils with random particle arrangement and remoulded samples will be isotropic.

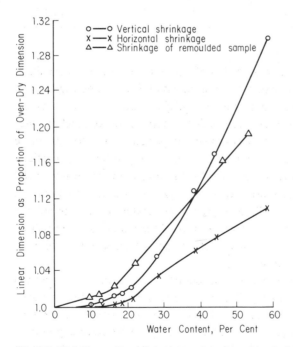

Fig.6.5. Shrinkage curves for a lacustrine clay with partial orientation of particles. (Warkentin and Bozozuk, 1961.)

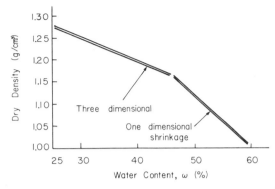

Fig.6.6. Change from uni-dimensional to three-dimensional shrinkage as clay soil dries. (Fox, 1964. *Soil Sci.*, 98, ©Williams and Wilkins Co., Baltimore.)

Shrinkage will not be three-dimensional if the soil is plastic and the overburden pressure affects the shrinking layer. One would expect shrinkage to be one-dimensional in a wet soil, changing gradually to three-dimensional in a dry soil. Fox (1964) has presented evidence that this change may be quite abrupt for the surface layer of a soil (Fig. 6.6).

Surface soils with a low clay content or with a well-developed crumb structure may show no range of normal shrinkage as the water content is decreased. In this case, air enters the samples as the water content decreases, and the total shrinkage is lower than it would be if normal shrinkage occurred (Fig. 6.7). There is no shrinkage limit (Fig. 6.8).

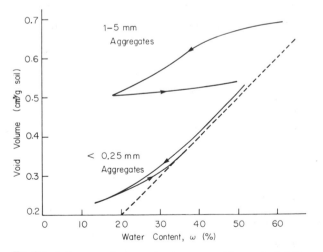

Fig.6.7. Influence of aggregate size on swelling and shrinking of soil. (Chang and Warkentin, 1968. *Soil Sci.*, 105, © Williams and Wilkins Co., Baltimore.)

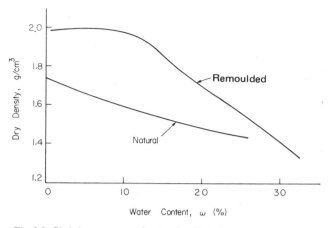

Fig.6.8. Shrinkage curves of natural soil clods and remoulded clods. (Gill, 1959.)

This is typical for soils with a wide range of void sizes. In a clay soil with a well-developed crumb structure there are large voids between crumbs in which water is held by capillary forces. The first water lost on drying the sample comes from these voids. The voids within the crumbs, where water is held by the forces associated with swelling, remain saturated up to the shrinkage limit. In such soils the first increments of water loss are accompanied by a volume decrease less than the volume of water lost. This has been termed structural shrinkage. As drying proceeds, normal shrinkage may occur when the water within the crumbs is lost. This is shown in Fig. 6.9.

The change in shape of the shrinkage curve on remoulding gives information on particle arrangement or fabric. When a surface soil with a crumb structure is

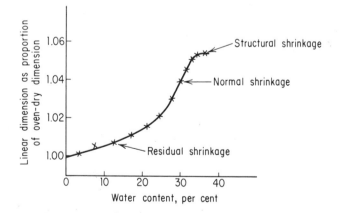

Fig.6.9. Measured shrinkage curve for soil with crumb structure.

·remoulded, the larger voids disappear. The range of normal shrinkage then increases while the water content at the shrinkage limit decreases (Fig. 6.8). For clay subsoils which do not contain large voids the change on remoulding may be very small. The change in the shrinkage curve will depend upon the original fabric. Fig. 6.5 illustrates the change in shrinkage on remoulding a soil with oriented particles.

A shrinkage curve for an allophane soil in Fig. 6.10 shows the high shrinkage limit and large total shrinkage typical of these soils at natural water content. Ovendrying before rewetting results in a shrinkage curve similar to that for soils with crystalline minerals.

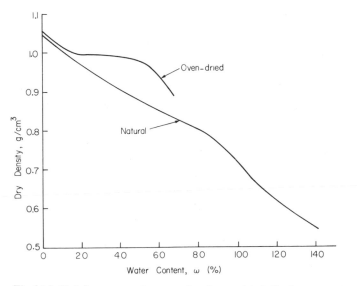

Fig.6.10. Shrinkage curves for natural and oven-dried allophane samples.

Crack formation during shrinking

Shrinkage cracks will form at the surface of clay soils on drying. One-dimensional vertical shrinkage might occur in a plastic soil at depth where the overburden consolidates the clay, but the usual case is approximately three-dimensional shrinkage, accompanied by cracking. Cracks form where the cohesion of the soil is lowest. Where drying is not uniform, cracks will form in the wetter soil. A change in particle orientation occurs at the crack surface, and on redrying after wetting, the cracks will appear in the same places if the soil has not been otherwise disturbed.

The number of cracks per unit area depends upon the clay minerals and the particle arrangement. A large number of cracks will form in flocculated clays, while in semioriented clay with higher cohesion a few, relatively large cracks will result. The cracks in a surface soil with a well-developed crumb structure may not be noticed because each crumb unit becomes separated by a small distance from the adjoining one. No crack is sufficiently large to be obvious.

The cracks formed in clay soils have an important role in water infiltration and movement. Water can be conducted rapidly in the cracks of an otherwise impermeable soil.

Measurement of shrinkage for samples

Shrinkage curves may be experimentally determined by measuring either linear dimensions or volume as a moist sample loses water. The average water content can be easily obtained by weighing the sample. There is the problem of ensuring that the sample has a uniform water content, since drying proceeds at the surface and water must move from inside the sample. This can be accomplished by slow drying of small samples or by allowing the samples to equilibrate in a saturated atmosphere after a short drying period. Accurate measurement of the volume is more difficult.

Haines (1923) in his classical work defining the characteristics of the shrinkage curve, measured volume by displacement of mercury in a special pycnometer bottle. If the samples have a plastic consistency, the required handling will deform the sample. At low water contents dry samples may break under the pressure. Another disadvantage is the health hazard in handling mercury extensively. However, despite these drawbacks, this remains the most accurate method of measuring volume unless samples of regular shape can be obtained.

Volume can be determined by displacement in water if the sample is coated to prevent water entry. Paraffin wax or films of plastic obtained by evaporating off the solvent are used. The disadvantage is that the sample can be used for only one measurement. Several samples must be used to define the shrinkage curve and variability of soil samples must be taken into account. Volume can also be measured if the voids are filled with a liquid such as kerosene which is immiscible with water.

Linear dimensions of the sample, from which volume shrinkage could be calculated, were measured in early studies of shrinking. More recently, Croney et al. (1958) measured the magnified image of a cube-shaped sample projected on a screen. Linear dimension changes have also been determined from the changing distance between two marks (either pins or marks scratched into the sample) on a sample face. The distance between two points on the surface of a small sample can be measured with a travelling microscope. If the sample has a thin wafer-shape, the water distribution within the sample on drying is more easily kept uniform.

6.3 SWELLING

Volume increase due to swelling does not always accompany water content increase on rewetting of a soil. A dry soil can take up water, with air in the voids being replaced by water, without a consequent increase in volume. This occurs typically for sandy and silty soils. Swelling requires a force of repulsion separating clay particles to increase the volume as the water content increases.

Soil characteristics affecting swelling

The amount of swelling depends upon the clay minerals and their arrangement or orientation in the clay soil, as well as upon physical-chemical properties such as valence of exchangeable cations, pore-water salt concentration, and cementing bonds between clay particles. Everything else being equal, swelling increases with increasing surface area of clay particles and with decreasing valence of the exchangeable cation. Specific differences between clays within a mineralogical group give rise to smaller differences in swelling.

Surface area of a clay depends more upon thickness of the tabular particles than upon the other dimensions. It decreases from the thin particles of montmorillonite to the much thicker kaolinite particles, as shown in Table 6.2.

Monovalent exchangeable cations such as sodium cause greater swelling than divalent calcium ions. This is explained by the greater extension of the diffuse ion-layer discussed in Chapter 2. Highly acidic clays have polyvalent aluminium as exchangeable ions, with a consequent low swelling. An increase in pore-water salt concentration decreases swelling of the high-swelling clays, especially if monovalent ions are present. Swelling and dispersion of soils with exchangeable sodium are reduced at higher salt concentrations. Usually swelling is undesirable in a soil, but for purposes such as lining irrigation canals swelling may be desirable. For example, montmorillonite with exchangeable sodium is used to make the surface impervious and reduce water loss from seepage.

TABLE 6.2

Size and swelling properties of clay minerals

Mineral	Approximate thickness (Å)	Maximum surface area (m²/g)	Observed volume change
Montmorillonite	20	800	high
Illite	200	80	medium
Kaolinite	1,000	15	low

Cementation between particles is a major factor in limiting volume increase of clays on swelling. Iron hydroxides, carbonates and various organic molecules are the cementing materials. It is not clear in many cases whether these materials bond between particles to form a restraint to swelling, or whether they affect the physical-chemical properties of the surface in such a way as to reduce the swelling force. Iron salts dried into a clay can markedly reduce swelling; this occurs in some "nonswelling" montmorillonites.

The decrease in swelling caused by cementing materials is greater for weathered clays which have undergone cycles of wetting and drying. Most of the mechanisms for cementation depend upon dehydration of colloidal material to form bonds between particles. Measurements of volume regain for two unweathered clays in Fig. 6.11 show that removal of iron oxides and carbonates resulted in only a small increase in swelling of the high-swelling clay and decreased the volume

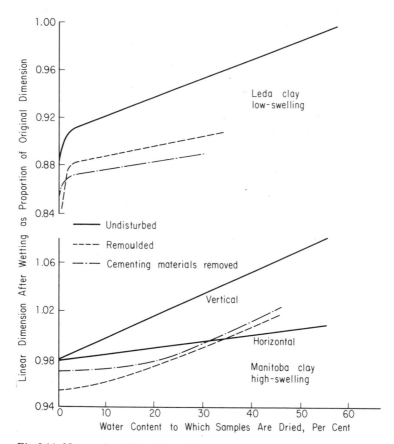

Fig.6.11. Measured swelling of two clays as influenced by remoulding and removal of cementing materials. (Warkentin and Bozozuk, 1961.)

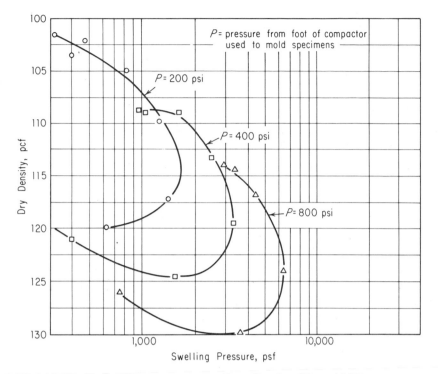

Fig.6.12. Measured swelling pressure curves for a silty clay. (Olson, 1964.)

regain of the low-swelling clay. This decrease was likely due to the more thorough remoulding during chemical removal of cementing materials. The change in particle arrangement caused by remoulding had the largest influence on swelling.

The influence of particle arrangement on swelling seems to depend upon the type of clay. Swelling against restraint is greatest for parallel particle orientation for high-swelling clays such as sodium montmorillonite. The swelling volume and swelling pressure reported by Seed and Chan (1959) is greatest for random or flocculated particle arrangement. Fig. 6.12 shows a higher swelling pressure with the higher compaction pressure which would produce more orientation. At the same dry density, swelling pressure depends upon whether the sample was compacted wet or dry of optimum. Fig. 6.11 shows the effect of remoulding on volume regain for the low-swelling Leda clay with random particle orientation and for a high-swelling, semioriented, Manitoba clay. Remoulding the Leda clay results in more parallel particle arrangement over short distances, and volume regain is decreased. The particle orientation in the high-swelling clay is changed by remoulding, but it is not known whether the remoulded orientation is more or less parallel from particle to particle. An explanation for the influence of particle arrangement on swelling is discussed in the next section on mechanism of swelling.

Swelling occurs primarily in the plane perpendicular to the flat surface of the clay particles, so swelling and shrinkage are anisotropic for clays with an overall orientation of particles. Samples of high-swelling montmorillonite oriented by pressure or by drying from dilute suspension show horizontal swelling which is only a few per cent of the vertical swelling. Anisotropic swelling of the Manitoba clay is shown in Fig. 6.11. Swelling in the horizontal direction does not reach the original dimension, but vertical swelling exceeds the original vertical dimension. Ward et al. (1959) found for the London clay, a greater swelling volume and a greater swelling pressure in the vertical direction compared with the horizontal.

Swelling pressure of soils

Swelling clays exert pressure against a confining load when water is available for a volume increase. This pressure exceeds usual loading, so that volume change is decreased but not prevented by the structure which the soil supports. The magnitude of these pressures can be seen in Table 6.3, where some measured pressures reported in the literature for soils are listed. Except for high-swelling clays, the swelling pressure decreases rapidly with small volume increases, and the amount of swelling under load is usually small.

TABLE 6.3

Some values of swelling pressure of soils reported in the literature

Soil	Swelling pressure (kg/cm^2)	Reference
London clay	2–9	Ward et al. (1959)
Black cotton soil	3	Palit (1953)
Bearpaw shale	6	Peterson and Peters (1963)
Sandy clay		
random	3	Seed and Chan (1959)
oriented	0.1	
Israeli clay	1–8	Kassiff et al. (1969)

The dependence of swelling pressure on volume change makes a precise measurement of swelling pressure difficult. Unless special precautions are taken in the measurement to prevent volume change, measured pressures will underestimate the swelling pressure. A pressure-measuring device is required which is actuated by very small volume changes. An ordinary oedometer (consolidometer) allows too much volume expansion. The normal rebound curve, i.e. unloading curve (see Chapter 7), therefore underestimates the swelling pressure at any void ratio.

A number of devices for measuring swelling pressure have been described in the literature. Typically they consist of a rigid cell confining the sample, with access to water through a porous stone or water-permeable membrane at one end and at the other end a solid or fluid piston which allows a pressure measurement. Swelling pressure is measured as the restraint which has to be applied to the piston to prevent movement. Such a device is shown in Fig. 6.13. Swelling pressure in the horizontal direction can be measured if the confining cell is fluid rather than a rigid wall, or if a measuring device is built into the wall.

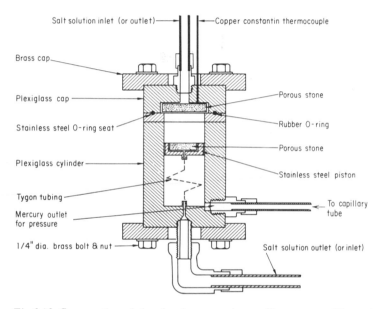

Fig.6.13. Cross-section of chamber for measuring swelling pressure. (Yong et al., 1963.)

The swelling pressure measured in these devices is not developed instantaneously. The results given by Palit (1953) show that the pressure increases approximately exponentially for a number of weeks. Two factors contribute to this increase. If an initially air-dry sample is being tested, the complete hydration of clay mineral surfaces on wetting will require time. The full swelling pressure will not be manifested until hydration is complete. Secondly, most pressure-measuring devices depend upon some volume change for actuation. This expansion of the sample requires that water moves into the clay. Since the permeability of swelling clays is low, the necessary distribution of water under small pressure gradients will require considerable time.

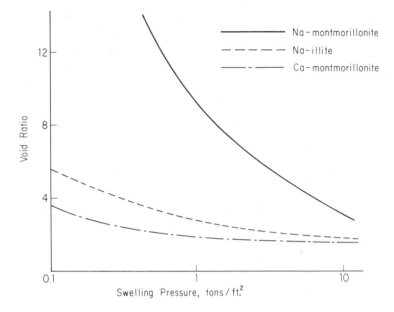

Fig.6.14. Measured swelling pressures of three clays. (Bolt, 1956.)

Measured swelling pressures of pure clays in Fig. 6.14 show the influence of surface area and valence of exchangeable cation. The sodium montmorillonite shows much higher swelling pressures than the other clays. The illite has a lower surface area, and the divalent calcium ion also decreases swelling. Fig. 6.14 also illustrates the decrease in swelling pressure with increasing void ratio.

Mechanism of swelling

Swelling, or a volume increase on wetting, requires an attraction of water to the clay to provide an effective repulsion between clay particles. From thermodynamic considerations, water will move into the soil as long as the free energy of the water in the soil is less than that of free water. Water fills the voids of a sand without a significant increase in volume, but for clays this wetting usually results in a volume increase. The reason is sought in the surface properties of clays. The different mechanisms suggested for the water adsorption and interparticle repulsion causing swelling have been discussed in Chapter 2.

It is of interest to calculate the amount of swelling predicted by the proposed mechanisms and to compare the values with measured swelling. The total amount of swelling is difficult to calculate because it depends upon the balance between forces of attraction and repulsion. Interparticle repulsion decreases as swelling volume increases until finally attraction balances repulsion. The final volume is sensitive to many outside influences. Swelling pressures at volumes less than completely swollen are more easily handled because they depend upon the larger net forces of repulsion.

The properties of bound water layers on clay surfaces cannot be calculated with present knowledge, so the swelling pressure cannot be predicted in this way. Osmotic swelling can be calculated with certain simplifying assumptions discussed in Chapter 2. These assumptions severely restrict the application of interparticle repulsion in predicting swelling of clay soils, but it will be seen that where the theoretical model is valid, the calculations can predict observed swelling pressures. The following discussion presumes that adsorbed water accounts for the first few water layers taken up on swelling and that extended swelling, especially with monovalent ions, results from osmotic pressure.

The swelling pressure is calculated as the osmotic pressure due to the difference in concentration of ions between clay particles and in the outside pore water. This gives only the forces of repulsion, and will be in error in predicting swelling by the amount of any forces of attraction. The model assumed is that of charged, tabular particles in parallel arrangement with diffuse layers of exchangeable ions which overlap, resulting in an ion concentration between particles higher than that away from their influence. The properties of the system which must be known to make this calculation are: specific surface area of the clay, kind of exchangeable cations, concentration of ions in the pore water, and exchange capacity or surface density of charge of the clay.

For water contents below 60% on a weight basis, and pore-water salt concentrations below 0.001 M, the concentration of anions between the clay particles can be neglected, and the calculation is simplified. The concentration of cations at the midpoint between two interacting plates cannot be measured. It can be calculated from the model, as shown in Appendix 2, and is given by:

$$C_c = \frac{\pi^2}{z^2 B(d + x_o)^2 \, 10^{-16}} \tag{6.1}$$

where: C_c = concentration of cations midway between two clay plates, in moles/litre; z = valence of exchangeable cations; d = half-distance between two clay plates, in Å; x_o = correction factor of 1–4 Å depending upon ion valence and charge density of clay; B = 10^{15} cm/mmole (this constant depends upon temperature and dielectric constant).

The swelling pressure can then be calculated from the Van't Hoff equation, which for monovalent ions is:

$$P = RT(C_c - 2C_o) \tag{6.2}$$

where: P = calculated swelling pressure; R = gas constant; T = absolute temperature; C_0 = concentration of salt in the pore water, moles/litre.

The relationship between water content and interparticle spacing is given by:

$$w = Sd/100 \tag{6.3}$$

where: w = water content in wt. %; S = surface area of clay, m^2/g for d in Å.

These calculations adequately predict measured swelling pressures for the high-swelling sodium montmorillonite at low salt concentrations (Fig. 6.15). At higher salt concentrations the measured pressures of sodium montmorillonite exceed calculated values. This is not due to flocculation, and probably results from the errors in using concentrations rather than activities of the exchangeable cations, and from neglecting the tactoid structure discussed in the next paragraph.

Calculations of swelling pressure from diffuse ion-layer for divalent ions are not substantiated experimentally. Divalent cations are present mainly in the Stern layer, not in the diffuse ion layer. For montmorillonite there is an uncertainty about the relevant surface area because the particles are arranged in packets or tactoids with extended swelling only between the tactoids. When the surface area is reduced to take this into account, theoretical and measured swelling pressures are in better agreement. Increasing salt concentration does not have the quantitatively

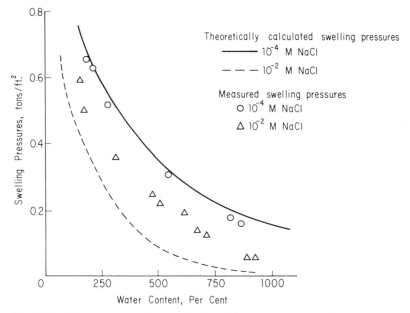

Fig.6.15. Comparison of calculated with measured swelling pressure for sodium montmorillonite at two salt concentrations. (Warkentin and Schofield, 1962.)

predicted effect in decreasing swelling pressure, although this may be the result of low swelling in water combined with the error in using concentrations rather than activities. Part of the decrease in swelling of calcium montmorillonite on adding salt is irreversible. This may result from irreversible formation of layer tactoids. Decreasing salt concentration does not provide enough energy to separate the tactoids. The packet structure is a less important factor for clays other than montmorillonite.

Parallel orientation of clay particles is assumed in the calculation of swelling pressure. For the random arrangement and the mixture with silt and sand-size particles found in soils, there will be a fraction of the pore space into which the diffuse ion layers do not extend. The calculation will consequently underestimate the water content at any swelling pressure. This dead volume is difficult to estimate, and presumably changes as the sample volume changes. As an estimate, the dead volume can be taken as the water content at the shrinkage limit. This is the point at which particle contact restricts shrinkage, and it is no longer the swelling pressure which keeps the particles apart. This correction can be used to calculate swelling pressures for natural clays at different water contents.

As an example we calculate the swelling pressure from the interparticle repulsion for the following conditions:

Water content	50%
Surface area	$150 \; m^2/g$
Pore water salt concentration	$0.0006 \; M \; CaCl_2$
Exchangeable ions	calcium of valence two

The half distance between particles d in Å for parallel particle arrangement, from eq. 6.3 is:

$$d = \frac{100w}{S} = \frac{100 \times 50}{150} = 33 \; Å$$

From eq. 6.1, using $x_o = 2 Å$:

$$C_c = \frac{3.14^2}{2^2 \cdot 10^{15} \times 35^2 \cdot 10^{-16}} = 0.020M$$

Substituting into eq. 6.2, using $3C_o$ because there are three ions per molecule of $CaCl_2$ rather than two for a monovalent salt, such as NaCl, and a temperature of 20° C:

$$P = 0.0848 \; \frac{kg \; litre}{cm^2 \, °K} \; 293 \; (°K) \; [0.020{-}0.0018]$$

$$= 0.0848 \times 293 \times 0.018 \; kg/cm^2 = 0.45 \; kg/cm^2$$

The swelling pressure for a montmorillonite clay with a tactoid structure must be calculated from the surface area of the outside of the tactoids, not the total surface area. Surface area measured by nitrogen absorption is approximately this outside area. If the number of particles per packet is known, the outside area can be calculated. As an example, if a swollen montmorillonite with exchangeable calcium ions has ten particles per packet, the surface area is reduced by ten times, neglecting the surface area of edges which is only a few per cent of the total. Only the two outer surfaces of the 20 particle surfaces are involved in swelling.

Within the packet, the particles are separated by 10-Å layers of water, which must be taken into account when calculating swelling pressure at any water content. For a total surface area of 800 m^2/g, if one-tenth is outside area, then 720 m^2/g is area inside the packet. This area has a water layer 5 Å thick (10 Å between particles). The water content then is 36%: 720 m^2/g $\cdot$ 10^4 cm^2/m^2 x 5 Å $\cdot$ 10^{-8} cm/Å = 0.36 cm^3/g.

The half distance d between effective swelling units for a water content of 85%, from eq. 6.3 is:

$$d = \frac{100\,(85 - 36)}{80} = 61 \text{ Å}$$

The calculation of swelling pressure is as above.

If the clay particles are in random arrangements, and other sizes of particles are present in the soil, there will be a "dead volume" where swelling forces are not effective. This must be subtracted from the total water content in calculating swelling pressure, or added to the water content calculated from the theoretical relationship between swelling pressure and water content. As an example, if the "dead volume" for the soil discussed above is 18%, the value for d is:

$$d = \frac{100\,(85 - 36 - 18)}{80} = 39 \text{ Å}$$

Summary of swelling

In summary, swelling can be visualized as the result of repulsion between adjacent clay particles over the area of approach. The initial small rebound on release of a load is probably elastic volume increase from unbending of particles. Adsorbed water layers form around the particles to cause some volume increase. This hydration of the surface and of the exchangeable cations can proceed under very high confining pressures. Elastic rebound and hydration account for some swelling at low water contents. Extended swelling is due to osmotic swelling pressure, dependent upon the ions and the salt concentration. The osmotic pressure decreases with increasing distance between particles, i.e., it decreases as swelling proceeds.

When particles are in random orientation, repulsion occurs at the points of closest approach, but considerable water is also held by capillary forces in pores between fabric units. For this reason, swelling volume is greater than for parallel arrangement if osmotic swelling is small. The amount of osmotic swelling and the swelling pressure are, however, dependent upon the amount of surface which is interacting, so for high-swelling clays the parallel particle orientation results in the greatest swelling.

A swelling clay deposited from fresh water has an oriented fabric arrangement with the flat faces lying horizontally. The particles or fabric units fall individually and settle in this minimum energy configuration. Repulsion between faces sustains the overburden. A sample prepared by one-dimensional consolidation will have the same characteristics, and will have a large average distance between edges in the direction perpendicular to consolidation. On drying, this sample will show considerable shrinkage in the horizontal direction as particle edges approach. Swelling, however, will occur predominantly in the vertical direction, from repulsion between faces.

A sample allowed to swell without being loaded will increase in volume until forces of attraction balance the forces of repulsion. These forces of attraction include Van der Waals forces, Coulomb forces between unlike charges, and organic and inorganic molecules which bond between particles.

6.4 VOLUME CHANGES IN THE FIELD

The soil characteristics affecting shrinking and swelling discussed in previous sections of this chapter have been studied on small samples in the laboratory. Volume changes of soils in situ may differ in degree. The only unconfined boundary occurs at the surface, so that aspects such as cracking must necessarily be studied in the field. In general, the volume changes occurring in situ will be lower than those measured for small samples. Climate determines the degree to which the potential volume change of a clay will be expressed. The amount of shrinking and swelling can be predicted if the clay properties and the climate are known.

Structures placed upon clay soils can change the water relationships of the soil beneath them by decreasing water loss through evaporation, or preventing water content increases from rainfall. Differential volume changes will then occur between exposed and unexposed soil. Deep-rooting plants decrease the water content of soils at depth. The increased shrinking around trees is well documented and readily observed on clay soils where reswelling is only a fraction of initial shrinkage. The volume changes may occur to a depth of 10–15 ft. Recent studies of naturally occuring volume changes, based upon a knowledge of shrinking and swelling mechanisms, have defined the important practical aspects of these volume changes.

In Fig. 6.17 vertical ground movements are shown in relation to soil water conditions. Volume changes accompanying water content change in natural soils can present major soil-engineering problems. Swelling or heaving occurs on water content increases, shrinkage may occur on loss of water. Since heaving and shrinking are responsive to changes in water content, they will vary seasonally as water content varies. Also, volume change will be restricted to the upper layers of soil, where the water content changes are largest. The engineering problems are therefore usually met in light structures bearing on the surface soil, in road pavements, etc. The influence of volume change on these structures then depends upon the water content of the soil at the time the structure is placed on it, and how the structure interferes with subsequent water content changes. A road pavement will prevent water loss by evaporation, and if it is placed on a dry soil, water content will increase and the soil will heave. Where drainage water goes is of critical importance for heaving soils.

The amount of heaving in response to wetting is determined by the nature of the clay minerals. Heaving is predominantly associated with montmorillonite clays. Irreversible shrinkage is associated with fabric changes, and occurs with clays having high initial water content due to the type of fabric. Reversible shrinkage is associated with active clay minerals such as montmorillonite.

The amount of swelling for a soil, and the swelling pressure, both decrease as water content increases (Fig. 6.15). The initial water content, therefore, is an important variable in amount of heave. Soil fabric and structure also have an important effect on swelling. The literature contains many studies on effect of type of compaction, of initial density, of moulding water content, and of stress history on swelling. The amount of air in the void space also influences swelling because some of the volume change can occur by expelling air.

Swelling is not instantaneous; the shape of the volume versus time curve depends upon the boundary conditions of the sample or in the field. The swelling rate curves show the same phenomena as the consolidation curves, with a "primary" swelling determined by rate of water movement into the clay and a "secondary" swelling due to small fabric rearrangement and small changes in free energy of bound water.

An example of heaving of pavement on swelling clay is shown in Fig. 6.16. In Fig. 6.17, vertical movements are shown in relation to soil water conditions. Many such measurements have been made. The problem of predicting the amount of heave from laboratory tests is discussed in the next section.

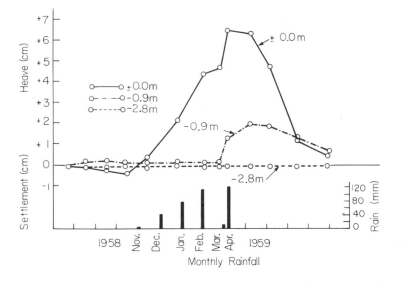

Fig.6.16. Seasonal vertical movements of an exposed clay. (Kassif et al., 1969.)

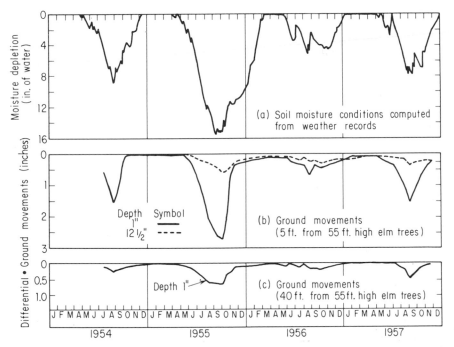

Fig.6.17. Seasonal variations in moisture conditions and vertical ground movements in Leda clay soil at Ottawa, Canada. (Bozozuk and Burn, 1960.)

Prediction of heave from soil properties

Several single index properties have been used to predict heave. Usually these properties cannot be used to predict the exact amount of heave, but they are very useful for identifying heaving soils and for qualitative predictions of the amount of heave. A mineral identification to check for amount of montmorillonite present is a useful index. The plasticity index increases as swelling potential increases. A high amount of shrinkage and a low shrinkage limit indicate a swelling soil. Simple "free-swell" tests, such as the final volume of a given quantity of soil poured into water, are also used.

A second group of tests is based upon measuring the amount of swelling which occurs under a small confining load. The specimen is confined in a cell, a vertical confining load is applied, water is added to the soil, and the increase in height or in volume of the specimen is measured. The confining stress can be applied by controlling soil suction on a pressure plate apparatus. This measurement has been used to calculate the coefficient of linear extensibility (Grossman et al., 1968). Some tests allow measurement of volume at several decreasing pressures. The swelling pressure of an undisturbed sample can be measured by adjusting the confining pressure so that no swelling occurs on wetting. Alternatively, by adjusting the confining pressure to the value of the overburden pressure, the amount of swelling to be expected in the field can be measured.

Prediction of heave from the Jennings double oedometer method (Jennings, 1965) involves measuring consolidation e-log p curves for a sample at the natural water content and for a duplicate sample which has been allowed to swell freely in water. Heave is predicted from the difference in void ratio between the curves at any overburden pressure.

A different approach has been suggested by Croney et al. (1958), which involves predicting the equilibrium water content from the equilibrium suction expected under a pavement. The relationship between water content and volume is determined separately on samples. Where the water table is within 5 m of the surface, equilibrium suction will be determined by the gravitational component of soil suction (Chapter 4). For greater depth to the water table, the suction under the pavement is assumed to become the same as the suction below the depth at which water content changes seasonally.

A more detailed discussion of prediction of heave can be found in specialized publications such as Aitchison (1965), or Kassiff et al. (1969).

The amount of swelling and the swelling pressure measured on samples in the laboratory usually exceeds that measured in the field. Under field conditions, suction often does not fall to zero, so the full swelling is not attained. In addition the depth of soil, natural bonding, non-uniformity, lateral constraints, etc. make

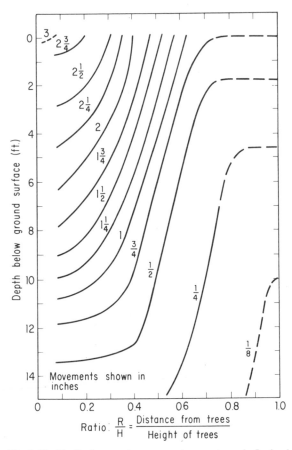

Fig.6.18. Vertical ground movements near trees in Leda clay soil at Ottawa, Canada. (Based on measurements near 55 ft. high elm trees in 1955 for a soil moisture depletion of 15.2" of water.) (Bozozuk and Burn, 1960.)

swelling in the field lower. Seasonal variation giving rise to changes in groundwater is not the only cause of ground movement. The water requirements of trees, for example, can be of such magnitude that moisture depletion in the subsoil would cause soil shrinkage. In Fig. 6.18 the vertical movements in Leda clay are plotted relative to distance and depth away from a large elm tree. In the case shown, subsidence of nearby buildings resulted because of this shrinkage. The obvious remedy in this case is to eliminate the tree or to locate superstructures far enough away from the influence of large trees.

In the following chapter we discuss compressibility and consolidation, at which time both the shrinkage and swelling influences on compressibility and consolidation will be examined.

6.5 SUMMARY

Shrinking and swelling accompanying drying and wetting are characteristic of clay soils. The amount of volume change depends upon the clay minerals, increasing with increasing surface area. Fabric also has a large effect on volume change, and shrinkage or swelling measured on remoulded samples will differ from that in the field. The cracks formed on shrinking of clay soils are important for water and air movement.

The swelling pressure exerted by a clay in contact with water is higher than the usual loading of a soil. The mechanism for the large amount of osmotic swelling which occurs in some clays is described.

The difficulties in prediction of swelling or heave in the field are discussed. Prediction requires a knowledge of the water-content changes which will occur due to rainfall and evaporation, and of the amount of volume change which will accompany a given change in water content.

Chapter 7

CONSOLIDATION AND COMPRESSION

7.1 INTRODUCTION

In the previous chapters we have dealt with the formation and structure of soils, water movement, and such fundamental properties as volume change, and consistency. The interaction of all of the above factors provide for the integrity of a soil mass. By this we mean that the integrity of a soil mass characterizes the resistance of the soil to deformation or distortion forces.

Under load application due to buildings and other kinds of super structures, physical deformation of the subsoil will occur. The nature and amount of deformation occurring is a function of not only the applied load, but also of the soil properties, and time. At least two kinds of mechanisms will be evoked in the soil in the deformation performance: (1) volume change due to extrusion of pore air and pore water; (2) shear distortion producing particle and fabric unit displacement phenomena with or without measurable pore-water extrusion and with development of slip planes.

It is likely that the two kinds of deformation mechanism, i.e., compression or volume change and distortion, occur concurrently with one or the other dominating — depending on the loading constraints. Where the superimposed load pattern does not exceed the yield strength of the soil, compression performance can be expected to dominate. Shear distortion will be small and can be ignored under general circumstances. Fig.7.1 illustrates the phenomenon schematically.

If the yield strength of the soil is exceeded by the loading, the resultant deformation will be due primarily to the shear-distortion mechanism. Initial deformation under these conditions will generally include compression volume change, but because of the excessive applied load, the shear distortion will override the initial volume change. Shear distortion and failure will be examined in detail in Chapters 8–10.

For the development and application of analytical theories to the problem of compression of soil, a proper modelling of the physics of compression volume change is required. This chapter deals with the phenomenon of compression volume change of clay soils due to applied loading. Similar considerations for cohesionless (granular) soils will be given in Chapter 9. Predictions of foundation settlements are not within the scope of this chapter.

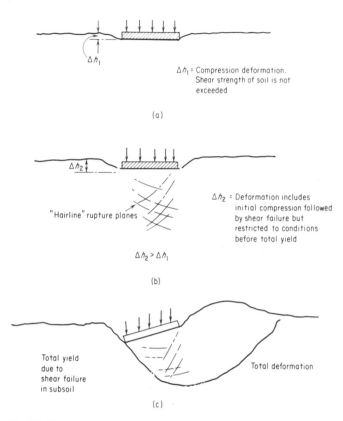

Fig.7.1. Various stages in deformation of soil, depending upon load and soil properties.

Compressibility, consolidation and creep

Whilst the term *compressibility* describes the volumetric response behaviour of a soil mass, recent and common usage of the term has restricted it to describing characteristics of compression behaviour not expressly or solely covered by the term *consolidation*. Fig.7.2 illustrates the region of compression behaviour fitting the description of "consolidation". The change in volume of a soil mass with time due to the extrusion of pore water is said to be a process of consolidation. The phenomenon covered by this definition yields a corresponding analytical model describing fluid flow through a "concerned" mass (see Section 7.2). The phenomenon of consolidation is one which depends upon time, soil and pore fluid properties, and load application. The overall load—volume change performance is identified as a stress—strain—time phenomenon and can be called *rheologic behaviour*.

Compressibility as a general term includes the overall performance of volume change (with or without some shear distortion) under load and is meant to describe the compression response characteristics along the lines of stress and strain. Thus, one may identify a soil as being more or less compressible, using *total deformation* and rigidity as a criterion for judgment. Under such circumstances, slight shear distortions are permitted. However, shear yield and failure are not included in overall performance characterization.

A common use for the term *compressibility* can be found in the description of compression behaviour of granular soils where physico-chemical properties are relatively insignificant and generally discounted. This is discussed in Chapters 8 and 9.

The term *creep* is used to describe stress—strain—time (rheologic) performance of materials. In the typical situation, material resistance to imposed loading is tested until yield or failure occurs (see Fig.7.3). Thus both volume change and shear distortion are involved in the deformation process. By and large, attention in creep analysis is focussed on providing working relationships applicable to description of stress—strain—time performance. Unlike consolidation in clays which refers specifically to the volume change phenomenon due to pore-water pressure dissipation, (see next section), "creep" in soils can be applied as an analysis

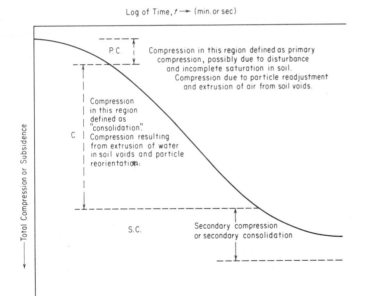

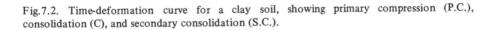

Fig.7.2. Time-deformation curve for a clay soil, showing primary compression (P.C.), consolidation (C), and secondary consolidation (S.C.).

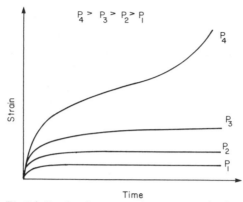

Fig.7.3. Strain–time curves from constant load application.

without considering the physical processes producing the creep curves shown in Fig.7.3. This will be examined in the last part of Section 7.5. Whilst this is not an uncommon procedure the recent developments in understanding of the mechanisms involved in particle interaction allow for the development of creep theories along the lines of actual soil behaviour.

Effective stresses

When external pressure is applied to a soil, internal resistance to the applied pressure consists of pressures developed in the pore fluid (water and air) and pressures (stresses) between individual soil particles. The stresses developed between soil particles due to stress transfer and contact are defined as effective stresses (i.e., intergranular stresses) and can best be shown in the following simple relationship for a fully saturated soil-water system:

$$\bar{\sigma} = \sigma - u \tag{7.1}$$

where: $\bar{\sigma}$ = effective stress; σ = applied total stress; u = pore pressure, i.e. pressure in the pore water.

Consider the saturated system shown in Fig.7.4. Particles A, B and C are rigid and incompressible. The fluid contained within the membrane in Fig.7.4 is also considered to be incompressible. If p represents the equal all-around pressure (i.e., hydrostatic pressure) acting on the membrane which encases the three-particle system, and if the particle-system geometry allows for movement of the particles within the membrane, the shape of the overall system shown in Fig.7.4a can distort. To illustrate the principle of effective stress, we will consider two separate cases involving the system shown in Fig.7.4.

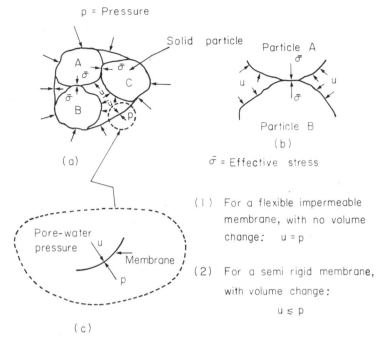

Fig.7.4. Simple effective-stress analogy. a. Ideal three-particle system with surrounding impermeable membrane. b. Effective stress between particle A and B. c. Pressures across the membrane.

For the first case, it is assumed that the impermeable membrane is flexible and possesses no membrane strength. Distortion of the system can occur easily without any volume change. Under the application of p, if no total volume change is allowed, $u = p$ (as shown in Fig.7.4c) at equilibrium. Thus, as shown in Fig.7.4b, and applying eq.7.1:

$$p = \sigma = \bar{\sigma} + u$$

Thus:

$$\bar{\sigma} = 0$$

This shows that where the pore-water pressure in a totally saturated soil is equal to the total applied pressure (i.e., total stress applied to a soil mass), the effective stress is zero. This indicates that the total applied load is carried in the fluid phase as hydrostatic pressure.

In case two, we will consider the impermeable membrane to be semiflexible, and to possess membrane strength. The soil particles and fluid are still considered as incompressible. If under the application of p, some fluid is allowed to escape, the system will compress and since the membrane will now develop stresses, $u < p$.

Without complicating the problem, we can assume that at the contact region between membrane and particle, the reaction between particle and membrane provides for a stress $\sigma = p$. Thus, the effective stress (stress between particles shown in Fig.7.4b) assumes some finite value, conditioned on the value of u as shown in eq.7.1. The smaller u becomes, the larger is the effective stress. At the extreme end, if the semiflexible membrane develops a leak, fluid will be extruded continuously under pressure p. Under such conditions, $u \rightarrow 0$ and $\bar{\sigma} = \sigma$. Thus, for the case where complete drainage is allowed (as in the case where $u \rightarrow 0$), the effective stress is equal to the total stress when the pore pressure dissipates to zero. This is the basis for the development of the theory of consolidation (see next section).

The principle of effective stresses is most basic to soil mechanics. It forms the basis for development of constitutive behaviour and relationships, and in essence states that the mechanical behaviour of soils is conditioned by relationships defined by the effective stresses. These conditions and response relationships will be examined in Chapters 8–10.

7.2 CONSOLIDATION OF CLAY

The volume-change performance of a clay under load is dependent on the relationship that is established between the effective stresses (i.e., stresses between particles) developed in the clay soil and the corresponding decrease in void ratio or porosity. For soils that are partially saturated, the decrease in void ratio is due to the extrusion of both air and water, a combination of a solution of air into the pore water and a partial extrusion of both water and air. At equilibrium, the air and water in the soil voids do not carry any stress, i.e., there is no pore pressure. In the equilibrium state under any effective stress regime, the stresses that must be transmitted for equilibrium to be maintained must be such that the particles will sustain the stress. This constitutes the basic structural matrix. Investigations to date have shown that this structural matrix may not be composed totally of solid particles in intimate contact but could be composed of solid particles surrounded by adsorbed water and contacting through the thin adsorbed water layers.

As defined in the previous section, for fully saturated clays, the mechanism by which compression occurs and during which the generated pore pressures are dissipated to reach an ultimate state of zero pore pressure is called *consolidation*. In essence consolidation is the compression that results when a load that is applied to a saturated clay gives rise to a compression, the magnitude of which is determined when the generated pore pressures are fully dissipated. Fig.7.2 shows that consolidation, however, is not the full extent of the compressibility of the clay under load since a further compression with time is generally recorded under that same load. This has been termed as *secondary compression* or *secondary consolidation* – to distinguish it from the *primary* mechanisms governing consolidation behaviour.

Review of the theory for one-dimensional consolidation

When a load is applied to a soil mass that is completely saturated, the load is initially sustained or carried by the fluid phase in the saturated soil. Since the volume change, ΔV, is initially zero due to the inability of water to move rapidly, the pore-water pressure in soil is initially equal or almost equal to the applied load. Fig.7.5 shows the dissipation of pore-water pressure with time as ΔV increases,

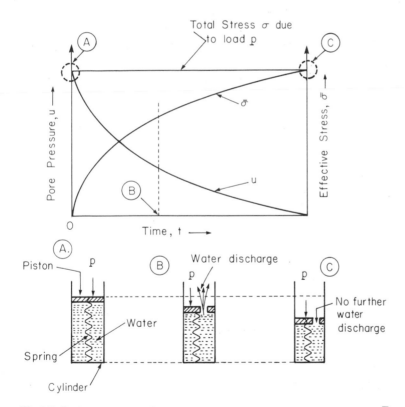

Fig.7.5. Pore pressure, u, dissipation and development of effective stress, $\bar{\sigma}$, in consolidation using the hydraulic piston and spring analogy.

together with a simple spring-dashpot analysis. The rate of discharge of the pore water, which controls the rate of volume change and the dissipation of pore-water pressure, u, depends on the soil structure and permeability. Obviously as ΔV increases (as shown in Fig.7.5) the density of the soil increases, and associated changes in soil structure and permeability will occur.

For simplicity in analysis, several conditions and assumptions are invoked. These are:

(1) Fully saturated clay-water system.

(2) Uni-directional flow of water.

(3) One-dimensional compression occurring in the direction of flow.

(4) A simple relationship between change in volume and applied pressure. (This is not unreasonable if the volume changes are small thus allowing us to apply linear small-strain theory.)

(5) The validity of Darcy's law, i.e., $\bar{v} = k\nabla h$, where: $\bar{v}$ = vector point function representing velocity; k = conductivity coefficient; h = total head; ∇ = vector operator $[i\,(\partial/\partial x) + j\,(\partial/\partial y) + k\,(\partial/\partial z)]$.

As a general case, we will begin by laying aside assumptions 2 and 3. These assumptions will be used later when we specialize the general relationship to the one-dimensional problem. Consider a fixed closed volume V with a surface area S. If $\bar{n}$ is the unit normal vector, the instantaneous volumetric flux rate through the surface S (Fig.7.6) is given as:

$$\iint_S \bar{V} \cdot \bar{n} \, dS + \iiint_V \Phi \, dV \qquad\qquad (7.2)$$

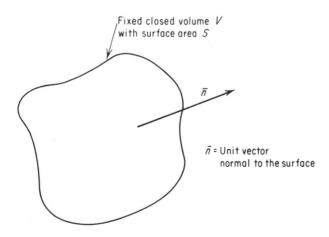

Fig.7.6. Fixed closed volume with unit normal vector, $\bar{n}$, emanating from surface — for use in derivation of consolidation equation.

where Φ represents the volume rate change associated with the internal forces in the system.

The divergence theorem by Gauss allows us to write:

$$\iint_S \bar{V} \cdot \bar{n} \, dS = \iiint_V \nabla \cdot \bar{V} \, dV \tag{7.3}$$

Substituting eq.7.3 in eq.7.2 and introducing Darcy's law for $\bar{v}$, we obtain the instantaneous volumetric flux rate through the surface S as:

$$\iiint_V [\nabla \cdot (k\nabla h) + \Phi] \, dV \tag{7.4}$$

The total change in volume resulting from the instantaneous volumetric flux rate is $\iiint (\partial V/\partial t) \, dV$. For a fully saturated system, the change in volume $(\partial V/\partial t) = (\partial V_v/\partial t)$ (i.e., change in total volume = change in volume of voids) if we assume the particles in the clay-water-system are incompressible and that volume decrement results directly from extrusion of water from the saturated pore spaces — as required by the use of the definition "consolidation".

Eq.7.4 may be combined with the total volume change since the rate of volumetric flux must be equivalent to the total volume change:

$$\iiint_V [\nabla \cdot (k\nabla h) + \Phi] \, dV = - \iiint_V \frac{\partial V_v}{\partial t} \, dV \tag{7.5}$$

Hence:

$$\iiint_V \left[\nabla \cdot (k\nabla h) + \Phi + \frac{\partial V_v}{\partial t} \right] dV = 0 \tag{7.6}$$

which, on the basis of the original consideration of an arbitrary volume V, allows us to write:

$$\nabla \cdot (k\nabla h) + \Phi + \frac{\partial V_v}{\partial t} = 0 \tag{7.7}$$

Expanding eq.7.7,

$$\nabla k \cdot \nabla h + k\nabla^2 h + \Phi + \frac{\partial V_v}{\partial t} = 0$$

To simplify eq.7.7, we recall that the initial assumption of constant permeability k

gives ∇k as zero. In addition Φ may reasonably be assumed to be zero on the basis of initial equilibrium conditions.

Hence, eq.7.7 is reduced to the following form:

$$k\nabla^2 h + \frac{\partial V_v}{\partial t} = 0 \tag{7.8}$$

By expressing h in terms of the pore pressure u [i.e., $h = (u/\gamma_w)$] and relating the change in volume of voids to the change in pore pressure [and this may be done by introducing the modulus of volume change m_v, where $m_v\,(\partial u/\partial t) = -\,(\partial V_v/\partial t)$], we may then rewrite eq.7.8 in the recognized form as presented by Terzaghi:

$$k\nabla^2 u = m_v\gamma_w\,\frac{\partial u}{\partial t} \tag{7.9}$$

Eq.7.9 is the three-dimensional form of consolidation developed with the assumptions shown previously. The proper three-dimensional relationship requires more rigorous considerations of coupled relationships describing stresses and volume change (Schiffman, et al., 1969). For one-dimensional compression, assumptions 2 and 3 stated previously are used. Thus we have:

$$\frac{\partial^2 u}{\partial z^2} = \frac{m_v\gamma_w}{k}\,\frac{\partial u}{\partial t} \tag{7.10}$$

or:

$$\frac{\partial u}{\partial t} = \frac{k}{m_v\gamma_w}\,\frac{\partial^2 u}{\partial z^2} = C_v\,\frac{\partial^2 u}{\partial z^2} \tag{7.11}$$

where: $C_v = k/m_v\gamma_w$ = coefficient of consolidation; m_v = modulus of volume change = $a_v/(1 + e)$; a_v = coefficient of primary or theoretical compressibility. If e = void ratio, for small pressure and void-ratio increments, $\partial e = a_v\partial u$, as in assumption 4.

The derivation of the consolidation equation depends upon certain assumptions other than those already mentioned. These are: that there is no change in permeability with both compression and time, and that the temperature does not affect the characteristics of the void ratio—pressure relationship, i.e., $e-p$ curve, that the load that is placed for compression is applied instantaneously, and that the resultant deformation occurs as a function of the instantaneous increase of load. It will be recognized that none of these will be absolutely true in actual field situations. The limitations of the laboratory test technique will not be discussed here in detail. However, it is necessary to point out that there are several factors that must be considered in actual laboratory tests which will affect the utility of the laboratory results for use in predicting field settlement. The major factors are side friction effects, load increment ratio, sample thickness and diameter (asperity ratio), thixotropic effects, duration of test, and creep effects. The influence of any or all of these factors on the actual consolidation, be it one-dimensional or three-dimensional, has not been evaluated completely. There are, however, several studies which have investigated single factors contributing to the consolidation phenomenon.

In practical situations, the time rate of pore-pressure dissipation, which reflects directly on the time rate of consolidation settlement, is of primary concern. If two adjacent surface loads produce identical total settlements (i.e., total consolidation), problems can arise if the rates of settlement of the two adjacent loads are different. Differential settlements will cause distress in superstructures if the differences in settlement at any one particular time exceed the structural design allowable limit, even though the final total settlements for the adjacent loads may be the same after a long time period. Thus, consolidation evaluation and solutions will also consider pore pressure response (under load) with time.

The solution of eq.7.11 will depend on the boundary conditions used. Solutions for the theoretical one-dimensional consolidation equation may be found in the standard texts on soil mechanics (Lambe and Whitman, 1969; Taylor, 1948). For example for double drainage, i.e., drainage at the top and bottom of a clay stratum of depth $2H$, the boundary conditions may be specified as follows:

$u = 0$ for top of clay layer $\quad z = 0$
$u = 0$ for bottom of clay layer $z = 2H$
$u = u_0$ for time $t = 0$

The solution for eq.7.11 takes the following form:

$$u = \sum_{n=1}^{n=\infty} \left(\frac{1}{H} \int_0^{2H} u_0 \sin \frac{n\pi z}{2H} \, dz \right) \left(\sin \frac{n\pi z}{2H} \right) e^{-n^2 \pi^2 C_v t / 4H^2} \qquad (7.12)$$

The following factors influencing consolidation may be deduced from an examination of the analytical model leading to development of the theory of consolidation:

(a) Change in hydraulic conductivity with applied pressure and time: this is evident if we associate water movement in soils with pore size, structure, and interaction characteristics of the soil-water system (Chapter 5). When soil compression occurs under applied load, the change in void ratio or porosity of the soil is reflected in corresponding changes in pore size, structure, tortuosity etc. Consequently, hydraulic conductivity must change in accordance with the changes occurring in the soil. Since k is related to porosity and other soil structure factors (Chapter 5), and if we assume that for small deformations the relationship between volume of voids and excess pore pressure is linear, we can write:

$$k = k_o - \beta(u_{max} - u)$$ (7.13)

where: β = coefficient concerned with variation in soil permeability; u_{max} = maximum pore-water pressure; k_o = conductivity coefficient at the beginning of consolidation.

From the above, when $u = 0$, k becomes, $k_{final} = k_o - \beta u_{max}$. The actual value of k which accounts individually for inter and intra fabric unit water movement (as described in Chapter 5) cannot really be defined. Thus eq.7.13 is approximate and is meant to illustrate a possible solution technique.

(b) Non-linear relationship between void ratio and applied pressure. In general, the relationship between effective stress, $\bar{\sigma}$, and void ratio, e, may be stated as follows:

$$\frac{\partial e}{\partial \bar{\sigma}} = f(\bar{\sigma})$$ (7.14)

Hence, assuming that the effective stress principle is valid:

$$\partial e = f(\bar{\sigma}) \, \partial \bar{\sigma}$$
$$= f(\bar{\sigma}) \, \partial(\sigma - u)$$

and:

$$\frac{\partial e}{\partial t} = f(\bar{\sigma}) \left(\frac{\partial \sigma}{\partial t} - \frac{\partial u}{\partial t} \right)$$ (7.15)

$f(\bar{\sigma})$ may be obtained from laboratory test results, or from non-linear stress–strain relationships — see Chapters 8–10.

(c) Loading rate: If P_o represents the final load placed on the test specimen and P is the load at any one time t, we may write an expression for pore-pressure response in terms of the load rate if we assume a relationship between load rate and corresponding pore-pressure response. The simplest assumption one might make is that the rate of imposition of pore pressure is the same as the rate of loading. The problem of loading rate and increments will be discussed later.

7.3 LABORATORY CONSOLIDATION TEST

The general procedure for laboratory evaluation of consolidation characteristics of clay soils involves a one-dimensional test. This is due not only to the difficulty of instrumentation for recording volume change and natural strains over and above axial strains, but also because of the complexities that will arise in the mathematical analysis of three dimensional consolidation behaviour. Recognizing the limitations of a one-dimensional analysis for consolidation, the accepted procedure has been to predict from the results of laboratory tests of one-dimensional consolidation, estimates of settlement of consolidation in the field. This may further be refined using correlations between triaxial consolidation and one-dimensional consolidation. The measurements generally taken for triaxial consolidation involve total volume change, from which it is possible to form correlations between one-dimensional and three-dimensional volume change. Correlation and validity of predictions have been discussed in detail by Schiffman et al. (1969).

The apparatus needed to perform a one-dimensional consolidation test is shown in Fig.7.7. This is referred to as an oedometer or consolidometer. A common method of load application is through a dead weight system applied in prescribed increments of load, and achieved by appropriate loading frames or mechanisms. This system of loading constitutes the constant total stress technique. The porous stones at the top and bottom of the sample provide drainage at top and bottom conforming to the boundary conditions of the solution (eq.7.12) for the consolidation equation given as eq.7.11.

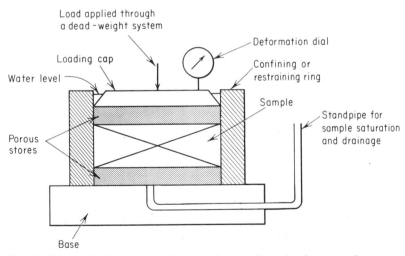

Fig.7.7. Schematic view of consolidometer for one-dimensional compression.

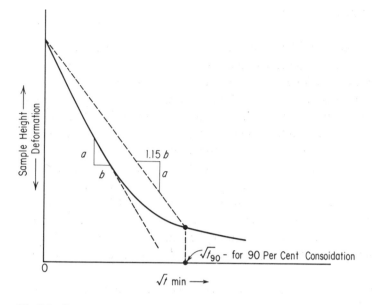

Fig.7.8. Square root time method for determination of 90% cent consolidation (Taylor method).

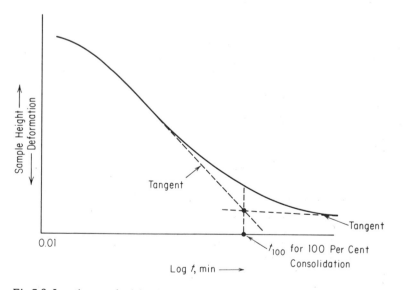

Fig.7.9. Log time method for determination of 100% consolidation (Casagrande method).

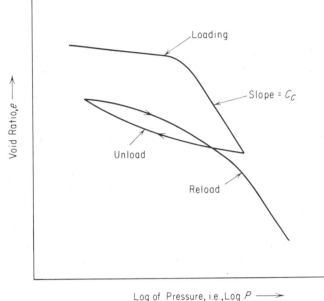

Fig.7.10. Typical e-log P curve from a consolidation test.

In the general constant total-stress test procedure, the sample in the consolidometer is subjected to an initial small loading. The measurements taken are time and deformation (shown by the deformation dial gauge). These deformation measurements are plotted either on a square root time base or a logarithm time base, shown in Fig.7.8 and Fig.7.9. Following the time taken to arrive at 90% consolidation in Fig.7.8 or 100% consolidation[1] in Fig.7.9 determined empirically, the next load (i.e., next increased level of constant stress) may be applied. For a fully saturated sample, the deformation may be related directly to void ratio change. Hence with a complete set of loadings and measurements of deformation, one may proceed to draw the e-log P curve. This is shown in Fig.7.10. From this, we can extract the slope of the curve C_c which would allow us to formulate a working model for prediction of total compression in the field.

There are at least two other methods of load application in consolidation tests; other than the constant total stress technique, these are:

---

[1] The 100% consolidation value obtained in this manner is not necessarily identical to that obtained from measurement of pore water pressures, i.e., 100% consolidation at $u = 0$. This problem of definition of "end of consolidation" is discussed later.

(1) Controlled strain-rate test: The idea here is to provide for a more rapid test by causing the sample to deform at some constant rate without undue distress arising from continued high pore-pressure response. Under such a system, results such as those in Fig.7.8–7.10 can still be obtained. They are, however, subject to severe interpretation, and forecasting limitations of the test sample which may not reach a state of near stress equilibrium – as in the constant total-stress technique. This situation would arise if the imposed strain-rate is high.

(2) Controlled gradient test: In view of the drawbacks or strict requirements for careful control on strain-rate with the strain-rate technique, a similar intent can be achieved by controlling the generated pore-water pressures. By controlling the pore pressure generated through application of total stress, the effective stress at any time is also controlled. Thus, if u = constant, and if the total stress σ has to be increased continuously throughout the test (since pore pressure, u, will dissipate in the test of σ = const.) to maintain u = constant. $\bar{\sigma}$, the effective stress will be maintained continuously. The advantage to this method of load application is in the fact that strain-rates are indirectly controlled and conditioned by the consolidation of the sample at any time (Lowe et al., 1969).

Estimation of total compression

To estimate the total compression (or consolidation) we can assume that the change in volume of a saturated soil ΔV is equal to the change in the volume of voids ΔV_v – conditional on the requirement of incompressibility of water and soil grains. Thus:

$$\Delta V = \Delta V_v = V_s \Delta e$$

or:

$$\Delta V = \frac{V \cdot \Delta e}{1 + e} \tag{7.16}$$

From Fig.7.10:

$$\frac{\Delta e}{\log P_2 - \log P_1} = C_c \tag{7.17}$$

We can write eq.7.17 as:

$$\Delta V = \frac{1}{1 + e} V C_c \log \frac{P_2}{P_1}$$

For one-dimensional compression $\Delta V = A\Delta h$, where: A = cross-sectional area of soil sample; h = height of sample. Thus:

$$A\Delta h = \frac{C_c}{1+e} Ah \log \frac{P_1 + \Delta P}{P_1},$$

and:

$$\Delta h = \frac{C_c}{1+e} h \log \frac{P_1 + \Delta P}{P_1} \qquad (7.18)$$

Eq.7.18 gives the total expected one-dimensional consolidation consistent with the inherent limitations of the test technique and the necessary assumptions. It is important to remember that the e-log P curve was obtained on the basis of a load increment shortly after estimated 100% consolidation. It follows that secondary consolidation is not accounted for in this analysis of total settlement. The interpretation of the time for 100% consolidation must be recognized to be an empirical division of the *total* time-compression phenomenon. Whether or not pore pressures do exist following this empirical division point is not certain. Under these conditions, P_1 represents the effective stress, $\bar{\sigma}_1$, at some initial point identified as 1. The increase in pressure, ΔP, will obviously produce an increase in effective stress, $\Delta\bar{\sigma}$, at the end of further consolidation due to $\Delta\bar{\sigma}$. Thus, eq.7.18 can be written as:

$$\Delta h = \frac{C_c}{1+e_1} h \log \frac{\bar{\sigma}_1 + \Delta\bar{\sigma}}{\bar{\sigma}_1}$$

$$= \frac{C_c}{1+e_1} h \log \frac{\bar{\sigma}_2}{\bar{\sigma}_1} \qquad (7.19)$$

where: $\bar{\sigma}_1 + \Delta\bar{\sigma} = \bar{\sigma}_2$ and e_1 = void ratio at the point taken for $\bar{\sigma}_1$.

For the case where the slope identified by C_c in Fig.7.10 is not constant, piecewise linearity techniques will have to be adopted. Eq.7.19 remains unaltered except for varying values of C_c consistent with the void ratio, e, and different thicknesses of h used for the test.

The modern theories associated with physico-chemical forces in clay-water systems and clay–water interaction do not emphasize the need for the division of the time-deformation curves into the three divisions of primary compression, consolidation, and secondary consolidation respectively. These theories and

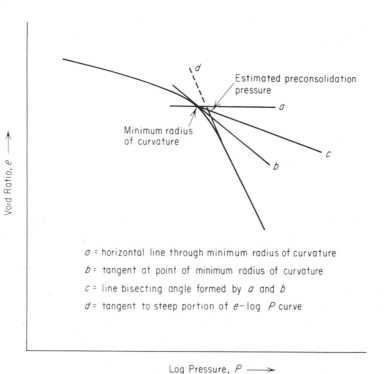

Void Ratio, e ⟶

Log Pressure, P ⟶

Fig.7.11. Graphical estimation of preconsolidation pressure. Intersection of lines c and d gives the estimated preconsolidation pressure (Casagrande method).

available consolidation research studies show too, that the definition of the pre-consolidation pressure is sensitive to the laboratory technique. The commonly used Casagrande technique for determination of preconsolidation pressure is given in Fig.7.11. This method predicts the normal preconsolidation pressure of the soil from the laboratory test assuming that the test constitutes a second (or thereabouts) loading cycle, i.e., extraction of sample from the ground is taken as an unloading cycle. The phenomenon of preconsolidation is obviously conditioned by previous load and time relationships in addition to loading patterns established in the evaluation of the consolidation characteristics of the material.

7.4   TIME AND LOAD-DEFORMATION CURVES

The time-deformation curves for a single loading are shown in Fig.7.12. The clays considered are a highly active or organic clay, a medium clay, and a silty clay. Based on the Casagrande method for determination of 100% consolidation, we can

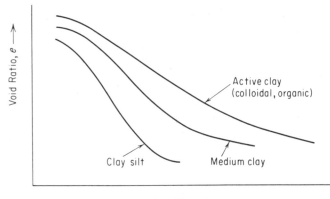

Fig.7.12. Comparison of time-deformation curves for three soils under a single load.

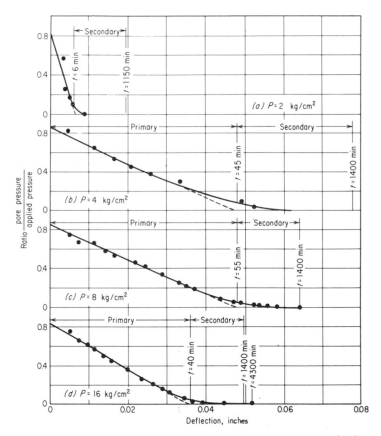

Fig.7.13. Relationship between pore pressure and deflection under incremental loads for a marine clay. (Crawford, 1964.)

establish the limit of consolidation in the silty clay, and the medium clay samples. However, for the highly active or organic clay, the time for 100% consolidation is not well defined. This is due in part to the soil structure (i.e., clay—water interaction and fabric) and also to the limitations in the definition of consolidation.

Accepting that the end of classical primary consolidation is that point where pore-water pressure is zero, a useful way of examining consolidation test results is one in which the measured pore-water pressure is expressed as a ratio of the applied pressure. This is shown in Fig.7.13. The test results for a marine clay (Crawford,

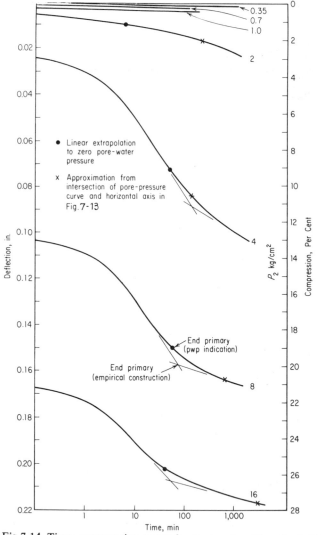

Fig.7.14. Time—compression curves for marine clay. (Crawford, 1964.)

1964), indicate that the relationship between the ratio of pore pressure to applied pressure and the deformation of the test specimen is approximately linear. Extrapolation of this linear relationship to the abcissa determines the end of pore-water-pressure registration. This intersection provides an empirical point for the end of classical primary consolidation. The extrapolated points can then be plotted on the time-compression curves, Fig.7.14. As a comparison, the empirical method for determination of end of primary consolidation based on the intersection between the two tangents is also shown in Fig.7.14.

While the empirical construction method and the extrapolated zero pore-water-pressure method do not give identical times for end of primary consolidation, this may be due in part to the sensitivity of the pore-water measuring device. In Fig.7.13 the end of actual pore-water registration is in all instances beyond the extrapolated point. Thus the two points shown in Fig.7.14, indicated by a solid circle or a cross, do not correspond. Taking the time of zero pore pressure as the

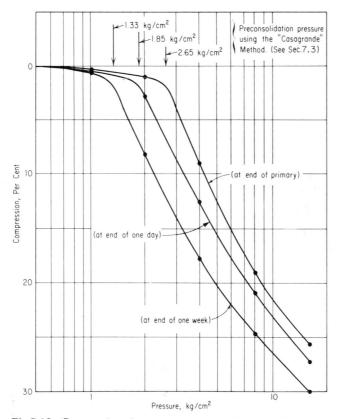

Fig.7.15. Compression–log pressure curves for normal and long-term incremental loading. (Crawford, 1964.)

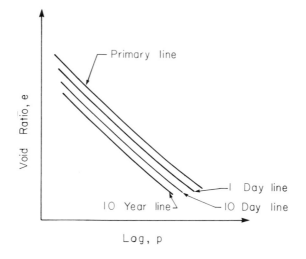

Fig.7.16. Time–load-compression curves.

point of intersection of the pore-pressure curve and the horizontal axis in Fig.7.13 and shown by the crosses in Fig.7.14, the results indicate that there is a measurable pore-water pressure beyond the time indicated by the empirical construction technique.

The point to be made here is that the definition of the end of primary consolidation on the basis of graphical interpolation techniques using laboratory results will provide empirical points. The significance of a proper choice representing the end of primary consolidation is shown in Fig.7.15 in terms of a compression and pressure relationship. This figure shows that it is possible to obtain three different values for the preconsolidation pressure dependent upon the choice of time for termination of primary consolidation. In the topmost curve the choice is made on the basis of the empirical construction (Fig.7.9) for the end of primary consolidation. The intermediate curve is derived on the basis of the total deflection after one complete day of applied loading (Fig.7.13). If total compression is taken after one week of applied loading for each incremental load, the compression–pressure relationship is again altered, as shown in the bottom curve in Fig.7.15 for a companion sample.

The results shown in Fig.7.15 reflect the same trends reported previously (Taylor, 1942). If one expresses the void ratio–pressure relationship in the form shown in Fig.7.16, it will be noted that the "one-day line", "ten-day line" and "ten year line" are essentially parallel, not unlike those shown in Fig.7.15. The longer one takes to reload, i.e., the greater the time interval between load applications, the lower would be the e-log p line shown in Fig.7.16. The lines below the "primary

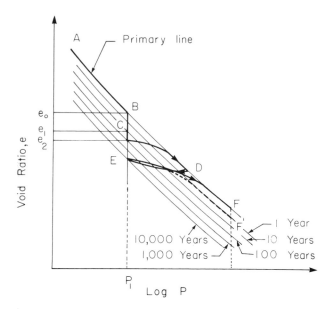

Fig.7.17. Load and unload sequence relative to time—load-compression curves.

line" account for the phenomenon of "secondary compression" since obviously the "primary line" denotes load increment after 100% consolidation. The greater the time period between loading, the larger is the "secondary compression". Bjerrum (1967) states that each of these lines represents the equilibrium void ratio for different values of effective overburden pressure at a specific time of sustained loading. As stated previously in Section 7.1 the total deformation indicated by the lines below the "primary line" must be considered as compression lines, as opposed to "consolidation lines", since the deformation includes more than that sustained by consolidation.

The total amount of secondary compression under any one sustained load can be obtained as in Fig.7.17. The figure shows that if primary loading is brought to P_1 and sustained at that level, the void ratio e_0 will continue to decrease with time. This is shown by the vertical line which cuts across the "time lines". The total amount of secondary compression after ten years is $e_0 - e_1$, and after 100 years is $e_0 - e_2$. The information contained in Fig.7.17 can be used to portray the following sequence of loading: load application as in a normal depositional load (line AB) followed by a sustained period of constant load (line BC), following which load application is made perhaps by glacial loading (line CD) and subsequent recession of the glacier causing unloading (line DE), and final loading through application of a building load (line EF). For primary loading, line EF must approach and superpose the primary line initially established by AB. If a delayed loading of one year is used, the line EF' will be obtained. Other time span delayed loadings (i.e. one day to n number of years) will give curves corresponding in shape to EF' and superposing on the appropriate "time line". From the above, it should be clear that the determination of an appropriate preconsolidation pressure must be sensitive to history of loading.

Load-increment ratio

The recommended standard constant total stress technique in laboratory tests for one-dimensional consolidation utilizes load-increment ratios of one, i.e. the loads or pressures are doubled for each loading $\Delta P/P = 1$. There is convincing evidence to show that load-increment or pressure-increment ratio is a very important consideration (Leonards and Ramiah, 1959). When the applied incremental loads are small, the ability for the individual particles and fabric units to readjust and to reattain equilibrium will control to a great extent the resultant

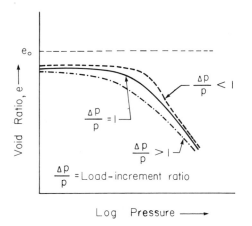

Fig.7.18. Effect of load-increment ratio on void ratio–log-pressure curve.

deformation. If this readjustment of the fabric is allowed to occur with minimal distortion and without generated pore pressures, the resultant compression is small. If, on the other hand, the succeeding load increment is high, the ability of the soil fabric to readjust to an equilibrium configuration would be affected since the applied loading is sufficiently large to create and cause shear distortions. Consequently, a greater degree of compression (see Fig.7.18) will occur.

7.5 SOIL STRUCTURE IN CONSOLIDATION AND COMPRESSION

In assumptions 2 and 3 in Section 7.2, the conditions for one-dimensional loading and fluid flow were stated as necessary requirements for the development of the one-dimensional theory of consolidation. The apparatus shown in Fig.7.7 indicates that drainage must be in the vertical direction, similar to the direction of load application. Fig.7.19 shows scanning micrographs of a natural soil following one-dimensional laboratory consolidation. The close-up views show: (1) that single particles are rarely seen — the fabric units are essentially peds; (2) that both large and small pore spaces exist between peds and that these pore spaces are quite continuous in the vertical direction (Fig.7.19B, D). Note that the flat face of the "micro" sample is in actual fact the side face of the consolidation sample. From the micrographs, we can assume that since the major pore channels are directed vertically — in the same direction of loading, one-dimensional theory can be applied. The views shown by the micrographs give an indication of the action of peds in compression behaviour. The pore spaces between peds and within peds, i.e. inter-ped and intra-ped pores, will provide for different fluid-flow characteristics.

In a typical soil structure where fabric units are common and indeed constitute the soil mass itself, compression behaviour can be classified in terms of dominant drainage characteristics. This is illustrated in Fig.7.20. The corresponding drainage control graph for Fig.7.20 is given in Fig.7.21.

The two kinds of flow or extrusion are not really separable. As time increases, the corresponding decrease in soil volume resulting from the decrease in inter-and intra-fabric unit pores will create different extrusion and flow characteristics. The terms "extrusion" and "flow" have been used to signify extrusion of pore water from individual fabric units and between fabric units (due to applied load), and subsequent flow of the extruded water. Undoubtedly in a soil mass consisting of a large number of fabric units, these two actions occur simultaneously and continuously throughout the test sample.

Fig.7.20 and 7.21 show that for successful predictions of consolidation, fluid flow characteristics appear to be governed by inter fabric pores, i.e. the diffusion equation derived using the development shown in Section 7.2 relies on flow through inter fabric unit pores. The values for hydraulic conductivity k must necessarily take into account flow characterization through inter and intra fabric

---

Fig.7.19. Scanning electron micrographs of a one-dimensionally consolidated sample of Bangkok clay. A. General view of test sample in specimen holder. B. Close-up of edge of sample shown in (A). Note ped formation and macropores. C. Top view of sample. D. Side view of sample. Note size of macropores. Orientation of macropore channels is in the same direction of fluid flow to drainage. (Sample for viewing obtained from Professor Za-Chieh Moh of Asian Institute of Technology.)

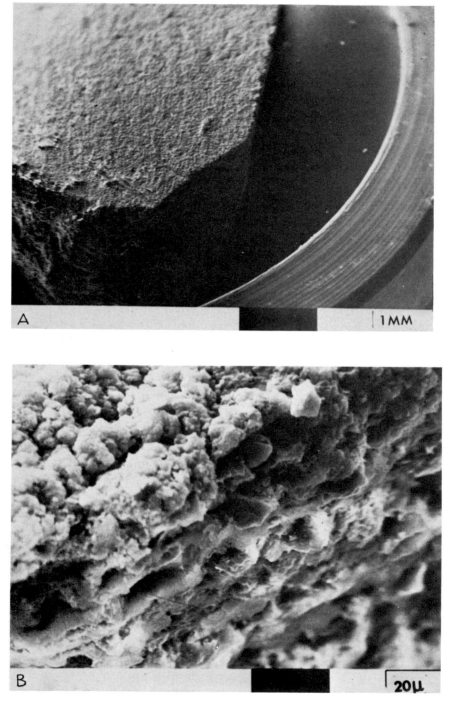

Fig. 7.19 (for legend see p. 247).

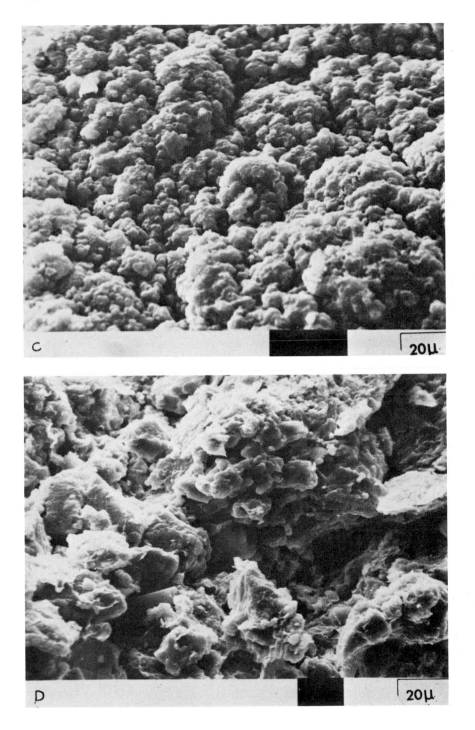

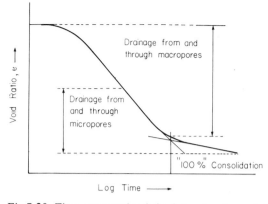

Fig.7.20. Time compression behaviour of a clay soil showing drainage sequence through macro and micropores.

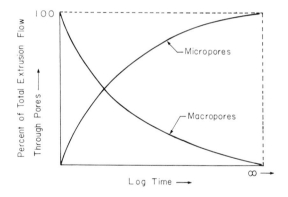

Fig.7.21. Drainage and water extrusion through macro and micropores.

unit pores. The measured value of k is in general a lumped parameter k, i.e. an effective k which samples all the kinds of flow involved and provides an average value.

In view of the soil structural behaviour aspect of the total compression problem it is difficult to rigidly classify the compression behaviour into primary and secondary consolidation components. The above reasoning based on prepondant flow would be appropriate as a guideline for classifying the phenomenon identified as "consolidation". When secondary compression or secondary consolidation occurs, without measurable pore pressures or fluid flow, it is likely that intra fabric unit flow occurs — thus providing not only the longer term deformation but also the insignificant quantities of water discharged.

Fabric changes in compression

To illustrate the fabric changes that occur during compression of a soil mass, we consider the typical situation of a soil composed of an infinite number of fabric units of various categories — i.e. peds, clusters and domains as shown in Chapter 4. In the initial state, the fabric may be considered to be random and uniform. From physical observations, e.g., Fig.7.19 and 7.23 using microscopy and X-ray diffraction methods, total fabric change in compression under external constraints will follow a sequential pattern as follows (Fig.7.22):

(a) Rearrangement and orientation of fabric units without any significant distortion of the units.

(b) Increased load application would cause a greater degree of orientation of fabric units. At this stage, some particle rearrangement within the fabric units can occur. Pseudo anisotropy (i.e., fabric unit anisotropy) effects may be observed.

(c) Further loading will cause more orientation of the fabric units and particles within the fabric units. Volume decrease in fabric units will also occur, leading to secondary compression effects with no measurable pore-water pressure.

In Fig.7.23 the X-ray peak height in counts per second is shown for two soil structures under consolidation. X-ray peak registration is low if particles are not

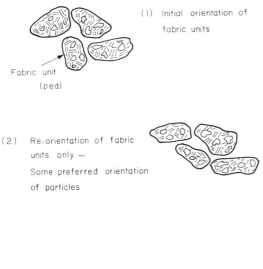

(1) Initial orientation of fabric units

Fabric unit (ped)

(2) Re-orientation of fabric units only — Some preferred orientation of particles

(3) Orientation of fabric units and particles in the fabric units

Fig.7.22. Fabric unit and particle orientation under load.

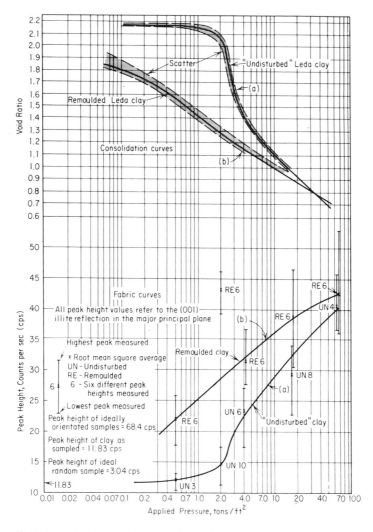

Fig.7.23. Orientation of clay particles produced by one-dimensional consolidation of undisturbed and remoulded samples of Leda clay. (Quigley and Thompson, 1966; by permission of the Nat. Res. Council of Canada, *Can. Geotech. J.*, 3(3), pp.61–73.)

oriented. The more nearly parallel the particles are, the higher is the X-ray peak intensity. Thus flocculated clays show low peak registration until breakdown and reorientation of fabric units and particles into an oriented fabric occurs. For curve *a,* the flocculated undisturbed clay, the X-ray peak intensity shows increasing parallel orientation of particles with increasing load after collapse of initial structures. The curve seems to mirror the e-log *p* curve shown in the top of the figure. Curves *b* are the consolidation and X-ray curves for the remoulded clay.

Beginning with a higher intensity due to orientation from remoulding, there is a gradual reorientation of fabric units and particles in the soil. As more load is applied, the soil fabric becomes more oriented as shown by higher peak intensity. The absence of structural collapse (compared with the flocculated structure) provides the basis for the relatively smooth curve obtained.

Silt inclusions in clay soils will tend to create "clay balls". These are clay coated silt particles. Under a compressive load, the sequence of fabric change is not dissimilar to that shown in Fig.7.22, except that stages two and three are combined. Clay-particle reorientation around the silt inclusions will now be confined to a tighter packing since the clay particles tend to be originally oriented around the inclusions. The rate of deformation will be dependent on the amount of silt inclusions within the soil mass.

Temperature effect

Laboratory consolidation tests are generally performed at room temperature. In the field, the subsoil temperature varies during the year. Because of seasonal and even daily fluctuations in temperature, it is necessary to examine the effects of temperature on consolidation of subsoil.

While a number of studies have been reported on the influence of temperature on both primary and secondary consolidation, there is still no general explanation which allows a prediction of the temperature effect for a particular soil. A number of papers on this topic can be found in *Highway Research Board Special Report* (1969). The general conclusions reached in these studies can be summarized as follows. Increasing temperature results in a lower line on the e-log P plot for most soils, i.e. as temperature is increased the void ratio is lower at a constant pressure. The magnitude of the difference depends upon the soil. The coefficient of consolidation, C_v, decreases slightly with increasing temperature for many soils, but is essentially independent of temperature for other soils. Secondary compression rates are increased with increasing temperature for many soils, but are independent of temperature in some soils. High-swelling soils increase in volume at a given pressure, with an increase in temperature.

Structure and creep

In Section 7.1 and Fig.7.3, we discussed very briefly the phenomenon of creep. We will not be concerned with the creep performance shown by the topmost curve in Fig.7.3, but will examine the behaviour pattern identified by the lower curves in Fig.7.3. The curves shown in Fig.7.24 illustrate both loading and unloading, where $P_3 > P_2 > P_1$.

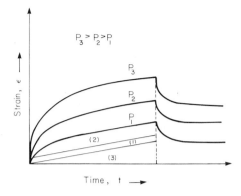

Fig.7.24. Load and unload curves in creep evaluation of a clay.

From the soil behaviour or soil structure point of view, a change in soil structure would produce different characteristics in the curves shown. The typical long-term time deformation relationship may be broken down into three separate parts, shown in Fig.7.24 as 1, 2 and 3. These are: (1) instantaneous deformation (strain), ε_e; (2) retarded deformation (strain), ε_r; (3) constant rate deformation (strain), ε_f. We can consider the deformation behaviour of the overall mass in terms of the deformation or relative displacement of a representative or basic group of fabric units in the soil mass. Thus, if the micro strain (i.e. fabric unit group strain) is given as ε, we will have:

$$\varepsilon = \varepsilon_e + \varepsilon_r + \varepsilon_f \tag{7.20}$$

Thus:

$$\varepsilon = a_i \xi_i + \alpha f_i(t)\xi_i + \beta b_i \xi_i t \tag{7.21}$$

where: a_i = the compliance of instantaneous deformation; $f_i(t)$ = the creep compliance for ith fabric unit; b_i = a viscous coefficient responsible for continuous flow; $\alpha = 0$, $\beta = 1$ when ξ_i is greater than the yield point of a basic group of fabric units; $\alpha = 1$, $\beta = 0$ when ξ_i is less than the yield point of a basic group of fabric units; ξ_i = microscopic stress acting on the basic group of fabric units.

The total deformation strain e is thus the integral effect of all the micro deformation strains ε, i.e.:

$$e = \Sigma a_i \xi_i + \sum_1^m f_i(t)\xi_i + \sum_1^n b_i t \, \xi_i \tag{7.22}$$

where: m = number of basic unit fabric groups whose yield strength has not been exceeded; n = number of basic fabric unit groups whose interaction strength been exceeded.

The basic consideration in terms of groups of fabric units interacting with each other to provide overall stability is not unrealistic. The micrographs shown in Fig.7.19, for example, depict individual peds and other fabric units forming larger groups. These have been identified as fabric unit groups, and it is this kind of group behaviour that ultimately reflects in overall mass behaviour. The total number of the units may be considered as semi-infinite. Thus: $n + m \rightarrow \infty$. The microscopic stresses ξ_i depend upon the arrangement of groups and individual fabric units, and the actual stress distribution within the sample. If ξ_i exceeds the basic strength of the ith group, the group will undergo continuous flow.

Eq.7.22 may be represented as a continuous function in the following form[1]:

$$e = A\sigma + \int_{-\infty}^{t} f(t-\tau)\sigma d\tau + Bt\sigma \qquad (7.23)$$

where: A = instantaneous compliance; B = flow parameter; σ = total stress; t = current time; τ = time variable.

In this instance the gross effect of the instantaneous deformation has been replaced by an overall instantaneous compliance A which may be easily evaluated from observations, and:

$$f(t) = \frac{dF(t)}{dt} \qquad (7.24)$$

This represents the rate of change of creep function $F(t)$ where $F(t)$ is of the form $\Sigma (1-e^{-t/\tau_i})$, in which τ_i is the retardation time[2] of the ith group. Thus the rate of retarded deformation decreases continuously, and this portion of deformation converges to a finite value under a given stress for long loading periods.

The pore-water pressures generated in the clay water system whenever loads are applied to the clay soil are generally not evenly distributed. Thus pressure gradients will exist, causing internal flow and redistribution of the pore fluid. The low permeability of clays, for example, causes the dissipation process to be

---

[1] The mathematical treatment and solution of problems of creep can be found in the modern textbooks on viscoelasticity, e.g., Christensen (1971). For our purpose, we are interested in obtaining an insight into the behaviour of typical soils (containing peds, etc.) under compression where longterm deformation is observed.

[2] The retardation time is that time which is required to produce $(1 - 1/e)$ of the full elastic deformation under a constant stress.

retarded, resulting in a time-retardation deformation phenomenon. Since different soil structures will provide different resistances to induced flow (Chapter 5), the retardation time τ can be related to the different fabric unit groups present in the soil. A change in rate of retarded deformation reflects the response of the fabric unit groups to the external load and the retardation time distribution can be regarded as an indication of the distribution of the fabric units and groups participating in the process of deformation.

In clay soils under sustained loading, the flow portion of the deformation may constitute a large portion of total deformation. This is the result of the continuous reorientation of fabric units and groups. Stress transfer takes place from one group to another due to the yielding or deformation of a group and in turn, over-stressing of other groups or fabric units beyond the limit equilibrium state will occur, thereby resulting in the observed constant rate of flow.

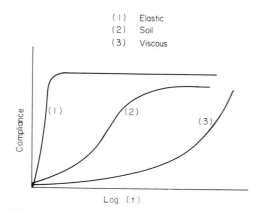

Fig.7.25. Compliance time relationships for different materials. (Yong and Chen, 1970.)

The creep function $F(t)$ (eq.7.24) can be obtained by plotting the compliance of the retarded portion of the deformation against the logarithm of time. A typical set is given in Fig.7.25. The function $F(t)$ is now specified (i.e. from Fig.7.25). The retardation time distribution is obtained as:

$$f(t) = \frac{dF(t)}{d(\log t)} \tag{7.25}$$

where actual slopes of the curves shown in Fig.7.25 may be evaluated. Plotting the slopes against the logarithm of time provides the retardation-time curve, from which the probability density function can be obtained by normalizing the retardation-time-distribution curves. Fig.7.26 shows the normalized curves for the same materials identified in Fig.7.25.

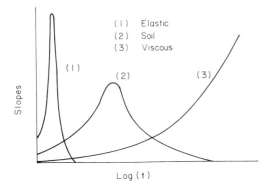

Fig.7.26. Normalized retardation-time-distribution curves for various materials. (Yong and Chen, 1970.)

The normalized retardation-time-distribution curves show that the mode and the mean values shift from the higher time decades to the lower ones where ultimately the bell-shape curve approaches that of a Gaussian distribution, i.e. mean and mode values are coincident. Note that for a material exhibiting ideal flow (i.e. curve *3* in Fig.7.25 and 7.26) and retardation time approaches ∞, while for a nearly perfect elastic material (i.e. curve *1*) this retardation time is close to zero. In tests on frozen soils for example, this value is in the range of 100 min. Increasing material rigidity will produce corresponding decreasing values in the retardation time.

Recognizing that the soil structure characteristics defining each fabric unit group in a soil mass depend on specific environmental constraints in regard to balance of energy, the probability of occurrence P_i of a group of fabric units at a particular level of integrity is a direct function of its energy state E_i, i.e. $P_i = f(E_i)$.

A canonical ensemble of volume $\bar{\Omega}$ can be constructed such that each fabric unit group of volume $\bar{V}$ constitutes an elemental system of the ensemble which is in weak thermal interaction with other groups forming the ensemble. If there are M systems in the ensemble $\bar{\Omega}$ each system together with the other $M-1$ systems constitutes a heat reservoir (i.e. the ensemble $\bar{\Omega}$ is a heat bath). It is apparent that $\bar{V}$ has many fewer degrees of freedom than $\bar{\Omega}$. In an equilibrium situation, the probability of occurrence of one system $\bar{V}$ in $\bar{\Omega}$ in a state i is proportional to the number of states accessible to $\bar{\Omega}$. Thus:

$$P_i = c\omega(E^{(o)} - E_i) \tag{7.26}$$

where: c = proportionate constant independent of i; $\omega(E^{(o)} - E_i)$ = number of energy states accessible to the systems remaining in $\bar{\Omega}$ in view of *one* system $\bar{V}$ being at state i. From the normalization procedure:

$$\Sigma P_i = 1 \tag{7.27}$$

where the summation includes all possible states of $\bar{V}$.

Eq.7.26 can be approximated by expanding the logarithm of $\omega(E')$ about the value $E' = E^\circ$ as follows:

$$\ln \omega(E^{(0)} - E_i) = \ln \omega(E^{(0)}) - \left[\frac{\partial \ln \omega}{\partial E'}\right]_0 E_i \tag{7.28}$$

where: $E' = E^{(0)} - E_i$ = energy of $M-1$ systems remaining in Ω if any *one* system has energy E_i.

It can be shown that since:

$$\left[\frac{\partial \ln \omega}{\partial E'}\right]_0 = \beta$$

is a constant if evaluated at the fixed energy $E' = E^{(0)}$ the dependency on E_i for system $\bar{V}$ does not exist. Thus:

$$\beta = 1/KT$$

where: K = positive constant with dimensions of energy; T = temperature.
Eq.7.28 can be written as:

$$\omega(E^{(0)} - E_i) = \omega(E^{(0)}) e^{-\beta E_i} \tag{7.29}$$

and can be substituted into eq.7.26 to give:

$$P_i = c\omega(E^{(0)}) e^{-\beta E_i}$$

Since $\omega(E^{(0)})$ is a constant independent of i, we obtain:

$$P_i = Ce^{-\beta E_i}$$

In view of eq.7.27:

$$\frac{1}{C} = \Sigma e^{-\beta E_i} = \text{partition function} \tag{7.30}$$

Hence:

$$P_i = \frac{e^{-\beta E_i}}{\Sigma e^{-\beta E_i}} \qquad\qquad (7.31)$$

The probability P_i of occurrence of energy state i which recognizes the availability of many fabric unit group forms g_i can be introduced into the creep performance evaluation in terms of the mathematical expectation of the total deformation $<\varepsilon(t)>$. Hence Eq.7.22 can be modified thus:

$$<\varepsilon(t)> = \Sigma a_k P_k \xi_k + \Sigma P_i f_i(t) \xi_i + \Sigma P_j \xi_j b_j t \qquad\qquad (7.32)$$
$$k = i, j$$

Correspondingly, we can modify eq.7.23 as follows:

$$<\varepsilon(t)> = A\sigma + \int_o^t \frac{P_i(\tau)}{g_i} f(t-\tau)\, \sigma d\tau + B\sigma t \qquad\qquad (7.33)$$

where: g_i = number of ped structural states at the same energy level.

The consequences in regard to eq.7.32 or 7.33 allow for microscopic instability (i.e. local yielding) without macroscopic failure or yield, and can be used to predict or analyze the performance curves shown in Fig.7.3 and 7.24 (Yong and Chen, 1970, 1972).

7.6 SUMMARY

Consolidation, as such, is a difficult phenomenon to analyze since the methods available for laboratory study, which must conform with physical reality, are limited in scope. The change in properties of soil with time under the influence of stresses is generally not properly accounted for in laboratory evaluation. Undoubtedly, as more is learned, more refinements to the simplified one-dimensional consolidation theory will be made.

The specific points covered in this chapter relate to the development of a realistic appreciation of the results of particle and fabric-unit interaction under external stressing. The particular role of pore water in the development of rates and total settlement is seen to be of utmost significance. A proper modelling of the physics of clay—water interaction under sustained loading is required for the production of analytical theories needed to predict performance. The mathematics of the problem which relate more to solution of the developed analytic formulations and predictions are treated in specialized publications on consolidation and in books on soil mechanics and foundations engineering.

Chapter 8

YIELD AND FAILURE

8.1 INTRODUCTION

To set the stage for a discussion of granular and cohesive soil strengths (Chapters 9 and 10) it is necessary to first establish concepts of yield and failure in soils.

In the determination of the shear strength of soils, it is relevant to bear in mind the fact that if a predetermined or prechosen yield or failure criterion is used, the parameters describing yield or failure must be realistically evaluated. Thus taking the Mohr-Coulomb failure theory as an example of accepted usage, the generally used parameters c and ϕ are a direct consequence of the application of the failure criterion which may not necessarily be applicable or admissible in the description of failure conditions in certain kinds of soils. The definitions associated with c and ϕ, such as cohesion and angle of internal friction pertain to the applied failure criterion and may not necessarily represent the actual mechanism being described in the shearing resistance of the soil. In other words, the requirement for compatibility between the physical behaviour of the soil in shear deformation (i.e., the "physical" model) and the "mathematical" model is implicit if proper application of yield and failure theories is to be made.

When a material specimen is stressed in simple tension or compression its mechanical behaviour is described by the stress–strain curve and a statement concerning its extensibility or compressibility. The actual shape of the stress–strain curve depends upon the material being tested. When the relationship between stress and plastic behaviour may be gradual as in Fig. 8.1a, or abrupt, as in Fig. 8.1b. In the is termed elastic. When the stress–strain relationship becomes nonlinear and irreversible, this is termed as plastic behaviour. The transition between the elastic and plastic behaviour may be gradual as in Fig. 8.1a, or abrupt, as in Fig. 8.1b. In the former case, the yield stress σ_y occurs at the end of elastic behaviour and subsequent material performance is defined as strain-hardening. Unlike the results shown in Fig. 8.1b where σ_y corresponds to a definite physical phenomenon, the yield stress point in Fig. 8.1a cannot be exactly defined.

The actual stress–strain curves may be approximated by a number of linear segments. Various ideal models are shown in Fig. 8.1c–f. Soils are not ideal materials and thus do not necessarily exhibit performance characteristics such as that shown in Fig. 8.1. The stress–strain behaviour and strength of soils will be discussed in Chapters 9 and 10. In this chapter we will be concerned with the concepts of yielding and failure of soils. Further detailed application of these theories will be developed in Chapters 9 and 10.

The concepts of yield and failure in soils

The terms *yield* and *failure* cannot be applied indiscriminately to soils. The failure of brittle materials, such as cast iron or rock, occurs as a fracture with little or no plastic yielding. Thus *fracture* can be readily identified with *failure,* The term "yield" in the field of plasticity is used to describe the onset of plastic deformation, or, conversely, the upper limit of elastic action, (see Fig. 8.1). The precise definition of yield in an actual material is related to the characteristics of the stress–strain curve of the material; only when there is a sharp "break" between the elastic portion (recoverable deformation) and the plastic portion (non-recover-

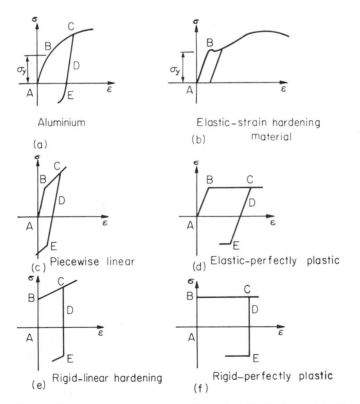

Fig.8.1. Typical stress–strain curves for various kinds of materials. The yield stress σ_y occurs at point *B* in a–f.

able deformation) of the stress–strain curve can yield be accurately defined. In Fig. 8.1 a perfectly plastic material exhibits a continued strain performance at the yield stress if the stress is sustained, as in Fig. 8. ld and f. The term *plastic* strain which s commonly used denotes irrecoverable strain performance as in curve *BC* in Fig. 8. .a. In this case, strain hardening is shown by *BC*.

The term *fracture* implies the appearance of distinct surfaces of separation in the body (Jaeger, 1962), whereas *yield* is used to describe the onset of plastic deformation with the resulting unrestricted plastic deformation defined as *flow*. "Failure" in a general sense includes both fracture and flow.

Several types of fracture commonly occur:

(a) Rupture occurs when a ductile material fails in tension. It is preceded by plastic deformation, causing "necking" and is usually terminated by a "cup and cone" fracture.

(b) Brittle fracture occurs for brittle materials in tension as a tensile or cleavage fracture which occurs on a plane perpendicular to the direction of the tension.

(c) Shear fracture occurs for brittle materials in compression. The resulting failure planes are approximately in the direction of greatest shear stress and always between this direction and the direction of the compressive stress.

Failure of granular soils and other particulate media can be adequately described by the term *fracture*. This is true when particle behaviour is controlled by gravity forces. At elevated temperatures and pressures however, brittle materials become ductile or plastic. Fortunately, for the range of temperatures and pressures encountered in most civil-engineering practice, the required extremes of elevated temperatures and pressures are seldom encountered, and the dominance of gravity forces for granular soils may be assumed.

As shown in Fig. 8. lb–f, yield can be clearly defined only when the material has a stress–strain curve consisting of straight lines (point *B* in the figures). When the stress–strain curves show that point *B* (in Fig. 8. ld and f) is the maximum stress point at which collapse occurs, then "yield" and "failure" are identical, i.e. the yield and failure stresses are the same. When the material does not have a linear stress–strain curve, the definition of yielding as the upper limit of elastic action loses its meaning. Such non-linear stress–strain curves are characteristic of soils. Further, in cohesive soils where interparticle forces and clay–water interaction control soil behaviour, the stress–strain curves are time dependent, and temperature dependent to an extent far greater than in granular soils.

Nadai (1950) defines yielding or fracture and rupture as follows:

"A state of stress which is just necessary to produce failure by plastic yielding or by fracture in an element may be described by the three principal stresses, σ_1, σ_2, σ_3. These three quantities may be represented by the rectangular coordinates of a point P. The totality of the points P representing different states of stress just necessary to produce sudden yielding or plastic deformation in a given material forms a surface $f_1 (\sigma_1, \sigma_2, \sigma_3) = 0$ which we may call the *limiting surface of yielding* of the material.

". . . . . solids under a uniform tension acting in all directions are able to resist only certain definite stresses. If the three principal stresses are equal tensile stresses solid materials break without preceding permanent deformation . . . . . in the region of tensile stresses there corresponds a totality of limiting states of stress $\sigma_1, \sigma_2, \sigma_3$ capable of rupturing a body which may be represented by a second surface $f_2 (\sigma_1, \sigma_2, \sigma_3) = 0$ which we may call the *limiting surface of rupture.*"

Newmark (1960) in pointing out another difficulty in the overall concept of failure states that:

"The definition of failure has not been given in general terms. We may mean the beginning of inelastic action, or we may mean actual rupture of the material. In cohesive soils the failure situation can be stated in various ways, some of them relatively arbitrary but in general accounting for the situation at the beginning of loss of shearing resistance or at a relatively advanced state in the loss of shearing resistance. It is difficult to be precise about the particular theories of failure which might be used if one is not precise about the description of the failure condition. The situation is further complicated by the fact that failure as measured by any of the commonly used concepts for cohesive soils may involve relatively large differences in the strains at which such failure occurs."

The problems associated with the study of yielding and failure of soils, together with a comprehensive review of results obtained from recent pertinent studies have been presented by Scott and Ko (1969) and Bishop (1972). The concept that yield refers to the onset of plastic deformation while *failure* signifies collapse of the material as a terminal stage of the process of yielding is not consistent with observed performance of many soils. In order that yield and failure in soils be properly accounted for, experimental and analytical techniques must provide a consistent and realistic evaluation of soil behaviour. The phenomenological coefficients or parameters must suitably represent material properties.

Principal stress space

A convenient way to examine the state of stress producing yield or failure in a material specimen is to plot the principal stress components $\sigma_1, \sigma_2,$ and σ_3 at yield or failure in principal stress space (Fig. 8.2). Point *1* in Fig. 8.2 represents a $\sigma_1, \sigma_2,$ and σ_3 combination producing yield in a material in a particular stressing situation. Similarly, point *2* represents another $\sigma_1, \sigma_2,$ and σ_3 principal stress combination obtained at yield for another stressing situation. By applying various stress situations, the line joining the points on a common octahedral plane (i.e. π plane) will define a surface which is called the yield surface. The function $f(\sigma_1, \sigma_2, \sigma_3)$ is thus called the yield function.

If points *1, 2* and *3* represent stress situations at failure, the surface defined is termed a failure surface, and the function $f(\sigma_1, \sigma_2, \sigma_3)$ will be called a failure function or failure theory. This will be examined in detail in a later section.

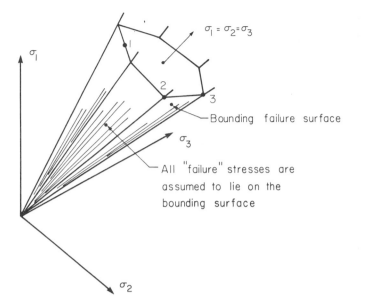

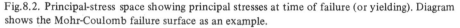

Fig.8.2. Principal-stress space showing principal stresses at time of failure (or yielding). Diagram shows the Mohr-Coulomb failure surface as an example.

8.2 YIELD CRITERIA

A few of the earlier theories describing the conditions causing failure of a material by plastic yielding or by fracture are first reviewed to show their relationship to description of mechanical soil behaviour. Whilst these are called yield theories, they also fulfill the condition of failure. The first three theories are contradictory at times whereas the subsequent two theories find a better fit to experimental data. The background for these may be found in any textbook on mechanics of materials.

The maximum-stress theory

The oldest theory of yielding and failure, sometimes known as Rankine's theory, postulates that the maximum principal stress in the material determines failure regardless of the magnitudes and senses of the other two principal stresses. This gives rise to its name "maximum-stress theory". Thus yielding in a stressed body in accordance with this theory begins when the absolute value of the maximum stress reaches the yield point stress of the material in simple tension or compression. Plotted in principal stress space the yield surface representing this theory is a cube as shown in the projected view of principal stress space (on to the

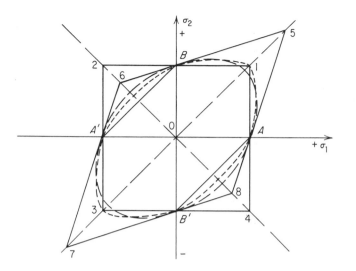

1234 = Maximum-stress theory
5678 = Maximum-strain theory
A1BA'3B'A = Maximum-shear theory
——— —— = Maximum-strain - energy theory
— — — — — = Maximum-distortional -energy theory
Material assumed to have same yield point in
tension and compression

Fig.8.3. Representation of maximum shear-stress equation. (Timoshenko, 1956, published by Van Nostrand Reinhold Co., ©1955 Litton Educational Publ. Inc.)

σ_1, σ_2 axes) in Fig. 8.3. The theory is contradicted in solid materials where three equal tensile or compressive stresses cannot produce a plastic but only an elastic deformation. For those materials in which hydrostatic compressive stresses do cause plastic deformation, the theory is contradicted by the fact that failure in simple tension in an isotropic material would be along inclined planes on which neither the tensile nor compressive stress is a maximum. However, there is some merit in the theory when one considers the strength of non-isotropic materials, particularly layered materials, where there is a pronounced difference in strength properties in different directions, e.g., a layered rock might have almost no tensile strength in the direction normal to the layers and would fail in tension by splitting along these layers. The theory has also found some use in a modified form to explain the cleavage fracture of crystals. With these few exceptions, the theory finds no application in modern practice.

The maximum elastic-strain theory

The maximum elastic-strain theory, attributed to St. Venant assumes that a ductile material begins to yield when either the maximum (elongation) strain equals the yield point strain in simple tension, i.e.:

$$\frac{\sigma_1}{E} - \frac{\mu}{E}(\sigma_2 + \sigma_3) = \frac{\sigma_y\,(\text{tensile})}{E} \qquad (8.1)$$

or the minimum (shortening) strain equals the yield point strain in simple compression,

$$\left|\frac{\sigma_3}{E} - \frac{\mu}{E}(\sigma_1 + \sigma_2)\right| = \frac{\sigma_y\,(\text{compressive})}{E} \qquad (8.2)$$

where the principal stresses σ_1, σ_2 and σ_3 are considered positive in tension, and are ordered such that $\sigma_1 > \sigma_2 > \sigma_3$ and μ, E and σ_y are Poisson's ratio, Young's Modulus and yield point stress respectively. In principal stress space, the yield surface corresponding to theory consists of two straight three-sided pyramids in inverted positions relative to each other, having equilateral triangles as sections normal to the axis which coincides with one of the space diagonals, e.g., $\sigma_1 = \sigma_2 = \sigma_3$ (see Fig. 8.3). The slopes of the sides of the pyramids would depend on Poisson's ratio. This theory is again contradicted by material behaviour under hydrostatic tensile or compressive stresses.

The constant elastic-strain energy theory

The quantity of strain energy per unit volume of the material is used as the basis for determining failure in the constant elastic-strain-energy theory. If we equate the strain energy for a given state of stress at failure to the energy stored at yield in simple tension the criterion may be written as:

$$\frac{(\sigma_y)^2}{2E} = \frac{1}{2E}(\sigma_1{}^2 + \sigma_2{}^2 + \sigma_3{}^2) - \frac{\mu}{E}(\sigma_1\sigma_2 + \sigma_2\sigma_3 + \sigma_3\sigma_1) \qquad (8.3)$$

Again, the performance of materials under hydrostatic stresses indicates that the elastic energy can have no significance as a limiting condition.

The maximum shear-stress theory

The maximum shear-stress theory assumes that yielding begins when the maximum shear stress in the material equals the maximum shear stress at the yield point in simple tension. The maximum shear stress in a material under some general state of stress $(\sigma_1 > \sigma_2 > \sigma_3)$ is $(\sigma_1 - \sigma_3)/2$ and the maximum shear stress in a tension test is equal to half the normal stress, $\sigma_y/2$. The condition for yielding is thus given as:

$$(\sigma_1 - \sigma_3) = \sigma_y \qquad (8.4)$$

This theory was advanced by Tresca in the period 1865 to 1870 and is generally attributed to him. It is a direct consequence of the Coulomb theory for a frictionless material. The maximum shear-stress theory (Coulomb theory) has been extended by Navier to account for pressures normal to the failure plane, which leads to its reference as the Coulomb-Navier theory. The concept of maximum shear stress to explain a fracture type failure in a cohesive soil appears in the work of Collin [1846]. Tresca's contribution to this theory appears to account for a yielding type of failure. The results shown by Guest [1900] supported this criterion and the theory which is thus sometimes referred to as Guest's Law.

In its most useful form the theory may be stated as follows:

$$\tau_{max} = \frac{\sigma_1 - \sigma_3}{2} = \text{constant} \qquad (8.5)$$

In uniaxial tension, $\sigma_1 = \sigma_0$, $\sigma_2 = \sigma_3 = 0$, and $\tau_{max} = \sigma_0/2$; in uniaxial compression, $\sigma_1 = \sigma_2 = 0$, $\sigma_3 = -\sigma_0$. Thus $\tau_{max} = -(\sigma_0/2)$. Hence the yield condition requires that:

$$\tau_{max} = \frac{1}{2}(\sigma_1 - \sigma_3) = \frac{\sigma_0}{2} \qquad (8.6)$$

Eq. 8.6 requires that the yield stress of the material in either simple tension or compression must be equal, which is approximately true in the case of mild steel. The "slip lines" (failure lines or planes) which appear at the onset of plastic flow should be inclined at an angle of $45°$ with respect to the directions of the principal stresses σ_1 and σ_3, that is, coincident with the directions of maximum shearing stress.

The condition of flow does not contain the intermediate principal stress, σ_2, which can have any value between σ_1 and σ_3. The flow condition in its most general form may be expressed by three equations:

$$\sigma_1 - \sigma_3 = \pm \sigma_y \ ; \qquad \sigma_2 - \sigma_1 = \pm \sigma_y \ ; \qquad \sigma_3 - \sigma_2 = \pm \sigma_y \qquad (8.7)$$

where σ_y is the absolute value of the yield stress in tension or compression. Thus the surface of yielding corresponding to the maximum shear stress theory consists of three sets of parallel planes which define a straight hexagonal prism in $\sigma_1, \sigma_2, \sigma_3$

space whose axis coincides with the space diagonal $\sigma_1 = \sigma_2 = \sigma_3$, i.e., in the positive
quadrant of the axes. Cross-sections of the prism are regular hexagons (Fig. 8.3).

The constant elastic strain-energy-of-distortion theory

This theory is also known as the constant octahedral shearing-stress theory.
The theory is variously attributed to Huber, Hencky and Von Mises, although it is
supposed to have been first mentioned by Maxwell in some private correspondence.
This theory states that plastic yielding begins when the strain energy of
distortion given by W_D, where:

$$W_D = \frac{1 + \mu}{6E} [(\sigma_1 - \sigma_2)^2 + (\sigma_2 - \sigma_3)^2 + (\sigma_3 - \sigma_1)^2] \tag{8.8}$$

reaches a critical value. For a material with a pronounced yield point in simple
tension, σ_y, we have $\sigma_1 = \sigma_y$ and $\sigma_2 = \sigma_3 = 0$. Substitution into the above formula
gives:

$$W_D = \frac{1 + \mu}{3E} (\sigma_y)^2$$

Thus the condition for yielding based on the distortion energy theory is:

$$(\sigma_1 - \sigma_2)^2 + (\sigma_2 - \sigma_3)^2 + (\sigma_3 - \sigma_1)^2 = 2(\sigma_y)^2 \tag{8.9}$$

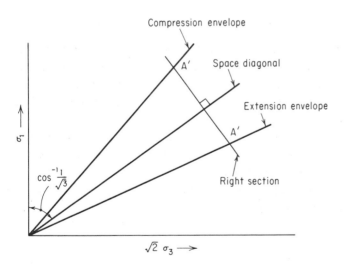

Fig.8.4. Failure envelopes.

A useful form of the theory is obtained by considering the octahedral shearing-stress. An octahedral plane is obtained by passing a plane through the unit points on the principal axes. Thus it is normal to a space diagonal (shown in Fig. 8.2 and 8.4) in $\sigma_1, \sigma_2, \sigma_3$ space i.e., principal stress space; there are thus eight such planes. The normal to each octahedral plane has the direction cos $(1/\sqrt{3})$ to each of the coordinate axes. Normal and shearing stresses on the octahedral plane are called "octahedral stresses". Thus the normal octahedral stress, σ_{oct}, is:

$$\sigma_{oct} = \frac{1}{3}(\sigma_1 + \sigma_2 + \sigma_3) = \frac{1}{3}J_1 \qquad (8.10)$$

where $J_1 = \sigma_1 + \sigma_2 + \sigma_3$ = first stress invariant. The octahedral shearing stress is:

$$\tau_{oct} = \frac{1}{3}[(\sigma_1 - \sigma_2)^2 + (\sigma_2 - \sigma_3)^2 + (\sigma_3 - \sigma_1)^2]^{\frac{1}{2}} \qquad (8.11)$$

Thus any state of stress consisting of three principal stresses may be resolved into two component states of stress, (a) a component consisting of equal tensile (or compressive) stresses acting in all directions, and (b) a component state of stress consisting of the eight octahedral shearing stresses.

Thus from eqs. 8.9 and 8.11:

$$9(\tau_{oct})^2 = 2(\sigma_y)^2$$

and hence:

$$\tau_{oct} = \frac{\sqrt{2}}{3}\sigma_y \qquad (8.12)$$

Eq. 8.12 is thus a statement of the maximum energy of distortion theory. The theory further shows that at the plastic limit the octahedral shearing stress in the material is constant, which depends on the yield point of the material in simple tension or compression. The yield stresses in simple tension and compression are thus assumed to be equal.

The yielding surface defined by this theory is a straight circular cylinder whose axis coincides with the space diagonal $\sigma_1 = \sigma_2 = \sigma_3$. Since planes normal to the axis of the cylinder are octahedral planes, the radius of the cylinder equals the octahedral shearing stress. The radius of the cylinder is therefore $\sqrt{2}/3\, \sigma_y$. This is similar to the Von Mises yield criterion.

8.3 FAILURE THEORIES

The failure theory proposed by Mohr (1900) followed the earlier work of Coulomb and Navier which considered the state of failure as a shear failure. As it turns out, both the Coulomb-Navier theory and the extended maximum shear-stress theory are special cases of the Mohr theory.

The theory which considers failure by both yielding and fracture (assuming slippage as a mode of failure) provides a functional relationship between normal and shear stresses on the failure plane, i.e.:

$$\tau = f(\sigma)$$

where: τ = shearing stress along the failure plane; σ = normal stress across the failure plane.

From the two-parameter nature of the theory the curve defined by this functional relationship may be plotted on the τ, σ-plane (Fig. 8.5).

Since changing the sign of τ merely changes the direction of failure but not the condition for it, the curve must be symmetrical about the σ-axis. The curve so obtained which is termed the Mohr rupture envelope, represents the locus of all points defining the limiting values of both components of stress (τ and σ) in the slip

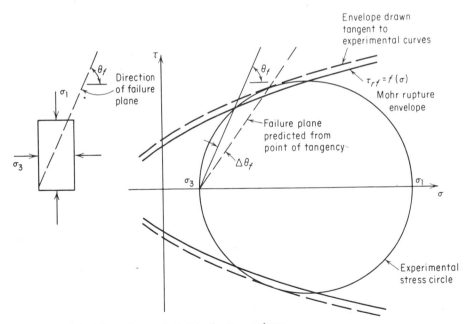

Fig.8.5. Mohr's failure theory plotted in the $\tau - \sigma$ plane.

planes under the different states of stress σ_1, σ_2, σ_3, to which the material may be subjected. The Mohr envelope thus reflects a property of the material which is independent of the stresses imposed on the material. The theory is attractive for use in studying the shear strength of soils since there is no requirement that the material obey Hooke's law or that Poisson's ratio be constant: also, the strength and stiffness of the material in tension and compression need not be equal.

The Coulomb equation, $\tau = c + \sigma \tan \phi$, represents a special case of the Mohr theory of strength in which the Mohr envelope is a straight line inclined to the normal stress axis at angle ϕ. The use of the Coulomb equation to represent the Mohr envelope in the Mohr diagram is called the Mohr-Coulomb theory.

Mohr's hypothesis states that failure depends upon the stresses on the slip planes and failure will take place when the obliquity of the resultant stress exceeds a certain maximum value. It is also stated that "the elastic limit and the ultimate strength of materials are dependent on the stresses acting on the slip planes".

The Mohr representation of stresses acting on the three principal planes is shown in Fig. 8.6. Stresses on any plane within the body must lie within the shaded area. The slope of the line joining the origin and point A gives the obliquity of stress. The maximum inclination of stress will be given by the tangents to the largest circle. Failure occurs on planes where stresses are represented by points B and C. These stresses act on planes which are parallel to the diameter of the intermediate principal stress. Therefore the diameter of the largest Mohr circle and the magnitude of the stresses at points B and C are independent of the intermediate principal stress, σ_2.

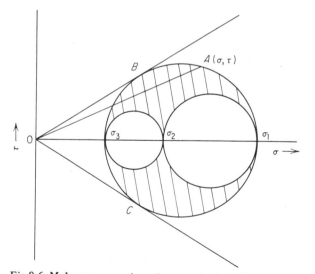

Fig.8.6. Mohr representation of stresses in three-dimensional system.

With the assumption that σ_2 is the intermediate principal stress, the largest of three circles representing the limiting state of stress will be of diameter $(\sigma_1 - \sigma_3)$ and centred at $(\sigma_1 + \sigma_3)/2$ along the σ-axis, as seen in Fig. 8.6, taking due account of the algebraic sign of the stresses. Since the two parallel sets of slip planes which occur when an isotropic specimen has been stressed slightly beyond the plastic limit by a state of homogeneous stress are symmetrically inclined with respect to the directions in which the major and minor principal stresses act, and the two plane systems intersect each other along the direction in which the intermediate principal stress acts, Mohr assumed that the intermediate principal stress is without influence on the failure of a material. Accordingly, some point on the perimeter of the circle of diameter $(\sigma_1 - \sigma_3)$ must represent the limiting stress condition. The theory thus affords a method of devising a failure theory for a specific material, i.e., establishing its Mohr rupture envelope, from actual test results.

In practice, a series of similar specimens is subjected to different stresses and brought to failure (as in the triaxial test described in the next section). The various Mohr's stress circles are plotted for ,the limiting states of stress and the unique failure stress on the failure plane for each test is taken as the point of common tangency between a smooth limiting curve (or envelope) and the various $(\sigma_1 - \sigma_3)$ circles. By taking the points of common tangency as representative of the σ and τ stresses on the failure plane, a state of homogeneous stress in an isotropic material is assumed. Due to experimental shortcomings, however, one may not necessarily obtain a state of homogeneous stress and thus the inferred stresses at the point of common tangency for the envelope, obtained experimentally, will not in all likelihood represent the actual stresses on the failure plane. It follows then that the predicted location of the failure plane, based on the common tangency points might be in error. The possible discrepancy between the actual Mohr rupture envelope and an experimentally obtained envelope is shown on Fig. 8.7.

Actual Mohr rupture envelopes are often curves. However, for soils, the curvature is usually not great and it has proved useful to approximate the envelope by a straight line, at least over a limited range of normal stress. The equation of a straight line in the τ, σ-plane is the Coulomb equation $\tau = s = c + \sigma \tan \phi$. From Fig. 8.7, the following formula may be derived:

$$R = \frac{\sigma_1 - \sigma_3}{2} = c \, \cos \phi + \left(\frac{\sigma_1 + \sigma_3}{2} \right) \sin \phi \qquad (8.13)$$

The parameters c and ϕ in eq. 8.13 are the analytical parameters of "cohesion" and "friction angle". They are a direct consequence of the application of the Mohr-Coulomb theory and need not bear any physical semblance to the real material properties of cohesion and friction of soil. When the physical conditions of failure in the test specimen are met, e.g., little or no volume change, development

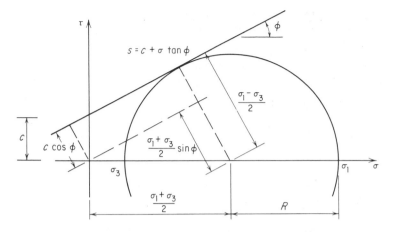

Notes : 1. σ_1 & σ_3 are limiting effective stresses at failure.
2. Compressive stresses considered positive in deriving Eq.(a-42).

Fig.8.7. Properties of a straight-line Mohr rupture envelope. σ_1 and σ_3 are limiting effective stresses at failure. Compressive stresses considered positive in deriving eq.8.13.

of failure plane, etc., the analytical parameters will more closely correspond and reflect the physical material (mechanistic) parameters.

The material shear-strength parameters c and ϕ which reflect the behaviour of the soil in shear response are not necessarily constant. They will vary with the void ratio or water content of the soil as well as the type of test used. The reasons for their variation are related to such factors as soil structure, fabric, stress history, particle interaction and clay-water forces. The mechanistic parameters will be discussed in detail in Chapters 9 and 10.

Eq. 8.13 may be manipulated in many ways to state the failure criterion in various forms. For example, by adding $\sigma_3 \sin \phi$ to both sides of the equation and by rearranging terms, we will obtain:

$$\frac{1}{2}(\sigma_1 - \sigma_3)(1 - \sin \phi) = c \, \cos \phi + \sigma_3 \sin \phi \tag{8.14}$$

Eq. 8.14 which was used by Skempton and Bishop gives straight lines when $(\sigma_1 - \sigma_3)/2$ is plotted against σ_3. By multiplying both sides of eq. 8.13 by two and rearranging terms, we get:

$$\sigma_1 (1 - \sin \phi) = 2c \, \cos \phi + \sigma_3 (1 + \sin \phi) \tag{8.15}$$

which gives straight lines when σ_1 is plotted against σ_3. This last equation has been used as a plotting method by Rendulic and more recently by Henkel (1959).

Expressed in its most general form, the failure surface corresponding to the Mohr-Coulomb condition of failure is:

$$[(\sigma_1 - \sigma_2)^2 - \{2c \, \cos \phi + (\sigma_1 + \sigma_2) \sin \phi\}^2] \times$$

$$[(\sigma_2 - \sigma_3)^2 - \{2c \, \cos \phi + (\sigma_2 + \sigma_3) \sin \phi\}^2] \times$$

$$[(\sigma_3 - \sigma_1)^2 - \{2c \, \cos \phi + (\sigma_3 + \sigma_1) \sin \phi\}^2] = 0 \qquad (8.16)$$

The failure surface defined by eq. 8.16 is a pyramid with the space diagonal $\sigma_1 = \sigma_2 = \sigma_3$ as axis and a cross-section which is an irregular hexagon with nonparallel sides of equal length (see Fig. 8.8). The projection of this irregular hexagon on the plane $\sigma_1 + \sigma_2 + \sigma_3 = $ constant (i.e. a plane at right angles to the space diagonal or an octahedral plane) is shown in Fig. 8.9. For reference, the circle and regular hexagon described by the extended Von Mises and extended Tresca criteria are also plotted on Fig. 8.9. The three criteria are seen to coincide for compressive tests but the strength in a tensile test is seen to be less for the Mohr-Coulomb failure theory.

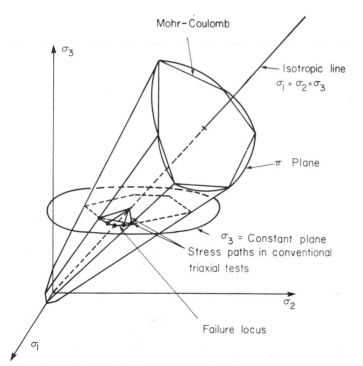

Fig.8.8. Mohr-Coulomb failure surface in principal stress space showing stress paths in conventional triaxial tests.

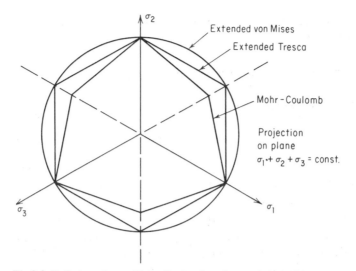

Fig.8.9. Failure surfaces—Mohr-Coulomb and extended yield criteria.

8.4 LABORATORY TRIAXIAL TEST TECHNIQUES FOR STRENGTH MEASUREMENT

Measured yield, failure and strength of a soil depend upon the method of test evaluation; both testing techniques and interpretation. The more common techniques used for determining soil strength are:

(1) Unconfined compression (uniaxial).

(2) Triaxial compression test: triaxial axisymmetric, ($\sigma_2 = \sigma_3$) compression — see Fig. 8.12, 8.13; triaxial unconsolidated undrained (UU); triaxial consolidated undrained (CU); triaxial consolidated drained (CD).

(3) True triaxial tests, $\sigma_1 \neq \sigma_2 \neq \sigma_3$ with similar considerations of UU, CU and CD.

The reasons for the control of drainage in sample preparation and during application of the shearing force are mainly to simulate field situations and specific requirements of the failure theory applied. The distinction between un-consolidated undrained (UU) consolidated undrained (CU) and consolidated drained (CD) tests in axisymmetric triaxial or true triaxial testing is illustrated as follows:

Consider an undisturbed soil sample under an all-around confining pressure of σ_3. For a fully saturated sample, no volume change can occur unless water is extruded from the soil voids (discussed in Chapter 7). The immediate pressure response in the sample is sustained by the pore water until some volume change occurs, i.e. the pore pressure $u = \sigma_3$ (see Fig. 8.10). The pore-water pressure, u, is significant since it affects the response behaviour of the sample under shear and will thus establish the operative shear strength parameters. For partially saturated soils, the measurements of pore-water pressure require a correction incorporating or accounting for pore-air pressure.

When volume change occurs in a fully saturated soil, there is a reduction in the volume of water if the total volume of the soil mass is decreased. Thus, a transfer of stress from hydrostatic to interparticle will take place. The amount of stress transfer is dependent on the restrictions placed on the decrease in fluid volume in the soil mass. The total stress within the soil mass resisting the external stress is carried by the soil particles if pore-water pressure is allowed to reduce to zero. Fig. 8.10 shows that when $u = 0, \bar{\sigma}$, the effective stress is equal to the total stress σ.

In the unconsolidated undrained (UU) triaxial test, the test sample is not allowed to consolidate prior to application of the shearing force, i.e. no drainage is allowed. Under a shearing force, no pore water is allowed to escape (see Fig. 8.11).

Consolidated undrained tests on the other hand require that the test samples be consolidated in the laboratory to some pre-established state defined generally in terms of the consolidation pressure. Thus the test sample no longer retains the same void ratio as the undisturbed field sample. The shearing process is performed after consolidation without allowing the pore water to drain under stress (see Fig. 8.11).

When pore water is allowed to escape during the shearing process, i.e. with drainage allowed, the test is referred to as a drained test. In the consolidated drained test, the test sample is consolidated in the triaxial cell and subsequent shear is performed with complete drainage. While the UU and CU shear tests are sometimes referred to as constant-volume shear tests, the CD test is a variable-volume test.

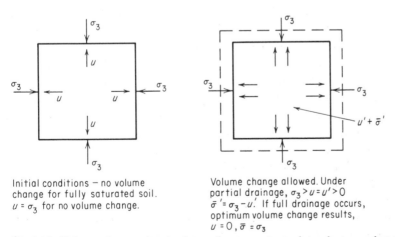

Initial conditions – no volume change for fully saturated soil. $u = \sigma_3$ for no volume change.

Volume change allowed. Under partial drainage, $\sigma_3 > u = u' > 0$. $\bar{\sigma}' = \sigma_3 - u'$. If full drainage occurs, optimum volume change results, $u = 0, \bar{\sigma} = \sigma_3$

Fig.8.10. Volume element showing internal response to σ_3 dependent on volume change due to allowance for drainage.

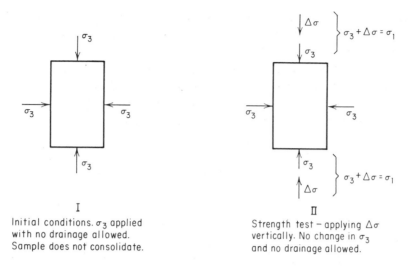

I

Initial conditions. σ_3 applied
with no drainage allowed.
Sample does not consolidate.

II

Strength test – applying $\Delta\sigma$
vertically. No change in σ_3
and no drainage allowed.

Fig.8.11. Sample conditions for UU test. (For CU test, stage I allows for drainage under σ_3, resulting in consolidation of sample. In stage II no drainage is allowed.)

Axisymmetric triaxial test

The commonly used $\sigma_2 = \sigma_3$ axisymmetric triaxial test is a cylindrical test, since the test sample is cylindrical in shape. Fig. 8.12 and 8.13 show the typical test system used. Fig. 8.14, which shows the stress plane investigated in the cylindrical test, demonstrates the influence of pore-pressure control on development of the stress value used.

We presume that a specimen is consolidated isotropically under a cell pressure equal to σ_c (u is zero at the end of consolidation). The point denoted by σ_c is shown on the isotropic line (space diagonal) where $\sigma_1 = \sigma_2 = \sigma_3$. If the piston load of p develops a stress increase of $\Delta\sigma$ on the specimen in the X_1 direction (corresponding to the direction in which σ_1 acts), the total stress σ_1 is shown to be $\Delta\sigma + \sigma_c$, where $\Delta\sigma$ is the vertical line rising from σ_c. Since the pore-water pressure u acts uniformly in all directions, its line of action must be parallel to the isotropic line. This is shown in Fig. 8.14 where the resultant effective stress path is also indicated. c^* is a function of c and the "friction" angle ϕ.

In Fig. 8.15, the general stress paths for certain types of cylindrical tests are shown. If point b on the isotropic line represents the stress state of the test sample following consolidation, ($u = 0$), and if the pore pressure u is maintained at zero throughout the test, the line bd representing the drained stress path will be obtained. This is defined as the consolidated drained (CD) test. On the other hand, if we start from point b and perform the cylindrical test in the undrained condition,

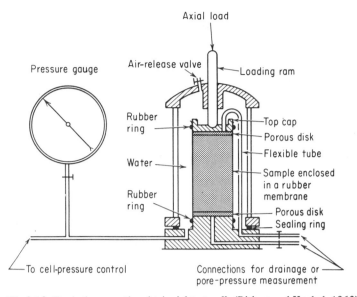

Fig.8.12. Typical conventional triaxial test cell. (Bishop and Henkel, 1962).

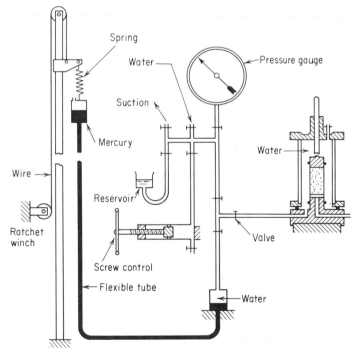

Fig.8.13. The basic layout of self-compensating mercury control. (Bishop and Henkel, 1962.)

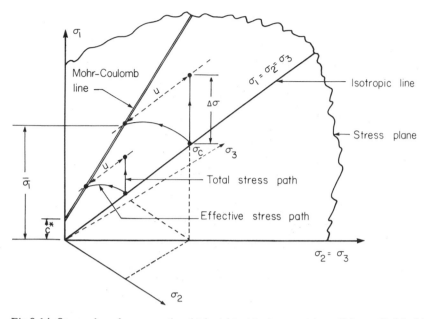

Fig.8.14. Stress plane for conventional triaxial test (axisymmetric, radial or cylindrical test).

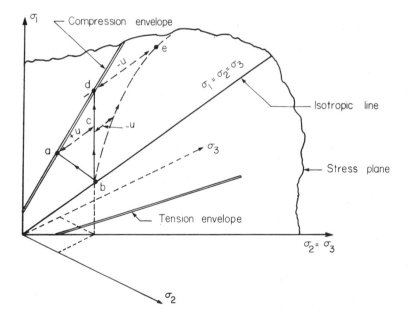

Fig.8.15. Stress paths in conventional triaxial test.

the application of the vertical stress increase $\Delta\sigma$ will produce the situation where $\Delta u/\Delta\sigma = 1$, as shown by the effective stress path ba. The total stress path for this is bc with ac representing the pore-water pressure generated at failure. All lines parallel to ac, between b and ac represent pore-water pressure paths.

In the shearing of overconsolidated materials where the tendency for dilatation exists, u can be negative. The effective stress path for such materials is shown by be. As before, de is parallel to ac. For illustrative purposes the compression test and tension test Mohr envelopes are also shown in Fig. 8.15. These can be replaced by other suitable envelopes as the situation demands.

If we examine the actual points of intersection of the two limiting envelopes (compression and tension) on a plane oriented perpendicular to the isotropic line (defined as the orthogonal plane), we will see only two points (see Fig. 8.16). From symmetry four other points corresponding to two compression and two extension points can be indicated on the projected $+\sigma_2$, $+\sigma_3$ and $-\sigma_2$, $-\sigma_3$ axes respectively. Using the extended Tresca, Von Mises or Mohr–Coulomb theories, straight or curved lines between these points can be drawn (see Fig. 8.9). As we observe from Fig. 8.16, however, no actual experimental information on intermediate principal stress effects on the limit trace is available, since no limit trace can be realistically defined by the two initial points d and f in Fig. 8.16. It is pertinent to point out at this stage that these are limit values and are generally accepted as failure values in the case of strength testing of soils. If suitable yield envelopes can be found, their projection or intersection on plane abc in Fig. 8.16 would also yield two points, similar to d and f.

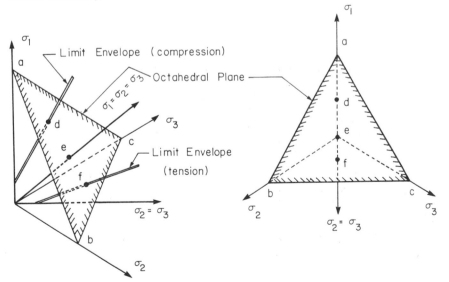

Fig.8.16. Point trace on octahedral plane for conventional triaxial test.

8.5 PRINCIPAL STRESS SPACE AND ADMISSIBLE YIELD OR FAILURE CRITERIA

As stated in Section 8.1, the examination of the stress behaviour or stress response of a test material at any state of stress or strain can be made with the aid of a principal stress space diagram. For example, the stresses at failure of any number of materials subjected to varying conditions of stress imposition may be examined in terms of the principal stresses existing at that particular failure strain. We have shown in Fig. 8.8 and Fig. 8.9, for example, that if these stress states are plotted in principal stress space, the failure locus traced on to an orthogonal plane will describe the failure criterion.

As an example, let us consider the results of triaxial (cylindrical) compression and extension tests of a material which conforms to the Mohr-Coulomb failure theory. The data will be treated in the manner used by Henkel (1959), based on eq. 8.15. Since there is radial symmetry in the tests, the three-dimensional stress space may be reduced to an equivalent two-dimensional space by considering the plane passing through the axial stress plane equidistant from the two radial stress planes — i.e. at 45° (see Fig. 8.17). The vector sum of the two radial stresses must be plotted along the radial-stress axis, as seen from the figure. The space diagonal in this plot defines its true angle with the axial stress axis, namely $\cos^{-1} 1/\sqrt{3}$. A right section to the space diagonal (an octahedral plane) can be drawn by drawing a line at 90° to the space diagonal (Fig. 8.18).

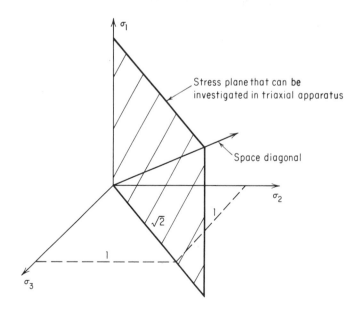

Fig.8.17. Equivalent two-dimensional space for radially symmetric stresses. (Henkel, 1959)

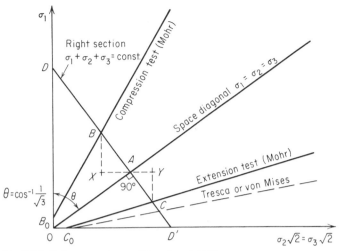

Fig.8.18. Failure envelopes. (Hvorslev, 1960.)

Consider the right section DD' which cuts the space diagonal at point A which corresponds to the all-round pressure σ_c. In the diagram, stress states leading to compressive failure give the limiting line B_oB, while limiting states in the extension test give the line C_oC. In the plane DD', the octahedral shear stresses at failure in compression and extension then are represented by points B and C respectively. Let us examine the ratio of these two strengths which must result if the Mohr-Coulomb theory is valid.

Consider the compression test first. If the stress at A is σ_c then the axial and radial components of the stress at B are:

Axial component

$$\sigma_A = \sigma_c + Bx$$

$$= \sigma_c + AB \sin \theta, \text{ where } \theta = \cos^{-1} 1/\sqrt{3}$$

$$= \sigma_c + \sqrt{2/3} \ AB$$

Radial component

$$\sigma_R = \sigma_c - \frac{Ax}{\sqrt{2}}$$

$$= \sigma_c - AB \cos \theta/\sqrt{2}$$

$$= \sigma_c - \frac{1}{\sqrt{6}} AB$$

Thus, for the compression test:

$$AB(\sqrt{2/3} + \sqrt{1/6}) = (\sigma_A - \sigma_D)$$

Now take the extension test where the coordinates of the point C are:

Axial component

$$\sigma_A = \sigma_c - YC$$

$$= \sigma_c - AC \sin \theta$$

$$= \sigma_c - \sqrt{2/3} \ AC$$

Radial component

$$\sigma_R = \sigma_c + \frac{AY}{\sqrt{2}}$$

$$= \sigma_c + \frac{AC}{\sqrt{2}} \cos \theta$$

$$= \sigma_c + \frac{AC}{\sqrt{6}}$$

And, for the extension test:

$$AC(\sqrt{2/3} + \sqrt{1/6}) = (\sigma_R - \sigma_A)$$

If the Mohr-Coulomb criterion holds, substitution of the axial and radial components in eq. 8.13 gives for compression:

$$(\sigma_A - \sigma_R) = 2c \ \cos \phi + (\sigma_A + \sigma_R) \sin \phi$$

$$(\sigma_c + \sqrt{2/3} \ AB - \sigma_c + \sqrt{1/6} \ AB) = 2c \cos \phi + (\sigma_c + \sqrt{2/3} \ AB + \sigma_c - $$

$$\sqrt{1/6} \ AB) \sin \phi$$

$$AB(\sqrt{2/3} + \sqrt{1/6}) = 2c \cos \phi + [2\sigma_c + AB(\sqrt{2/3} - \sqrt{1/6})] \sin \phi$$

$$AB[(\sqrt{2/3} + \sqrt{1/6}) - (\sqrt{2/3} - \sqrt{1/6}) \sin \phi] = 2c \cos \phi + 2 \sigma_c \sin \phi \quad (8.17)$$

and for extension:

$$(\sigma_R - \sigma_A) = 2c \cos \phi + (\sigma_R + \sigma_A) \sin \phi$$

$$(\sigma_c + \sqrt{1/6} \, AC - \sigma_c + \sqrt{2/3} \, AC) = 2c \cos \phi + (\sigma_c + \sqrt{1/6} \, AC + \sigma_c -$$

$$\sqrt{2/3} \, AC) \sin \phi \tag{8.18}$$

$$AC(\sqrt{2/3} + \sqrt{1/6}) = 2c \cos \phi + [2\sigma_c - AC(\sqrt{2/3} - \sqrt{1/6})] \sin \phi$$

$$AC[(\sqrt{2/3} + \sqrt{1/6}) + (\sqrt{2/3} - \sqrt{1/6}) \sin \phi] = 2c \cos \phi + 2\sigma_c \sin \phi$$

Equating (8.17) and (8.18) and multiplying by $2(\sqrt{2/3} - \sqrt{1/6})$ gives:

$$AB \left(1 - \frac{1}{3} \sin \phi\right) = AC \left(1 + \frac{1}{3} \sin \phi\right)$$

Thus the ratio:

$$\frac{AC}{AB} = \frac{1 - \frac{1}{3} \sin \phi}{1 + \frac{1}{3} \sin \phi} \tag{8.19}$$

must apply for the Mohr-Coulomb failure theory. For $\phi = 0$ this ratio is 1.0, and thus it will be seen that the extended Tresca criterion applies. Typical values of the ratio AC/AB for $\phi = 30°$ and $40°$ are 0.715 and 0.647.

The extended Tresca criterion can be written as:

$$[(\sigma_1 - \sigma_2)^2 - \{c + \frac{k}{3}(\sigma_1 + \sigma_2 + \sigma_3)\}^2] \times$$

$$[(\sigma_2 - \sigma_3)^2 - \{c + \frac{k}{3}(\sigma_1 + \sigma_2 + \sigma_3)\}^2] \times$$

$$[(\sigma_3 - \sigma_1)^2 - \{c + \frac{k}{3}(\sigma_1 + \sigma_2 + \sigma_3)\}^2] = 0 \tag{8.20}$$

where k = constant related to $\sin \phi$; c = cohesion. The extended Von Mises condition is given as:

$$(\sigma_1 - \sigma_2)^2 + (\sigma_2 - \sigma_3)^2 + (\sigma_3 - \sigma_1)^2 = [c + \frac{k}{3}(\sigma_1 + \sigma_2 + \sigma_3)]^2 \tag{8.21}$$

Scott (1963) shows various Mohr-Coulomb envelopes plotted on an octahedral plane for ϕ values of $30°$, $40°$ and $50°$ (reproduced as Fig. 8.19) showing

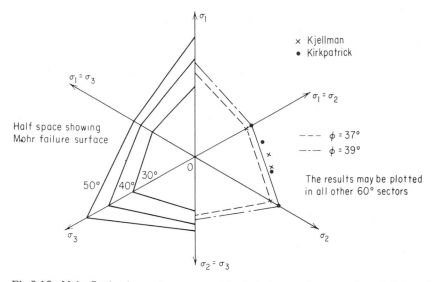

Fig.8.19. Mohr-Coulomb envelopes on octahedral plane and test results of Kirkpatrick and Kjellman. (Scott, 1963.)

too the results of tests on sands under varying stress conditions obtained by Kirkpatrick (1957) and Kjellman (1936).

The plastic potential

If we plot total principal strains in principal strain space similar in concept to the principal stress space discussed in Section 8.1, and if the principal strain axes are aligned with the principal stress axes shown in Fig. 8.2, under conditions of perfect plastic yielding the plastic-strain increment vector is normal to the yield surface (Drucker, 1967). The relationship between the strain rates and stress will be given by an associated flow rule (eq. 8.23) where the connection between the flow rule and yield condition is known as the theory of the plastic potential. In physical terms, one might associate this with the limit of isotropic plastic behaviour.

Consider a set of generalized stress $Q_1 \ldots Q_n$, acting on a body with corresponding generalized strains $q_1 \ldots q_n$ developed such that we can define the condition that:

$$dW = Q_1 \, dq_1 + \ldots \ldots + Q_n \, dq_n$$

is the work that the stresses do on infinitesimal increments of strain.

For an elastic, perfectly plastic material such as that shown by the stress–strain curve in Fig. 8.1d, each strain change dq_i generally has an elastic (recoverable) and a plastic (nonrecoverable) component:

$$dq_i = dq_i^{(e)} + dq_i^{(p)} \qquad\qquad (i = 1, \ldots . n)$$

where the superscripts (e) and (p) refer to elastic and plastic respectively.

If $Q_1 \ldots Q_n$ specify a state of stress at the yield limit, and the stress increments $dQ_1 \ldots dQ_n$ that are imposed will create a neighbouring state of stress at the yield limit, both former and new states of stress must satisfy the yield criterion. Hence:

$$\Phi \, (Q_1 \ldots Q_n) = 0$$

$$d \, \Phi = \frac{\partial \Phi}{\partial Q_1} \, dQ_1 + \ldots + \frac{\partial \Phi}{\partial Q_n} \, dQ_n = 0 \qquad\qquad (8.22)$$

Eq. 8.22 describes the orthogonality of the vectors $(dQ_1 \ldots dQ_n)$ and $[(\partial\Phi/\partial Q_1) \ldots (\partial\Phi/\partial Q_n)]$, with the first of these representing stress increments which joins the neighbouring points of the yield locus, and is tangential to this locus. The vector with the components $\partial\Phi/\partial Q_1 \ldots \partial\Phi/\partial Q_n$ is normal to the yield locus at the considered stress point. Since Φ changes from negative to positive values as one crosses the yield locus coming from the interior, this vector has the direction of the exterior normal (see Fig. 8.20).

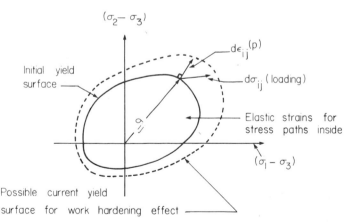

Fig.8.20. Yield surface on plane $(\sigma_1 - \sigma_3)$, $(\sigma_2 - \sigma_3)$, when the plastic potential surface coincides with the yield surface, the plastic strain-increment vector $d\epsilon_{ij}$ (P) assumes the outward normal direction in regard to the yield surface. In actual practice, normality between $d\epsilon_{ij}$(P) and the plastic potential surface can be obtained and decisions in regard to yield surfaces vis-a-vis plastic potential surface made according to evaluation of physical behaviour (see Section 10.7 and Fig.10.51 and 10.52).

For each state of stress at the yield limit, the flow rule which specifies the components of the plastic strain increment may be written as follows:

$$dq_1^{(p)} = \lambda \frac{\partial \Phi}{\partial Q_1} \quad \ldots \quad dq_n^{(p)} = \lambda \frac{\partial \Phi}{\partial Q_n}$$

or:

$$dq^{(p)} = \lambda \frac{\partial \Phi}{\partial Q} \tag{8.23}$$

where λ is a proportionality constant. The plastic strain increment $dq^{(p)}$ is normal to the surface defined by Φ since $\partial\Phi/\partial Q$ is normal to the surface, as shown above. The surface defined by Φ is called the plastic potential and is assumed to coincide with the yield surface for the elastic perfectly plastic model.

The constant λ is restricted by the condition that plastic flow always involves dissipation of energy:

$$Q_1 dq_1^{(p)} + \ldots + Q_n dq_n^{(p)} = \lambda \left(Q_1 \frac{\partial \Phi}{\partial Q_1} + \ldots + Q_n \frac{\partial \Phi}{\partial Q_n} \right) > 0$$

As $Q(\partial\Phi/\partial Q_1)$ is a scalar product of the radius vector of the stress point under consideration on the yield surface and a vector that has the direction of the exterior normal at this point (and is positive), λ must be positive.

In a material which is not perfectly plastic, e.g., a strain hardening material, the yield criterion may be written as follows:

$$f(J_2', J_3') = C \tag{8.24}$$

where J_2' and J_3' are the second and third invariants of the stress deviation tensor, i.e.:

$$J_2' = \frac{1}{2} S_{ij} S_{ij}$$

$$J_3' = S_1 S_2 S_3.$$

$$S_{ij} = \sigma_{ij} - [(\sigma_1 + \sigma_2 + \sigma_3)/3] \, \delta_{ij} = \text{stress deviation}$$

$S_1, S_2, S_3 = $ Principal components of stress deviation tensor

$\delta_{ij} = $ kronecker delta

The function f does not depend on the strain history, which enters only through the parameter C. In a neutral change $d\sigma_{ij}$ where the stress point remains on yield-surface:

$$df = \frac{\partial f}{\partial \sigma_{ij}} d\sigma_{ij} = \frac{\partial f}{\partial J_2'} dJ_2' + \frac{\partial f}{\partial J_3'} dJ_3' = 0$$

The condition that $d\varepsilon_{ij}^p$ is zero for a neutral change of stress is satisfied by assuming:

$$d\varepsilon_{ij}^p = G_{ij} df$$

where G_{ij} is a symmetric tensor and is a function of the stress components and of the previous strain history. It is *not* dependent on the gradient of the stress increment, and does not need to coincide with the yield surface.

Hill (1950) states that G_{ij} must satisfy:

(1) $G_{ii} = 0$ so that zero plastic volume change is obtained

$$d\varepsilon_{ii}^p = G_{ii} df = (0)df = 0$$

(2) Principle axes of the plastic strain-increment tensor, G_{ij} must coincide with the principle stress axes, since the element is isotropic. This means $\partial G_{ij}/\partial \sigma_{ij}$ can exist.

For a strain hardening or plastic body to be stable, the plastic work done by a probing stress increment $d\sigma_{ij}$, at the yield limit, must be non-negative (Drucker, 1967). Mathematically:

$$d\sigma_{ij} d\varepsilon_{ij}^p \geqslant 0 \qquad\qquad (8.25)$$

In the limit, the final state of stress lies near to the initial state of stress. The plastic strain-increment vector is taken to be perpendicular to the stress-increment vector, and as the stress-increment vector is tangential to the yield curve at the initial state of stress, it is apparent that the plastic strain-increment vector is an outward normal as shown in Fig. 8.20. This shows that the plastic potential coincides with the yield surface.

The application of yielding and failure criteria, together with examination of plastic potential and normality concepts to soils will be given in Chapter 10 where the strength and performance of clays are examined.

8.6 SUMMARY

While the determination of the shearing resistance of soils can be obtained through some form of laboratory test, the fundamental factors associated with the inherent shearing strength of soils are difficult to evaluate. Some of the extrinsic and intrinsic factors, other than soil type, are: type of test system, strain or stress-controlled load application, stress and strain history, temperature, soil fabric, density, saturation, water content, etc. Variations in any of these and other factors will provide resultant changes in stress–strain behaviour and computed shear-strength parameters. It becomes difficult therefore to establish meaningful criteria for soil strength and pertinent parameters without paying particular attention to at least some if not all of the significant factors involved. It is evident that for those factors that cannot be controlled, a proper evaluation of their significance and influence is needed.

Recognizing that the currertly used Mohr-Coulomb criterion does not account for the effects of intermediate principal stresses – quite apart from inattention to inertia terms and the necessary requirement for development of limit values, the problem of shear strength determination as performed in the laboratory and used in practice can be stated quite simply as follows:

(1) What do the standard strength tests examine?

(2) If the Mohr-Coulomb theory is fundamentally an empirical-analytical theory, how does this represent actual soil response behaviour? How does this correspond to the mechanistic interpretation of c and ϕ?

(3) Can we offer a more realistic means for evaluation of soil strength which would account for the properties of the soil and the test system used?

The simplest distinction that can be used to distinguish between failure and yield lies in the definition of the overall state of the soil both during and as a consequence of straining under load. In brittle materials for example, fracture with little or no plastic yielding occurs. In many soils, however, fracture does not occur. The term fracture which can readily be identified with failure implies the appearance of distinct surfaces of separation in the material undergoing straining. On the other hand, yield as a phenomenon describes the onset of plastic deformation with resulting successive deformation which may be defined as flow. The concept of failure in the general sense includes both fracture and flow.

In granular soils, plastic yielding and flow constitute a small portion of the failure of the material. Failure occurs due to rupture, and hence we assume that the prediction of the overall state of granular soils under load may be suitably described by a "failure" theory. The state of failure in many cohesive soils is arrived at as a consequence of yielding and flow. At large strains following successive plastic yielding, slippage occurs thus giving the overall impression of a surface of separation. This is not to be confused with a rupture plane associated with failure during sample straining, but is in actual fact a slippage plane arising as a consequence of large sample straining.

Strength theories describing failure and yield of materials are useful in that they provide material parameters needed for design considerations. In soils, the

Mohr-Coulomb theory has found wide usage. The shear strength parameters "cohesion" c, and "friction angle" ϕ are obtainable from the Mohr-Coulomb diagram. Misunderstanding arises when we attempt to relate the parameters to actual mechanistic behaviour. The values obtained from the Mohr-Coulomb diagram do not necessarily reflect actual cohesion and friction mechanisms operative in the demonstration of shearing strength of the soil. A distinction must therefore be made between analytic shear strength parameters such as those obtained from the application of a failure theory, and mechanistic parameters which describe contributing mechanisms generating shear resistance. From recent knowledge of clay-water systems, it is known that physical grain-to-grain contact does not exist in many clay soils. Contact between particles in such cases is achieved through layers of water which may describe a quasi friction quality under the conditions specified in shear. However, their dependency on many other factors makes it difficult to prescribe a rigorous relationship between thickness of water layers and friction. One could attempt a solution based on the determination of the nature and magnitude of the forces holding water to soil particles, however, it will soon become evident that this approach will not be fruitful. A separation between the generating mechanisms for cohesion and friction for clays with distinct clay-water forces is not feasible at the present time. It becomes important therefore to understand the need for distinguishing between analytic shear strength parameters which provide quantitative values, and mechanistic parameters which will provide an understanding of the generation mechanisms in the study of soil strength.

Chapter 9

GRANULAR SOIL STRENGTH

9.1 INTRODUCTION

It was shown in Chapter 3, Section 3.3, that granular soils consist of discrete particles and are not readily influenced by surface forces. The behaviour of granular soil is governed by gravitational and mechanical forces. Resistance to shear deformation and distortion will be described by particle contact and packing relationships. The Mohr-Coulomb theory of failure described in Chapter 8, Section 8.3, can be satisfactorily applied to granular soil systems provided that proper accounting is made or given to the parasitic energy components developed.

In the study of granular soil strength, we will confine ourselves to soil-water systems where no measurable "cohesion" exists. We will assume that behaviour under load of these granular soil-water systems will be influenced solely by the friction properties of the soil. If $\bar{\sigma}$ = effective stress, then the shear strength τ is given by:

$$\tau = \bar{\sigma} \tan \phi \qquad (9.1)$$

where ϕ = angle of internal friction.

9.2 FRICTION PROPERTIES

The frictional property of granular soils is the primary physical component governing the development of shear strength. It is intimately involved in the movement of one particle relative to another. The physical processes contributing to motion include both sliding of particles and removal or displacement of particles from interlocking action between adjacent particles.

Sliding friction consists of microscopic interlocking due to surface roughness of contacting particle surfaces. There is no significant volume dilatation associated with this type of action. Interlocking friction consists of physical restraint to relative particle translation afforded by adjacent particles. Fig.9.1 shows the difference between the two. Since there is obstruction to relative particle translation, individual particles must be plucked from their interlocking seats and

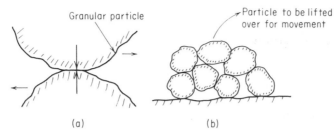

Fig.9.1. Friction between soil grains. a. Microscopic interlocking (surface friction). b. Macroscopic interlocking (mechanical locking).

made to ride over the adjacent particles to provide translation. Hence, a volume increase is generally associated with breakdown of interlocking. The friction property which is responsible for shear resistance, and which one measures in a direct shear or triaxial test includes both types of friction. These are not easily separable in a granular soil mass since innumerable particles are involved. Under an applied shear, these two types of friction resistance must occur simultaneously. While some particles are interlocked and require a plucking action for initiation of particle translation, others may be undergoing translation.

Rolling friction may be included in the category of sliding friction in that little or no volume change is expected. We have tended to ignore this since it is difficult to separate rolling resistance from the phenomenon associated with particles overriding one another in release from their mechanical locking by adjacent particles.

Apparent friction parameter, ϕ

It was stated in Chapter 8 that the shear strength parameters obtained through application of a yield or failure theory, were analytical parameters. In the preceding discussion, we note that the frictional property of granular materials is made up of two components:

(1) Sliding friction — the corresponding parameter is ϕ_s which denotes the angle of friction developed from sliding resistance. The analogy of a solid block resting on another is used to describe this phenomenon. The resistance to displacement between the two blocks is a direct function of the surface friction and normal pressure.

(2) Interlocking friction — the parameter ϕ_i is used to denote the "friction angle" developed due to packing of articles.

The parameter ϕ_i is not measured or computed directly. It is obtained by deduction and analysis through an evaluation of developed shear strengths and corresponding densities.

Let us consider the various methods of stress application on a sample of soil as, for example, in the standard axisymmetric triaxial test performed in the laboratory, where the intermediate and minor principal stresses σ_2 and σ_3 respectively are the same, i.e. $\sigma_2 = \sigma_3 = \sigma_r$. The axial stress–strain relationship shown in Fig.9.2 for a drained test is as follows: Curve 1 depicts the situation where the sample is compressed hydrostatically, i.e. $\sigma_1 = \sigma_r$. With greater volume changes, the soil becomes denser, hence the resistance to deformation becomes larger, resulting in a concave upward stress–strain curve. Curve 2 which is also concave upwards is the result obtained from a one-dimensional constrained compression test. With the restriction that $\varepsilon_r = 0$, σ_r increases because of the lateral or radial restriction, but not necessarily equal to the increase in σ_1, i.e. $\Delta\sigma_1 \neq \Delta\sigma_r$. In Curve 3, we have the more normal stress–strain relationship identified with the standard axisymmetric triaxial testing method (see Chapter 8). In this instance, σ_r is kept constant while σ_1 increases.

The relationship between stress–strain of the type shown as Curve 3 in Fig.9.2 and volume change for two states of packing is shown in Fig.9.3. The degree of confinement afforded the test sample will affect both the volume change, ΔV, and stress. This is shown in Fig.9.4 in terms of both void ratio and confining

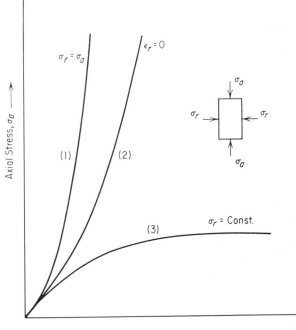

Fig.9.2. Axial stress–strain curves for soil under various states of stress.

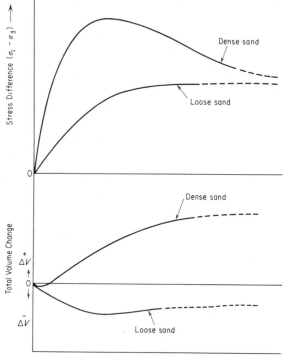

Fig.9.3. Stress–strain–volume relationships for sand in a drained shear test.

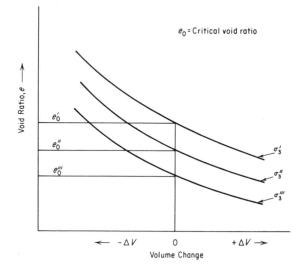

Fig.9.4. Critical void ratio as defined at failure in drained shear test on the basis of zero volume change.

pressure where the volume change ΔV at failure is seen to be a function of initial void ratio e and confining pressure σ_3. The curves are obtained by conducting tests on samples at various initial void ratios and at various confining pressures. By measuring the volume change at failure for all the test samples, the curves may be drawn and the point of $\Delta V = 0$ defined. For a particular state of confinement, there exists an ideal initial void ratio where no resultant volume change occurs.

Consider the energy applied to a sample during a conventional axisymmetric triaxial test. At and near failure, since the stress curve is at the peak stress point (see Fig.9.3), we may reasonably assume that $\Delta\sigma_1/\Delta\varepsilon_1 = 0$. With a constant confining pressure, $\sigma_2 = \sigma_3 = $ constant. Denoting ε_1, ε_2 and ε_3 as compressional strains in the same directions indicated by σ_1, σ_2 and σ_3, respectively, the work terms associated with differential strains may be expressed as follows:

Total work input in application of shearing force must be the sum total of (a) the work required to overcome the internal shear strength w_i and (b) the work associated with volume change:

$$\sigma_1 \Delta\varepsilon_1 = w_i + \sigma_3 (- 2\Delta\varepsilon_3) \tag{9.2}$$

where the negative sign associated with $\Delta\varepsilon_3$ indicates extensional strain. Hence:

$$(\sigma_3 + \Delta\sigma)\Delta\varepsilon_1 = w_i + \sigma_3 (- 2\Delta\varepsilon_3)$$

or:

$$\Delta\sigma \cdot \Delta\varepsilon_1 = w_i + \sigma_3 (- \Delta\varepsilon_1 - 2\Delta\varepsilon_3)$$

Specifying $(- \Delta\varepsilon_1 - 2\Delta\varepsilon_3)$ as $\Delta\nu$ the differential volumetric strain (where the positive sign associated with $\Delta\nu$ indicates volume increase):

$$\Delta\sigma = \frac{w_i}{\Delta\varepsilon_1} + \sigma_3 \frac{\Delta\nu}{\Delta\varepsilon_1} \tag{9.3}$$

To separate friction components in development of shear resistance, we consider eq.9.3 in terms of work and energy associated with the stresses.

$$\Delta\sigma \cdot \Delta\varepsilon_1 = \text{work put into the specimen}$$
$$w_i = \text{intrinsic energy of the specimen}$$
$$\sigma_3 \Delta\nu = \text{work done by the specimen}$$

i.e.:

$$\Delta\sigma \cdot \Delta\varepsilon_1 = w_i + \sigma_3 \Delta\nu$$

Hence, actual shear strength derived from friction resistance along a rupture plane must be given by w_i. The work input $\Delta\sigma\Delta\varepsilon_1$ is made up of two parts, $\sigma_3\Delta\upsilon$ representing mechanical work of the system (no entropy increase), and w_i representing the conversion of work into heat energy because of friction resistance across the rupture plane (entropy increase). The requirement for failure to occur in rupture is implicit in this development. This is not unreasonable for granular soils.

Consider the Mohr's circle diagram shown in Fig.9.5 for a drained test on a granular soil. If the applicability of the Mohr-Coulomb relationship is accepted, we can describe the relationship between dissipative energy (energy used in overcoming friction along the rupture plane) and the shear strength of the specimen by assigning the w_i component to the shearing resistance across the rupture plane. We note that since $(\sigma_1 - \sigma_3)/2 = \Delta\sigma/2$, eq.9.3 can be divided by two for the following development:

From the diagram, if τ represents the shearing stress at failure and is assumed to act on the developed rupture surface, we will obtain:

$$\tau = \frac{w_i}{2\Delta\varepsilon_1}$$

The stress associated with work output is given by:

$$\frac{\Delta\sigma}{2} = \tau + \frac{\sigma_3}{2}\frac{\Delta\upsilon}{\Delta\varepsilon_1}$$

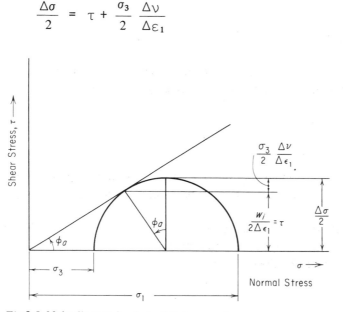

Fig.9.5. Mohr diagram for drained test on sand.

By physically measuring the quantity $(\Delta\sigma/2 - \tau)$ from Fig.9.5, we can quantify stresses associated with volume expansion without measuring Δv and $\Delta\varepsilon_1$ although these may be routine measurements in themselves. The limitations in this method of separation of energy associated with volume change or no volume change are inherent in the initial formulation of the relationship. By assigning $w_i/2\Delta\varepsilon_1$ to τ one assumes that mechanistically, the separation may be made without influencing the normal relationships.

Designating ϕ as the apparent friction angle, i.e. the angle subtended by the rupture envelope, and assuming that linear superposition of friction parameters is permitted:

$$\phi = \phi_s + \phi_i \tag{9.4}$$

From Fig.9.5, we can state by analogy:

$$\sigma_3 \frac{\Delta v}{\Delta\varepsilon_1} = \Delta\sigma(1 - \cos\phi) \tag{9.5}$$

Noting that $\Delta\sigma = \sigma_1 - \sigma_3$:

$$\frac{\Delta v}{\Delta\varepsilon_1} = \left(\frac{\sigma_1}{\sigma_3} - 1\right)(1 - \cos\phi)$$

From Mohr's circle:

$$\frac{\sigma_1}{\sigma_3} = \frac{1 + \sin\phi}{1 - \sin\phi}$$

This gives us:

$$\Delta v = 2\sin\phi\left(\frac{1 - \cos\phi}{1 - \sin\phi}\right)\Delta\varepsilon_1 \tag{9.6}$$

We note from eq.9.6 that $\Delta v/\Delta\varepsilon_1$ is completely describable in terms of the apparent friction angle ϕ. In effect, we may write: $\Delta v = \Delta\varepsilon_1 f(\phi)$.

The important and necessary conditions are that failure is occurring throughout the test samples, and that the expressions derived are limited to the failure state.

We note that the stresses associated with volume change can be uniquely described by the parameter ϕ. This allows us to predict the volume change

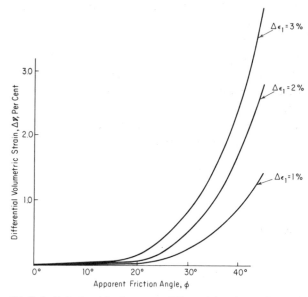

Fig.9.6. Relationship between differential volumetric strain and apparent friction angle for varying values of differential axial strain $\Delta \varepsilon_1$ at failure.

characteristics of the material based upon a knowledge of ϕ (eq.9.5 and 9.6). The greater ϕ is, the larger is $\Delta v / \Delta \varepsilon_1$ — indicative of a densely packed sample.

The differential volumetric strain at failure may be suitably expressed in the form of a graph (Fig.9.6), relating $\Delta v / \Delta \varepsilon_1$ to ϕ for any particular cohesionless soil.

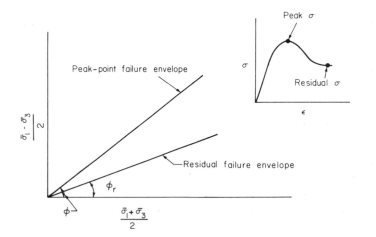

Fig.9.7. Peak stress and residual stress failure envelopes. Note that the residual stress obtained in this post peak performance of a granular material may be called the asymptotic stress at large strains.

Sliding and interlocking friction

In Fig.9.3, we show that if axial straining is allowed to continue, we would ultimately obtain an asymptotic stress value for the granular material under test, regardless of initial density. This demonstrates the fact that with continued straining beyond the failure point, the incremental volume change $d(\Delta V/V)$ becomes zero as noted in the total volume-change curve in Fig.9.3. At this stage, the void ratio remains constant. This is the critical state defined by Hvorslev (1937) which shows that continued straining of the test sample will not cause any further change in the shearing resistance or the void ratio.

The asymptotic stress state which has been defined as the residual stress state (Skempton, 1964) for clays can also be applied to granular soils. At this stage, since the void ratio remains constant with continued straining, and since the shearing resistance remains constant, it can be assumed that the friction property operative is sliding friction beause localized interlocking is minimal. Thus, ϕ_r can be measured directly from tests. The practicalities and difficulties involved, however, will require careful interpretation of test results (Bishop, 1972).

In Fig.9.7 the two failure envelopes describe the condition arrived at by taking peak point values of stress from the stress—strain curve shown, and the residual values obtained in continued straining after failure. Angle ϕ_r which denotes the angle subtended by the residual strength envelope may be taken to be equal to ϕ_s the sliding friction angle. Thus, if we accept super position, the interlocking friction angle ϕ_i which contributes to the strength increase in a granular soil can be obtained by subtracting ϕ_s (i.e. ϕ_r) from ϕ as in eq.9.4. Thus: $\phi_i = \phi - \phi_r$, and assuming that $\phi_r = \phi_s$:

$$\phi_i = \phi - \phi_s$$

The assumption that $\phi_r = \phi_s$ requires that the strain conditions at residual or asymptotic stress do not create undue distortion in the test sample. In addition, the stress and strain fields (distributions) must be continuous except at the zone of failure. When these conditions are not easily met, it is necessary to establish ϕ_s through a method similar to that used for determining the critical void ratio, shown in Fig.9.4. By plotting the rate of volume change at failure $[\,d\,(\Delta V/V)/d\epsilon_1\,]_F$ against the friction angle at failure ϕ_f for various samples under various test conditions, extrapolation of the relationship to the condition of $[\,d\,(\Delta V/V)/d\epsilon_1\,]_F = 0$ would yield the value of ϕ_f at the critical state. This can be assumed to be ϕ_s. There are some constraints to this method of determination of the sliding friction angle (Bishop, 1972).

9.3   LABORATORY MEASUREMENT OF GRANULAR SOIL STRENGTH

The common techniques used for determining granular soil strength are:

(1) *Axisymmetric triaxial test.* This is the conventional triaxial test where σ_2 = σ_3 = σ_r and is commonly referred to as the triaxial test. When reference is made to the "triaxial test", we will mean the axisymmetric triaxial test. The procedure for testing can either be conducted on dry, wet or drained samples, similar to the overall procedure described in Chapter 8.

The triaxial tests are thus (a) fully saturated undrained tests with measurements of pore-water pressure, (b) drained tests where pore pressures do not develop, and (c) tests on dry samples. The more common procedure however is to conduct drained tests.

(2) *Direct or simple shear tests.* These are best conducted as drained tests.

(3) *True triaxial tests.* To distinguish this from the (axisymmetric) triaxial test, true triaxial tests refer to the situation where equipment capability can produce $\sigma_1 \neq \sigma_2 \neq \sigma_3$. Test samples are either rectangular or cubic.

(4) *Plane strain tests.* These tests are meant to simulate two dimensional situations. In laboratory testing, rectangular specimens are used and the condition of $\varepsilon_2 = 0$ is imposed.

Measurement of friction angle, ϕ

In theory, if one measures the generated pore-water pressure u in an undrained triaxial test, the effective stress $\bar{\sigma}$, can be computed by subtracting u from the total stress σ. However, because of volume change and contact

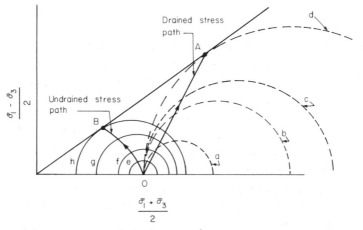

Fig.9.8. Drained and undrained stress paths.

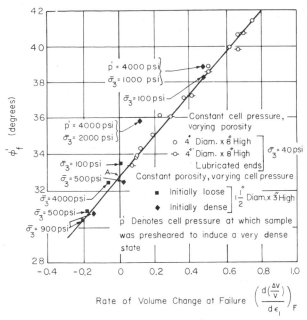

Fig.9.9. Experimental determination of ϕ_s using measurements of ΔV and ε. (Bishop, 1972.)

relationships, the different stress paths followed by drained and undrained triaxial tests may not produce similar results for ϕ. Fig.8.15 (p.280) shows the stress paths followed by the two tests. We can examine these paths in the conventional Mohr-Coulomb graphical representation shown in Fig.9.8. Mohr circles a,b,c and d represent drained test results whilst e, f and g represent undrained test results. The stress paths OA and OB indicate drained and undrained stress paths. The closer OB is to OA, the better is the likelihood of obtaining identical values for ϕ. Problems of boundary and dilatation energy corrections become increasingly significant with major differences in stress paths OA and OB.

Undrained tests on granular materials represent special interest cases for situations where liquefaction or other generated pore-water pressures are anticipated. Because of the case of pore-water drainage in field situations, stressing of a granular soil mass will in general be realized without creating significant pore-water pressures. Thus, the drained test is more generally used.

The procedures for determining the appropriate friction angle for the granular soil will depend on anticipated boundary constraints and on volume change. The apparent friction angle ϕ measured or determined from the Mohr-Coulomb diagram must be distinguished from ϕ_r as in Fig.9.7. Alternatively, ϕ_r or ϕ_s can be computed with measurements of ΔV and $\Delta\varepsilon_1$ using the graphical method of projection to zero for $d[(\Delta V/V)/d\varepsilon_1]_F$ as in Fig.9.9. Point A in Fig.9.9 is the

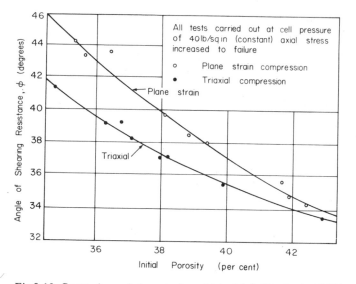

Fig.9.10. Comparison of plane strain and triaxial ϕ. (Conforth, 1964.)

critical state (i.e. critical void ratio) ϕ and may be considered identical to ϕ_r the residual friction angle. Bishop (1972) suggests that it is necessary to distinguish between pure and simple shear in granular soils since the critical states defined in shear testing will be sensitive to the mechanism of failure.

The considerations and care in ascertaining similar modes and mechanisms of failure cannot be lightly dismissed. If plane strain test results are compared with triaxial tests, it is generally noted that unless volume change (dilatation) energy and boundary energy corrections are made, the values for ϕ will not correspond. Measured values of the apparent friction angle ϕ will consistently show that the plane strain $\phi >$ triaxial ϕ (Fig.9.10). The difference between plane strain ϕ and triaxial ϕ becomes greater at lower porosities because of the greater amount of energy input required for volumetric expansion (dilatation) during stressing of the test samples to reach failure. When diligent measurements and corrections are made to account for ϕ_i, the values for ϕ_s are generally relatively close. Where ϕ plane strain is 5° greater than ϕ triaxial, ϕ_s plane strain may be about 1° greater than ϕ_s triaxial. The different failure modes or mechanisms may be responsible for this reduced difference.

Stress and strain

The stress–strain relationships obtained in triaxial testing will depend on the initial density of the material. Since particle packing is the critical consideration, the denser the packing, the greater is the need for energy input for shear distortion and volume change.

The range in stress–strain curves is shown in Fig.9.11. The stress–strain performance of the dense granular soil can be approximated by elastic strain softening theories whilst loose granular soil performance would be identified with work hardening concepts. In cyclic loadings, the stress–strain performance is

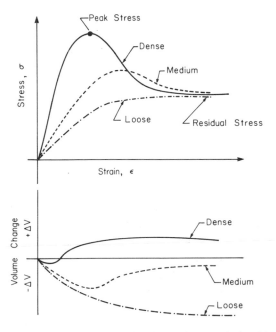

Fig.9.11. Stress–strain and volume change relationships as a functional of initial density. Note that the residual stress is the asymptotic stress at large strains. This may be also identified as large-strain stress.

approximated by elastic work hardening theories. The curves obtained are shown in Fig.9.12. Since the cyclic loading serves to densify the sample, continued loading beyond the last cyclic application would generally produce a stress–strain curve identified with dense granular soils. Thus strain softening is experienced beyond the peak stress point.

The concepts of peak and residual stresses can be easily applied when stress–strain curves demonstrate peak values as in the dense soil shown in Fig.9.11. Peak values can also be seen in medium dense soils. However, in initially loose granular soils, no peak stress is achieved. The residual stress is the peak stress. Thus, if interpretations for ϕ are to be made, it will be apparent that where peak stress is the same as residual stress, $\phi = \phi_r = \phi_s$.

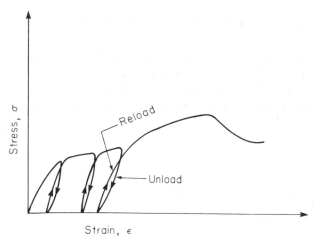

Fig.9.12. Cyclic loading and stress–strain performance.

9.4 THE INTRINSIC FRICTION ANGLE

Skempton (1960) presents the angle of intrinsic friction as a formal concept defining the relationship between strength and pressure. How does this differ from ϕ the apparent friction angle? In tests on cohesionless soils, at very high confining pressures, the Mohr rupture envelope does not describe a unique relationship (Fig.9.13). The angle ψ is defined as shown in Fig.9.13 and its intercept on the τ axis is defined as k in magnitude. The shear strength of solid substances τ_i is given in terms of k and ψ:

$$\tau_i = k + \sigma \tan \psi \qquad (9.7)$$

Based upon the analysis by Skempton, the effective stress is given by:

$$\bar{\sigma} = \sigma - \left(1 - \frac{a \tan \psi}{\tan \phi}\right) u \qquad (9.8)$$

where: $\underline{a} = A_c/A$; A_c = contact area between particles on a plane normal to the applied stresses; A = total cross-sectional area (including void spaces).

If $\underline{a}$ is negligible, eq.9.8 degenerates into the classical equation of: $\bar{\sigma} = \sigma - u$.

We note that in the range of pressures normally encountered $\underline{a}$ is small and that the rupture envelope is still a straight line. When $\underline{a}$ becomes significant, ψ also becomes important, since by definition ψ concerns itself with totally solid substances.

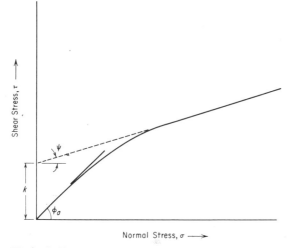

Fig.9.13. Shear and normal stress relationship at high confining pressures.

The value for ψ given for a sand containing quartz particles is $13°$ (Skempton, 1960). $\tan \psi / \tan \phi$ is of the order of 0.3. However, a is probably 0.1, hence:

$$\bar{\sigma} = \sigma - 0.97\ u$$

Under high pressures, crushing of soil particles will occur, causing degradation of the granular soil. This explains the curvature seen in Fig.9.13. When degradation occurs, A_c becomes larger and thus a becomes significant. The correction factor involved in eq.9.7 for modification of the pore-water pressure becomes correspondingly important. In drained tests, however, since $u = 0$, the need for correction does not exist. The effect of crushing of particles in drained tests will be suitably accounted for if proper attention is paid to the determination of residual stresses.

9.5 VOLUMETRIC STRAIN

A theoretical model may be used to provide an analytical description of the mechanical behaviour of an assemblage of discrete particles. Several simplifying assumptions can be made to facilitate the analysis. The simplest case considers an arrangement of two discrete spherical particles in elastic contact with one another (Fig.9.14). The theory according to Hertz (Timoshenko and Goodier, 1951, pp.372–382) predicts that the constant plane between the particles is circular with a contact radius of r. If d is the diameter of the like and equal spheres, $k =$

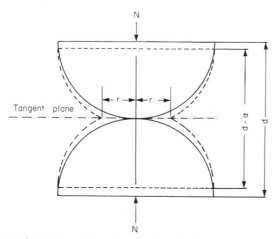

Fig.9.14. Two spheres in normal contact.

$(1 - \mu^2) / \pi E$, where μ and E are Poisson's ratio and Young's modulus of the solid material respectively. The decrease in distance between centres of two spheres in contact under a normal force N will be given as:

$$\delta d \ = \ \sqrt[3]{\frac{9\pi^2 \ k^2 \ N^2}{d}} \tag{9.9}$$

The elastic macroscopic volumetric strain ν_e of an assemblage of spheres will be:

$$\nu_e \ = \ 3 < \frac{\delta d}{d} >$$

$$= \ 3 \ \frac{(9\pi^2 \ k^2)^{1/3}}{d^{4/3}} \ <N^{2/3}> \tag{9.10}$$

where $< \ >$ denote the mathematical expectation of the random variable contained within the carats.

Under hydrostatic loading and unloading the geometry of packing of the spheres comprising the assemblage will not change, except for a possible increase in the number of particle contacts. The relationship between elastic volumetric strain ν_e and σ can be obtained by considering a model of a random array of spheres, where normal and tangential forces exist at constant points. A space structure whose members represent forces at contact points with both direction and magnitude, can be generated as seen in Fig.9.15.

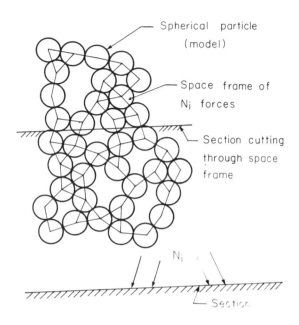

Fig.9.15. Simulated force space structure obtained from forces acting in the system. The bottom section shows the N_i forces acting on the cut section depicted above. (Wong, 1972.)

If a section through the space structure is made as shown in Fig.9.15, an analysis using the method of sections shows that:

$$\sigma = \frac{c}{3} <Nd>$$

where c is the total number of contact forces per unit volume.

For equal spheres since d = common for all spheres:

$$\sigma = \frac{c}{3} d <N> \tag{9.11}$$

With eq.9.10 and 9.11 we will obtain:

$$\nu_e = \frac{G'}{c^{2/3} d^2} R \sigma^{2/3} \tag{9.12}$$

where:

$$G' = 3(81 \pi^2 k^2)^{1/3}$$

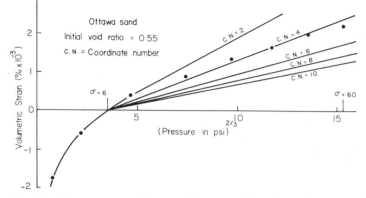

Fig.9.16. Elastic volumetric strain of dense sand in hydrostatic compression. (Yong and Wong, 1972. *Proc., 3rd Southeast Asian Conf. Soil Eng., Hong Kong.*)

and:

$$R = <N^{2/3}> / <N>^{2/3} \tag{9.13}$$

For a regular packing of spheres, eq.9.12 becomes (Yong and Wong, 1972):

$$\nu_e = \frac{3^{5/3} \, 2^{4/3}}{c^{2/3} \, d^2} \, w^{2/3} \, \sigma^{2/3} \tag{9.14}$$

where:

$$w = \frac{3}{4} \, [(1 - \nu^2)/E]$$

For a simple cubic packing with "holes" (similar in philosophy to Ko and Scott, 1967) it can be shown that (Wong, 1972):

$$\nu_e = 3 \, (16 \, w\sigma)^{2/3} \tag{9.15}$$

Additionally, for a face centre cubic packing with "holes":

$$\nu_e = 3 \, (2 \sqrt{2} \, w\sigma)^{2/3} \tag{9.16}$$

Fig.9.16 and 9.17 show test results on Ottawa sand compared with the computed results using the above relationships for ν_e. The coordinate numbers *C.N.* shown in the Figures are obtained from Table 3.1 (p.75) and are associated with

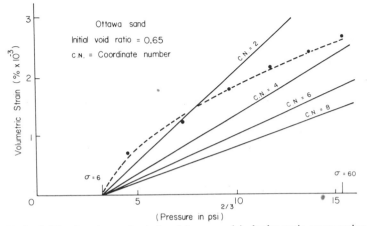

Fig.9.17. Elastic volumetric strain of loose sand in hydrostatic compression. (Yong and Wong, 1972. *Proc., 3rd Southeast Asian Conf. Soil Eng.*)

eq.9.14–9.16 for computation of ν_e. The Figures show that as the pressure is increased, the slopes of the experimental curves become smaller and can be related to the variation in *C.N.* i.e., if the slope of the experimental curve is identical with the slope of the line for *C.N.* = 8, this indicates that the coordinate number for the test sample in the pressure range examined is 8. Taking Fig.9.16 as an example, we note that at low pressures, the slope of the experimental curve is approximately similar to that shown for the line representing *C.N.* = 6. From Table 3.1, this represents a simple cubic packing. As the pressure is increased, more particle contacts are established between adjacent particles in view of deformation of the particles. The experimental curve shows that the slope at higher pressures is similar to the *C.N.* = 10 line. Thus, from Table 4.1, the likely ideal configuration for a coordination number of 10 is tetragonal – sphenoidal. Since real systems do not possess complete spherical particles, the comparisons with ideal packing arrangements shown in Table 4.1 must be made only on the basis of qualitative evaluations.

The volumetric strain can be written as:

$$\nu_e = \frac{G}{(1-n_o)^{2/3} \, a_n^{2/3}} \, R\sigma^{2/3}$$

$$= K_v \sigma^{2/3} \qquad\qquad (9.17)$$

where:

$$G = 2^{2/3} \, G' \, (\frac{\pi}{6})^{2/3}$$

and:

$$K_v = \frac{G}{(1-n_o)^{2/3} \, a_n^{2/3}}$$

a_n = average number of active contacts per sphere i.e. average coordination number.

It is apparent from the Figures that a_n must be varied continuously if the proper predictions of v_e are to be made. The increase in coordination number, or number of contacts may be determined from probability theory. If a_i and a_t are the average coordinate number of the assemblage at hydrostatic pressure $\sigma = 0$ and $\sigma = \infty$ respectively, the probability of a new contact being made as a result of $\Delta\sigma$ will be given by:

$$\frac{m}{E^{2/3}} \, \Delta\sigma$$

where E is the elastic modulus of the solid material and m is a material parameter of the assemblage. The two-thirds power of the elastic modulus is a direct outcome of the amount of volumetric compression which varies with the two-thirds power of the modulus of the material in accord with Hertzian theory.

The probability that a new contact will not be made when σ increases by $\Delta\sigma$ is $(1-m/E^{2/3} \, \Delta\sigma)$. As the pressure increases continuously from zero to σ we will obtain:

$$\lim_{\Delta\sigma \to 0} \left(1 - \frac{m}{E^{2/3}} \, \Delta\sigma\right)^{(\sigma/\Delta\sigma)} = e^{-(m/E^{2/3})\sigma} \tag{9.18}$$

This demonstrates that:

$$(a_t - a_i) \, e^{-(m/E^{2/3})\,\sigma}$$

potential contact points per grain will not be activated at σ. The number of active contacts per grain, a at a hydrostatic pressure σ will be obtained as:

$$a = a_t - (a_t - a_i) \, e^{-(m/E^{2/3})\sigma^r} \tag{9.19}$$

Eq.9.17 may be written as:

$$v_e^{3/2} = \frac{G^{3/2} \, R^{3/2}}{(1-n_o) \, a_n} \, \sigma \tag{9.20}$$

and incremental changes may be considered in conjunction with eq.9.19 to yield:

$$dv_e^{3/2} = \frac{G^{3/2} R^{3/2}}{(1 - n_o) a_n} d\sigma$$

Thus, it can be shown that (Yong and Wong, 1972; Wong, 1972):

$$v_e = \frac{GR}{(1 - n_o)^{2/3} a_t^{2/3}} \left[\sigma + \frac{E^{2/3}}{m} \log_e \frac{a_t - (a_t - a_i) e^{-(m^{2/3}/E^{2/3})}}{a_i} \right]$$

(9.21)

Eq.9.21 represents the stress–strain relationship in hydrostatic compression for a granular soil. The elastic volumetric strain is expressed as a function of the hydrostatic stress σ and constitutive parameters G, P, a_t, a_i, m and R. m is a packing factor describing the rate of increase of contact points of an assemblage of discs in plane strain. The average value of m has been found from actual visual tests by Wong (1972) to be 431. This describes the rate of increase of the average coordinate number with mean stress. The initial (at zero pressure) and final (at infinite pressure) values of the average coordinate numbers, a_i and a_t must be known before the coordinate number at any particular pressure can be calculated. Assuming a random packing of identical spheres of porosity n to be composed of clusters of simple cubic and face-centre cubic arrays of spheres, the following equation for the average coordinate number may be obtained:

$$a_t = 26.48 - \frac{10.73}{1 - n}$$

(9.22)

Under an infinitesimal pressure, the "average" minimum number of contact points

TABLE 9.1

Physical Properties of Ottawa sand

Average size	0.016 inch
Average radius of curvature	0.005 inch
Specific gravity	2.65
Hardness	6
Modulus of elasticity	$1.25 \cdot 10^7$ psi
Poisson's ratio	0.17

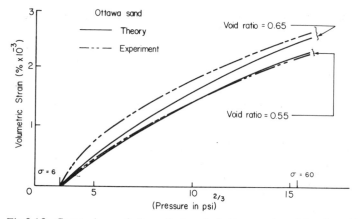

Fig.9.18. Comparison of theoretical and elastic experimental volumetric strain obtained by Wong (1972).

is 1.74 per sphere. This is the average coordinate number of a simple cubic array with holes, and is the minimum value for all stable regular packings of spheres.

The values of m, a_i a_t and the properties of Ottawa sand given in Table 9.1, have been used to obtain the theoretical curves shown in Fig.9.18. These are plotted with the experimental curves obtained by Wong (1972). The agreement between theory and experiment is very good for the dense sand. For the loose sand the theory seems to underpredict the volumetric strain at low stresses.

9.6 SUMMARY

We show in Chapter 8, Section 8.5, that granular soil failure can be adequately described by the Mohr-Coulomb theory of failure. The importance of separation of the two physical friction resistance properties, ϕ_i and ϕ_s becomes significant if proper application of the Mohr-Coulomb theory is to be made. The irregular hexagon shown in Fig.8.8 and 8.9 describe the principal stress states at failure for the condition of total energy dissipation along a rupture plane. In effect, one should expect that ϕ_s can be adequately described by the failure theory.

Where volume changes are large, irregularities in the Mohr-Coulomb failure surface become evident. This, plus other factors, is demonstrated by the points in Fig.8.19 (p.286) Kirkpatrick's experiments on hollow cylindrical samples allowed for varying values of the intermediate principal stress, thus accounting for the points between the σ_2 and $\sigma_1 = \sigma_2$ axes in the diagram. The results show some deviation from the Mohr-Coulomb criterion when the intermediate principal stress is allowed to vary.

The shearing strength of granular soils is highly dependent on the degree of particle packing. Since at least two components of physical friction are operative in shear resistance, the need for a proper accounting becomes important if the Mohr-Coulomb theory is to be applied. The results show that this failure theory is admissible if due attention is given to the requirements for material performance in conformity with the principle of compatibility between theoretical and physical models.

Chapter 10

COHESIVE SOIL STRENGTH

10.1 INTRODUCTION

In a particulate system where particles are easily discernible and where gravity forces are dominant, friction forces constitute the primary control on the response of the system to loading. The concepts of yield and failure in soils, and strength of granular soils have been discussed in the previous chapter. These show that for granular soils, the assumption of the Mohr-Coulomb parameter $c = 0$ is valid and that granular soils do behave as cohesionless materials.

In this chapter, we will examine the strength of cohesive soils. We describe a cohesive soil as one which possesses physical properties of friction and cohesion. The two properties exist and participate in the demonstration of shearing resistance. The constituents and composition of cohesive soils have been discussed in Chapters 2, 3 and 4. These show that the relationships established by the clay minerals with water and with each other are responsible for the integrity of the cohesive soil.

Analytical and physical strength parameters

The general procedure for obtaining or studying the strength of cohesive soils is to undertake physical measurements of strength using a suitable laboratory test, e.g., the triaxial test discussed in Chapter 8. In conjunction with an appropriate yield or failure theory, one will obtain the analytical shear strength parameters needed for engineering design or analysis. Alternatively, one may examine the mechanisms involved in soil deformation under applied loading and deduce therefrom the physical or material parameters responsible for the strength of the soil.

The difference between the two approaches or procedures lies in the definition of the strength parameters. In the first instance where an analytical yield or failure theory is used, the strength parameters obtained are a direct consequence of the theory used and are dependent on the validity and admissibility of the theory. The strength parameters obtained are identified as analytical strength parameters. As pointed out in Chapter 8, the admissibility of a suitable strength theory is rigorously defined within strict boundaries encompassing response behaviour.

A proper modelling of the physics of deformation behaviour will lead to a direct correspondence between analytical and physical models. Such a method of behaviour analysis requires that the generalization necessary to cover the variability in soils be performed with the proper insight into material behaviour. This is identified as a basic approach since it requires an examination of the interaction between soil and water in the development of the physical mechanism for failure or yield. The analytical model can be formulated therefrom.

Unlike many other materials, the strength of cohesive soils cannot be defined or categorized in terms of any one specific property such as ultimate, fracture or yield strength. The nature and form of particle and fabric unit interaction renders it difficult to establish a definite strength quantity except in terms of certain governing conditions and limitations. We have shown in previous chapters that the results of clay—water interaction cannot easily be described in "cause-and-effect" terms. Since a clay soil is not ideally elastic or plastic in nature or form, it becomes relatively difficult to apply conventional theories of mechanics to explain and evaluate cohesive soil strength.

10.2 PORE-WATER PRESSURE

In Chapter 7, Section 7.1, we discussed the concept of effective stress, showing through a simple model in Fig.7.4 that the intergranular stress is obtained by subtracting the pore-water pressure from the total applied pressure. This is given as eq.7.1. The role of pore-water pressure is particularly significant in cohesive soils because of the low permeability of the soils. The time taken to dissipate pore-water pressures is directly dependent on the permeability of the material. In field situations, evaluation of stability and support capacity of cohesive soils must include considerations of pore-pressure development and dissipation under applied loading. Fig.10.1 illustrates the change in overall system stability as a result of a corresponding change in pore-water pressures developed.

The pore-water pressure in a cohesive soil will often be negative if a soil suction exists as described in Chapter 4. The effective-stress equation may still be used, and the result of the suction is that the effective stress is now greater by the amount of the negative pore-water pressure. This negative pore-water pressure or suction can exist in a saturated soil, i.e. a soil in which no air is present in the voids. This can come about as a result of shrinkage or consolidation.

Components of pore-water pressure – fully saturated soil

Unlike granular soils where solid particle contacts can be established, the

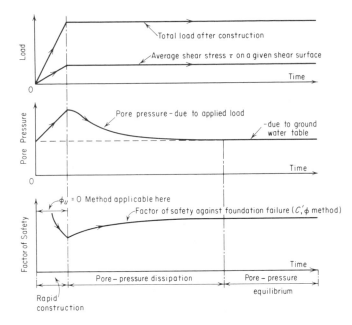

Fig.10.1. Variation of time of applied load, pore pressure and factor of safety for saturated clay foundation beneath a structure. (Bishop and Bjerrum, 1960.)

presence of surface and interparticle forces (Chapters 2 and 3) in cohesive soils require a close examination of the nature of the pore-water pressure generated in the soil as a result of applied load. If total pore-water pressure = u, then the individual components of pore-water pressure may be identified as follows:

u_p = pore-water pressure resulting from reaction to applied external pressure.

u_g = pore-water pressure arising due to the position of the soil mass in the gravitational field.

u_π = the pore-water pressure arising from osmotic effects of the dissolved solutes in the fluid phase, i.e. excluding osmotic pressure arising from the exchangeable ions associated with the clay.

u_m = pressure in the pore water arising from forces originating in the particles. This includes adsorptive forces holding the first few layers of water, surface tension forces holding water in coarse-grained soils, and osmotic forces from exchangeable cations in swelling fine-grained soils.

These component pore-water pressures are represented pictorially in Fig.10.2. They are equivalent to the components of soil water potential in Chapter 4. In many instances it is not possible to distinguish between all these component pore-water pressures nor to evaluate the separate components precisely.

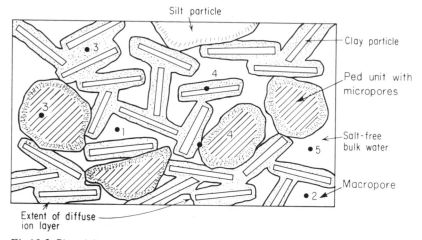

Fig.10.2. Pictorial representation of components of pore-water pressure in a saturated soil-water system. Note that only larger ped units are shown. Smaller particles shown could also be smaller fabric units. (Legend points *1–5*, see text.)

The pore spaces between particles in the fabric units are extremely small in comparison to the pore spaces between individual fabric units (Chapter 3). Intra fabric unit pores are identified as micro pores and inter fabric unit pores as macro pores as in Chapter 3. Points *1–5* in Fig.10.2 represent u_p, u_p, u_π, u_m, and u_g respectively. Application of external pressure will modify or alter the magnitude of the individual components represented by points *1, 3, 4* and *5* because of rearrangement and deformation of fabric units. Points *3* and *4* lie to a greater extent within the micro pores than in the macro pores. Specific location for points *1, 2* and *5* lie in the macro pores and in the bulk water portion of the pore fluid. The component pore-water pressures are important contributors to the total pore-water pressure and can be summed to give:

$$u = u_p + u_g + u_\pi + u_m \tag{10.1}$$

In high-swelling soils, the pore-pressure component u_m is related to the swelling pressure of the test sample and is thus a negative quantity. The amount of pressure required to sustain constant volume for a high-swelling soil in the presence of available water represents the force or pressure that must be placed on the available water to keep it from being drawn into the soil sample. In the absence of other pore-pressure components, the effective stress in a high-swelling soil under no external pressure excepting that which is required to maintain constant volume is given as follows:

$$\bar{\sigma} = \sigma - u$$

But since $u = u_m$, which is a negative number, then:

$$\bar{\sigma} = \sigma + u_m \tag{10.2}$$

In non-swelling or very low-swelling clays, generated pore pressures equal applied all around pressure if samples are completely saturated and if constant sample volume is maintained. This is not true for saturated high-swelling clays. In high-swelling clays, if the applied confining pressure is less than the osmotic or swelling pressure, negative pore pressures will be measured, i.e. the tendency for swelling occurs. If the sample is allowed to swell, i.e. if it is allowed to take in water, it will swell until the measured pore pressure u is zero, if access to water is maintained.

Under constant volume conditions, in high-swelling soils, if the applied external hydrostatic confining pressure is greater than the swelling pressure of the test sample, the net pore pressure generated will be the difference between the positively generated pore pressure and the swelling pressure. Thus the net pore pressure will be less than the applied confining pressure. In high-swelling clays, if consolidation is incomplete, residual consolidation pore pressures can be set up which would not be immediately apparent. These will add to the pore pressure generated as a result of further applied loading.

Pore pressures in a partly saturated clay

Consider the general case of a partly saturated clay-water system with random arrangement of particles as shown in Fig.10.3. Assuming that all points are essentially at the same elevation, the gravitational component can be neglected.

If point *1* is located in the free water region, the pore-pressure components are:

$$u_1 = u_{\pi_1} + u_{m_1} + u_p \tag{10.3}$$

where: u_1 = total pressure; u_{π_1} = osmotic pressure = RTC, where C is the solute concentration.

Point *2* represents a point on a curved air–water interface beyond the range of the bound water layers surrounding the clay particles. Analysis of static equilibrium across the interior shows that if the surface tension T is considered positive:

$$u_2 = u_{m_2} = \left(u_{air} - \frac{2T}{r_2} \right) \tag{10.4}$$

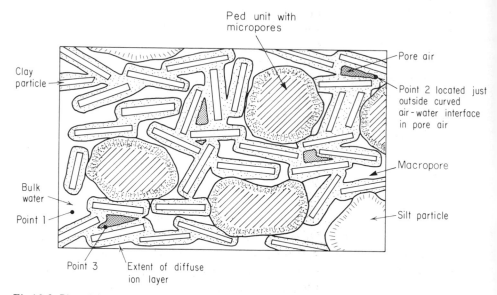

Fig.10.3. Pictorial representation of a partly-saturated soil-water system. Only larger ped units are shown. The presence of smaller fabric units can be assumed. (Legend points *1–3*, see text.)

where: u_{air} = air pressure in the void; T = surface tension; r_2 = radius of curvature of the meniscus at point 2.

Consider a point which lies just on the curved air—water interface but is within the diffuse ion-layer, as shown by point *3* in Fig.10.3.

$$u_3 = u_{air} - \frac{2T}{r_3} + u_{m_3} \qquad (10.5)$$

The curvature and ion concentration at any point are governed by a number of factors such as arrangement of particles, size of pore spaces, degree of saturation, structure, etc. as discussed and shown in the previous chapters. The free energy of the system at equilibrium is at a minimum. For partly saturated soils, the capillary and osmotic pressure effects will, in all probability, vary in equal but unlike manner from one point to another in the entire system in order to maintain equilibrium.

Practical considerations in pore-water measurements

From the preceding theoretical discussion on pore-water pressure components it is apparent that no method is available for physically measuring all the individual component pore-water pressures.

Practical considerations do not necessarily require anything more than a measure of the generated pore-water pressure. The effective stress that is obtained by subtracting the measured pore pressure from the total stress does not imply direct intergranular stress (i.e. grain-to-grain contact stress).

The commonly used pore-pressure measuring systems for saturated soils (without air) rely on the null-point method and do not require volume change to measure the pressure in the pore water. The details of the test system may be found in laboratory test manuals or specialized texts, e.g., Bishop and Bjerrum (1960), Bishop and Henkel (1962).

In partially saturated soils, both air and water occupy the soil voids. The preceding theoretical treatment of this problem may not be easily applied because of its complexities and uncertainties. Since the value of valid pore-water pressure measurements lies in the use of the effective stress equation, it is necessary to obtain either a value for pore-air pressure together with pore-water pressure, or to modify the measured pore-water pressures to account for both air and water. Bishop and Blight (1963) showed from laboratory tests that the effective stress equation for soils with both air and water might take the form of:

$$\bar{\sigma} = \sigma - u_a + \chi(u_a - u_w) \qquad (10.6)$$

where: u_a = pore-air pressure; u_w = pore-water pressure; χ = parameter dependent on sample saturation.

Typical values for the parameter χ for four compacted and partly saturated soils are shown in Fig.10.4. The values which were determined from axisymmetric triaxial compression tests by Bishop and Blight (1963) show that the relationship between χ and per cent saturation is different for each soil. However, the curves for the four different clays seem to follow similar trends.

Pore-pressure coefficients

The development and characterization of pore-water pressures may sometimes be better expressed in terms of pore-pressure coefficients (Skempton, 1954). The following development of pore-pressure coefficients is based on Skempton's treatment of the subject.

Consider an element of soil at equilibrium under an initial isotropic effective stress, p. At this stage, initial pore water pressure is zero. The application of pressures $\Delta\sigma_1$ and $\Delta\sigma_3$ occurs in two stages (Fig.10.5). The pore-pressure changes corresponding to the increased pressures of $\Delta\sigma_1$ and $\Delta\sigma_3$ are Δu_1 and Δu_3 respectively. The total pore-pressure change is:

$$\Delta u = \Delta u_3 + \Delta u_1 \qquad (10.7)$$

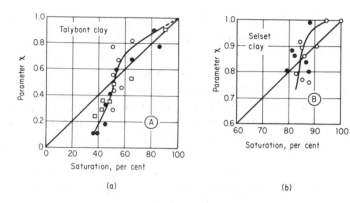

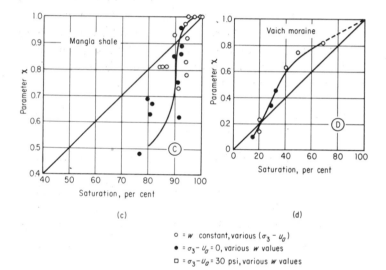

o = w constant, various $(\sigma_3 - u_a)$
● = $\sigma_3 - u_a = 0$, various w values
□ = $\sigma_3 - u_a = 30$ psi, various w values

Fig.10.4. Variation of parameter χ with degree of saturation for partly-saturated soils from triaxial compression tests. (Bishop and Blight, 1963.)

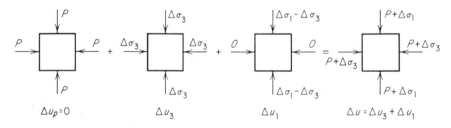

Fig.10.5. Representation of soil element subject to stress increments and associated pore-pressure response.

When the element is subjected to an all around cell pressure $\Delta\sigma_3$ as in the triaxial test, the increase in effective stress $\Delta\bar{\sigma}_3$ is given by:

$$\Delta\bar{\sigma}_3 = \Delta\sigma_3 - \Delta u_3$$

If C_c = compressibility of the soil structure, the corresponding volume change is:

$$\Delta V = - C_c \cdot V(\Delta\sigma_3 - \Delta u_3)$$

$$\Delta V_v = - C_v \, n \, V \cdot \Delta u_3$$

where: V = original volume of the sample; C_v = compressibility of the fluid (air and water) in the voids; n = porosity of the soil; V_v = volume of void space. These volume changes must be identical:

$$- C_c \cdot V(\Delta\sigma_3 - \Delta u_3) = - C_v \, n \, V \cdot \Delta u_3$$

Hence:

$$\frac{\Delta u_3}{\Delta\sigma_3} = B = \frac{1}{1 + \dfrac{nC_v}{C_c}} \tag{10.8}$$

In saturated soils, since water is less compressible than the mineral structure, $C_v/C_c = 0$. Consequently, $B = 1$, i.e. the degree of saturation = 1. For dry soils, C_v approaches infinity, hence $B = 0$.

The application of a stress difference $(\Delta\sigma_1 - \Delta\sigma_3)$, gives us:

$$\Delta\bar{\sigma}_1 = (\Delta\sigma_1 - \Delta\sigma_3) - \Delta u_1$$

$$\Delta\bar{\sigma}_3 = - \Delta u_1$$

As a first approximation, we assume that the soil skeleton is elastic. Hence, from elastic theory, the volume change of the mineral soil structure is given as:

$$V_c' = - C_c \, V \frac{1}{3} (\Delta\bar{\sigma}_1 + 2\Delta\bar{\sigma}_3)$$

The corresponding volume change in the void space is:

$$\Delta V_v = - C_v \, n \, V \, \Delta u_1$$

As before, the two volume changes must be equal:

$$\Delta u_1 = \frac{1}{1 + \dfrac{nC_v}{C_c}} \cdot \frac{1}{3} (\Delta\sigma_1 - \Delta\sigma_3)$$

$$= B \cdot \frac{1}{3} (\Delta\sigma_1 - \Delta\sigma_3)$$

Refining our approximation, and now accounting for the non elastic nature of the soil structure, we will replace the 1/3 factor by A.
Hence:

$$\Delta u_1 = B A (\Delta\sigma_1 - \Delta\sigma_3)$$

Referring back to eq.10.7 we will have:

$$\Delta u = B [\Delta\sigma_3 + A (\Delta\sigma_1 - \Delta\sigma_3)]$$

$$= B \Delta\sigma_3 + \bar{A} (\Delta\sigma_1 - \Delta\sigma_3) \qquad (10.9)$$

where:

$$B = \frac{\Delta u_3}{\Delta\sigma_3} , \qquad \bar{A} = \frac{\Delta u_1}{(\Delta\sigma_1 - \Delta\sigma_3)}$$

Note that $\bar{A} = A$ if $B = 1$. Hence we see that for a fully saturated soil, with $B = 1$

$$\Delta u = \Delta\sigma_3 + A (\Delta\sigma_1 - \Delta\sigma_3) \qquad (10.10)$$

The variation in A will allow us to evaluate the equilibrium state of the soil. The values given by Skempton for A at failure (with $B = 1$) are shown in Table 10.1.

The amount of volume change occurring under the stress difference $(\sigma_1 - \sigma_3)$ accounts for the variation in A. With highly compressible clays, the pore-pressure response is high. The value of A greater than unity indicates: (a) pore-pressure augmentation from initial trapped pore pressure, released due to breakdown of group fabric units under load; and (b) pore-pressure gain due to presence of residual consolidation pore pressure, i.e. the sample is not fully consolidated under the terminal cell confining pressure.

TABLE 10.1

A values for various clays at failure
(From Skempton, 1954)

Type of clay	A		
Clays of high sensitivity	$+\frac{3}{4}$	to	$1\frac{1}{2}$
Normally consolidated clays	$+\frac{1}{2}$	to	$+1$
Compacted sandy clay	$+\frac{1}{4}$	to	$+\frac{3}{4}$
Lightly overconsolidated clays	0	to	$+\frac{1}{2}$
Compacted clay gravel	$-\frac{1}{4}$	to	$+\frac{1}{4}$
Heavily overconsolidated clays	$-\frac{1}{2}$	to	0

In overconsolidated clays, the application of a stress difference will cause a volume expansion. This dilatation will create a negative pore pressure. Thus, the pore-pressure coefficient A will be negative. The greater the dilatation, the more negative is A, i.e. the greater will be the soil suction.

Laboratory determination of pore-pressure coefficients

The pore-pressure coefficients $\bar{A}$ and B are determined from measurements of generated pore pressures under $\Delta\sigma_1$ and $\Delta\sigma_3$, respectively. From eq.10.8: $\Delta u_3/\Delta\sigma_3 = B$.

If a normally consolidated soil sample is at equilibrium at some initial stage of cell confinement, say at σ_c, where u_c = pore pressure under cell confinement of σ_c, the change in pore pressure Δu_3 measured under an increase in cell pressure $\Delta\sigma_3$ is expressed as follows:

$$u_c + \Delta u_3 = u$$

$$\sigma_c + \Delta\sigma_3 = \sigma_3$$

Since u_c pertains to equilibrium conditions under σ_c, we should examine Δu_3 and $\Delta\sigma_3$. Thus the ratio of the increase in pore pressure to the increase in cell confinement which is given as B is readily determined from measurement of Δu_3. Within the limitations of laboratory experimentation, a B value of 0.95 and above may be taken to represent complete saturation of the test sample.

The determination of $\bar{A}$ is performed in a similar manner. In the consolidated undrained CU axisymmetric triaxial test, $u_c = 0$ immediately prior to the application of the axial pressure (represented as $\Delta\sigma$ where $\sigma_1 - \sigma_3 = \Delta\sigma$). Thus, generated pore-pressures under $\Delta\sigma$ represented as Δu_1 may be measured and expressed as a ratio:

$$\frac{\Delta u_1}{\Delta\sigma} = \frac{\Delta u_1}{(\sigma_1 - \sigma_3)} = \bar{A}$$

A typical relationship between pore pressure and strain is shown in Fig.10.6 together with the corresponding stress–strain relationship. Δu_1 increases continuously, and in many instances, this increase still occurs beyond peak stress. Under such circumstances, it is reasonable to assume that the rate of load application exceeded the time rate for equilibrating pore pressures. Typically, the ideal situation is represented by curve *1* where Δu and $\Delta\sigma$ peak at the same strain. Curve *2* represents conditions where insufficient time is allowed for pore pressures to equilibrate during application of $\Delta\sigma$.

In high-swelling soils, if consolidation under cell pressure is incomplete, residual consolidation pore pressure will add to Δu_1 since $u_c \neq 0$. This is less likely

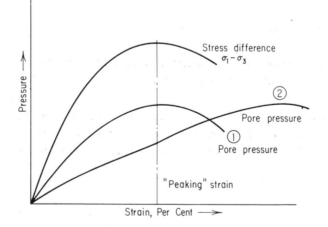

Fig.10.6. Stress–strain and pore pressure–strain relationships, showing lag in pore pressure due to non-equilibrium.

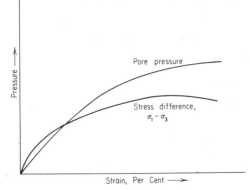

Fig.10.7. Typical stress—strain curve for bentonite for $A > 1$.

to happen in non swelling soils since the time taken for u_c to become zero under σ_c is not as long. Thus, measured values for pore pressure under $\Delta\sigma$ may exceed the applied pressure resulting in $\bar{A}$ values of greater than $1\frac{1}{2}$; i.e.:

$$\frac{\Delta u_1 + \Delta u'_c}{\Delta\sigma} > 1\frac{1}{2}$$

where: $\Delta u'_c$ = augmented pore pressure from trapped and residual consolidation pore pressure.

Fig.10.7 shows a typical test on a bentonite soil where $\bar{A} > 2$. $\Delta u_1 + \Delta u_c$ cannot exceed $\sigma_3 + \Delta\sigma$.

10.3   ANALYTICAL SHEAR STRENGTH PARAMETERS (FROM MOHR-COULOMB FAILURE THEORY)

In its simplest form the Mohr-Coulomb law developed in Chapter 8 provides a relationship between the strength τ of a material in terms of two soil strength parameters known as "cohesion" c and "friction" angle ϕ. In applying this relationship to soils we presume that the conditions necessary for application of the relationship are met. This aspect of the problem will be discussed in a later section. For the present discussion on analytical strength parameters, we will accept direct application of the relationship. Thus:

$$\tau_f = c + \sigma_n \tan \phi \tag{10.11}$$

where: σ_n = normal stress; τ_f = shearing stress on failure plane at failure.

To simplify the terminology, we will designate the conventional axisymmetric triaxial test $\sigma_2 = \sigma_3$ as the "triaxial test", and when the occasion demands, true triaxial tests on prismatic samples where $\sigma_1 \neq \sigma_2 \neq \sigma_3$ will be called "true triaxial tests". In unconsolidated undrained triaxial tests on saturated cohesive soils, the compression strength (i.e. the stress difference $\sigma_1 - \sigma_3 = \tau_f / 2$) is independent of the confining pressure used to test the sample. σ_1 and σ_3 denote the major and minor principal stresses, respectively. Typical Mohr stress circles obtained from UU tests (i.e. unconsolidated undrained tests) are shown in Fig.10.8. Since they are unconsolidated, the test samples possess the same void ratios as one might expect from natural soil. Hence, in the fully-saturated state, the applied cell pressure (i.e. confining pressure) is reflected by the sample pore-water pressure of equal magnitude. Thus, if $\sigma_3 = 30$ psi, $u = 30$ psi. The Mohr stress circles (solid lines) shown in Fig.10.8 are total-stress circles since they represent the total applied stresses. The soil strength parameters are c_{uu}, the apparent cohesion and ϕ_{uu} the apparent angle of internal friction, which in this instance is zero. The subscripts uu associated with c and ϕ denote unconsolidated undrained conditions.

$$\tau_f = c_{uu} = \tfrac{1}{2}(\sigma_1 - \sigma_3)_f$$

where the subscript f denotes stresses at failure. The analytical parameter is c_{uu}.

If pore pressures are measured in the UU test and the values of the pore pressures subtracted from the total stresses, the effective stresses obtained can be plotted on the same diagram. However, if the samples are fully saturated $B = 1$, i.e. $\Delta u_3 / \Delta \sigma_3 = 1$. Thus, as UU conditions theoretically demand that constant volume is maintained throughout the test, the circles shown as *1*, *2*, and *3* in Fig.10.8 would all collapse into one circle with $\bar{\sigma}_3 = 0$. The diameter of the circle will remain

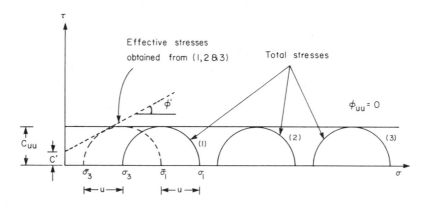

Fig.10.8. Unconsoildated undrained tests on saturated soil showing total and effective stress circles. In actual fact, $\bar{\sigma}_3$ will in all likelihood be zero.

unaltered. In essence, effective-stress computations for UU tests are not useful since UU conditions provide for all test samples to be at the same density and void ratio regardless of cell confining pressure σ_3.

Correlations between soil performance and soil properties show that $\bar{\sigma}$ may be used successfully. Much depends on the accurate determination of the pore-water pressure u. By measuring the pore-water pressure u at failure, the effective stresses at failure may be determined. These are shown in Fig.10.9 for consolidated undrained tests, i.e. samples consolidated under different cell-confining pressures thus resulting in samples with varying void ratios or densities. The analytical soil strength parameters derived from the effective-stress analysis c and ϕ' are defined as effective-stress parameters.

The consolidated drained triaxial test presumes that the shearing test is performed at a sufficiently slow rate such that no pore-water pressures are generated. In the absence of pore pressures the stresses in the soil sample are thus effective stresses. Because the consolidated undrained test is a constant-volume test, and the consolidated drained test is a variable-volume test, the physical soil strength parameters need not necessarily be similar since actual performances between the

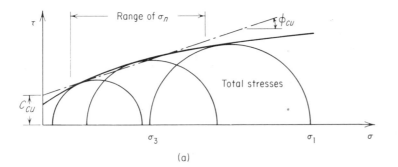

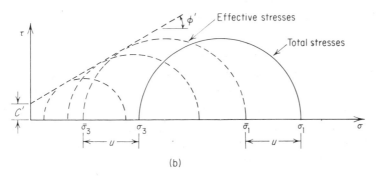

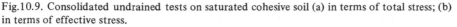

Fig.10.9. Consolidated undrained tests on saturated cohesive soil (a) in terms of total stress; (b) in terms of effective stress.

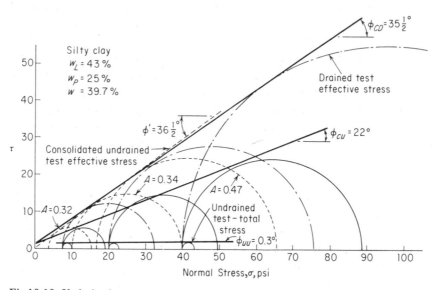

Fig.10.10. Undrained consolidated-undrained and drained tests on undisturbed samples of silty clay, with maximum stress difference. (Bishop and Bjerrum, 1960.)

two kinds of tests will not be similar. The consolidated drained test analytical parameters are given as c_{CD} and ϕ_{CD}. The comparison between these parameters, i.e. c_{uu}, ϕ_{uu}, c'', ϕ', c_{CD} and ϕ_{CD}, is given in Fig.10.10. The reasons for the differences will be evident from the discussions to follow in the later sections of this Chapter.

Observed failure characteristics in cohesive soils present a complex picture. For initial considerations the drained shear test can be used to illustrate certain phenomena in laboratory test on clays subjected to various stress histories. The following nomenclature will be used in discussing these properties:

$\bar{\sigma}_c$ = consolidating pressure = effective consolidating stress

σ_c = σ_{cf} at failure

σ_f = normal stress on failure plane at failure.

In the consolidated drained shear test, if the test sample is fully consolidated prior to shear, and if the shearing process is conducted slowly so that pore pressures are not generated, the stresses developed in the soil are effective stresses.

For no stress history of any significance, i.e. for any condition of $\bar{\sigma}_c$ > previous load history, the relationship between shear stress τ_f and consolidation load at failure, $\bar{\sigma}_{cf}$, is essentially linear (Fig.10.11).

The importance of load history in the development of shearing strength of cohesive soils can be demonstrated by subjecting overconsolidated test samples to shear testing at various confining pressures. We define an overconsolidated sample as one where the previous consolidation load is greater than the present consolidating load. This is accomplished in the laboratory by applying a high consolidating load and then allowing the sample to rebound later to some lower consolidating load. A normally consolidated sample is one which has never been subjected to a consolidation pressure larger than the existing one.

The drained shear strength of samples A, B and C, normally consolidated as shown in Fig.10.11 at varying consolidation loads indicates a linear relationship

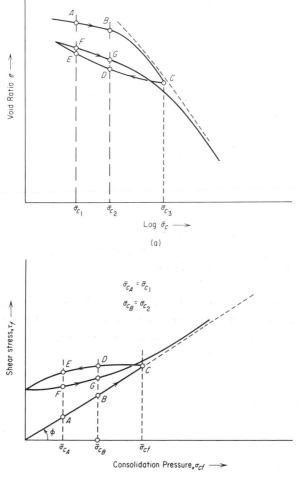

Fig.10.11. a. e-Log $\bar{\sigma}_c$ curve for samples consolidated for drained shear test. b. Drained shear test results for the consolidated samples.

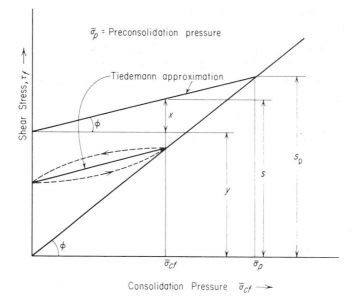

Fig.10.12. Tiedemann approximation of rebound shear curves.

between strength and preconsolidation pressure. If rebound to a lower consolidating or confining pressure is allowed after preconsolidation to a terminal pressure, the drained shear strength is much higher than that of a companion sample sheared without preloading. This is demonstrated in the shear strengths of samples E and D.

For simplicity analysis, the hysteresis loops in Fig.10.11b are approximated as straight lines (see Fig.10.12). The strength parameters defined from the graph in Fig.10.12 are c' and ϕ' — the effective-stress parameters:

$$\tan \phi' = \frac{S_p - y}{\bar{\sigma}_p}$$

or:

$$y = \bar{\sigma}_p \left(\frac{S_p}{\bar{\sigma}_p} - \tan \phi' \right)$$

$$= \bar{\sigma}_p \left(\tan \phi - \tan \phi' \right)$$

From Fig.10.12, since:

$$x = \bar{\sigma}_{cf} \tan \phi'$$

and:

$$\tau = x + y$$

we will obtain:

$$\tau = \bar{\sigma}_p (\tan \phi - \tan \phi') + \bar{\sigma}_{cf} \tan \phi'$$

Hence:

$$\tau = \kappa \bar{\sigma}_p + \bar{\sigma}_{cp} \tan \phi' \qquad\qquad (10.12)$$

where $\kappa = \tan \phi - \tan \phi'$.

The equation given for shear strength τ shows that the cohesion term is dependent on the stress history and preconsolidation pressure of the clay soil. This recognizes the need for rendering compatibility between the analytical strength parameters obtained through application of the Mohr-Coulomb law and actual physical performance parameters. We will discuss the significance of these and other parameters in the latter portion of this chapter.

10.4 MECHANISMS FOR DEVELOPMENT OF SHEAR STRENGTH

Since observed behaviour of clay soil in shear tests shows different values for ϕ and c depending upon test techniques and methods used for test data evaluation, it is necessary to determine what constitutes the significant effective stress parameters. To do so, shear strength analysis may be viewed in the light of two basic components contributing to the soil strength, (a) physical and (b) physico-chemical. These components describe in part, the mechanism governing soil behaviour under shearing forces. The actual measurement of strength parameters must still depend on the use of suitable laboratory test techniques.

Physical and physico-chemical components of friction and cohesion

The physical components of shear strength in cohesive soils arise primarily from the resistance to relative movement of sliding of one particle on another and to interlocking between particles, similar to the mechanical phenomenon discussed in Chapter 9 for granular soils.

Rosenqvist (1959) divides the interlocking phenomenon into two distinct types:

(1) Macroscopic interlocking of particles requiring appreciable movement of particles normal to the failure plane thereby resulting in an increase in volume prior to failure, and

(2) Microscopic interlocking because of surface roughness of particles resulting in small movements of particles normal to the failure plane.

The magnitude of these physical phenomena are proportional to the effective normal stress on the failure plane and are characteristic of granular materials (see Chapter 9). To what extent these phenomena are valid and operative in cohesive soils depends on:

(1) Initial bonding – cementation, organic, other types (described in Chapter 3).

(2) Soil fabric – number, size and distribution of fabric units.

(3) Inclusion of non-clay minerals and particles,

(4) The interaction of the interparticle forces of attraction and repulsion associated with the clay-water system.

For a perfectly elastic mineral the shearing resistance is a function of the effective stress and is expressed by:

$$\tau_f = \bar{\sigma} \tan \phi$$

where $\tan \phi$ in this instance is the ratio of the shearing resistance developed by the adhesive bonds to the normal effective stress.

In dense materials where particles are packed in close configuration, it is necessary for particles to move over adjacent particles for shear displacement. As shown in Chapter 9 for granular materials, the interlocking that occurs will thus cause resistance to motion and consequently volumetric expansion must occur to accommodate particle displacement. In undrained triaxial tests on saturated soils, volumetric expansion causes a decrease in pore-water pressure and an increase in effective stress, thus resulting in an increase in measured shear strength.

Frictional resistance may sometimes be considered as a physico-chemical phenomenon. Because surfaces of particles are not absolutely smooth, when two particles are brought into contact with each other under stress (and since it is not possible to have infinite stress at a point) the contact points will deform elastically or plastically to an amount sufficient to sustain the applied effective stress. The close proximity of contacting areas results in adhesion due to the electrical forces of attraction. The adhesive forces are responsible for shear resistance and must be overcome if sliding between contacting particles and fabric units is to occur. The shear resistance is proportional to the strength of the adhesive bond, which is a function of the distance between the atoms and molecules in the contacting surfaces and the composition of the material. Assuming that the spacing between oppositely charged atoms for a given mineral is constant, it then follows that the strength of the bond must be proportional to the total contact area. Further, it may be assumed that at failure, the contact pressure at each point of contact is constant and constitutes the strength contribution from the solids.

The study by Mitchell et al. (1969) using the rate process model results in a similar conclusion regarding bond strength development. The compressive strength for several soils correlated with the number of bonds computed on the basis of the model assumed is shown in Fig.10.13. The separation, in this instance, between physical and physico-chemical components in shear resistance is not made. The model assumes that the seat of shear resistance lies solely in the bonds developed at the contacting points between flow units. This assumption may be less valid with active clays or clays with high water-holding capacities.

In a clay-water system, the interaction of clay particles and fabric units with water is such that there may be little actual particle-to-particle contact. Thus the definition of friction must be altered. Friction can no longer be accepted as that property derived from physical solid particle contact (including interlocking) if the interaction of clay particles and fabric units is through the adsorbed water layer and the layers of exchangeable ions. The demonstration of a friction component of

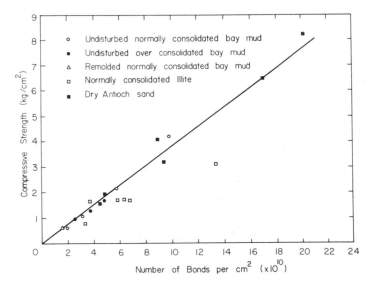

Fig.10.13. Compressive strength as a function of number of bonds. (Mitchell et al., 1969.)

shear strength of clay soils will depend on the constraints imposed in the test technique.

The component of shear strength generally defined as cohesion is thought to be a direct function of the physico-chemical properties of the clay soil. Cohesion may be defined as that property arising from physico-chemical forces of an interatomic, intermolecular and interparticle nature, as discussed in Chapters 2 and 3. Because of the many inseparable interactions that occur between the interparticle forces, direct measurement of actual cohesion as a physical property quite apart from the parameter derived in conjunction with a failure theory, is difficult.

The interaction of clay particles occurs through the layer of adsorbed water, through diffuse layers of exchangeable cations, and through direct particle contact. In Chapter 4, soil structure is seen to depend on this interaction of clay particles and fabric units. To quantify cohesion as a particular property, we must be able to separate individual effects arising from specific particle and fabric-unit interaction. The analytical cohesion parameter c is not to be confused with the property of cohesion. As noted previously, the analytical parameter is obtained from laboratory tests coupled with the application of a failure or strength theory.

Fabric units and bonding in shear strength

We have shown previously (Fig.10.11b) that prestress history affects shear strength. This may be examined on the basis of interaction of fabric units in shear and of their importance in the development of the ultimate strength of the soil sample. Note that individual clay particles are also fabric units. The result of prestressing or preloading is to reorient the initial arrangement of fabric units into a different final configuration. The development of shear resistance within the soil depends on the interaction of the fabric units. If there is increased resistance to displacement and distortion of these soil fabric units under the action of an applied shear, then greater strength may be found. It is possible on this basis, to formulate a working hypothesis which would provide the mechanism for shear and answer some questions concerning pore pressure, interparticle forces and particle interaction.

Consider first the case of pure clay where all the soil fabric units will be oriented in some particular configuration, e.g., flocculated or random.

The bonding that can occur as a result of interparticle forces and clay–water interaction may be classified as follows:

(a) edge-to-surface (es) bonds derived from edge-to-surface arrangement,

(b) edge-to-edge (ee) bonds derived from edge-to-edge arrangement,

(c) surface-to-surface (ss) bonds derived from parallel orientation.

The edge-to-edge and edge-to-surface bonds can include both (a) cementing bonds resulting from oxides of iron, silicon and aluminium, carbonates and organic matter precipitated between particles and fabric units, and (b) those resulting from interparticle forces. Denoting these as es, ee, and ss bonds, in the order given above (in order of decreasing intensity of bonding), the ss bonds will tend to be the weakest for the same effective particle spacing. The differentiation between ee- and es-bonding strength must depend upon the type of minerals present, the cementing agents, and the specific charges on the minerals themselves. We note that whilst flocculated fabrics can exist in bonded and unbonded clays, the strength of the fabric is also dependent on the cementing bonds formed. The term "bonded" clay denotes strong edge-to-edge and edge-to-surface bonding through cementing, carbonate or organic bonds between particles and fabric units (e.g. peds) providing for a semi-rigid fabric. This type of bonding is likely to be found in marine clays — e.g., Champlain clays of eastern Canada.

When the cementing bonds are strongly developed, the mineral framework which forms the fabric of clays becomes strongly tied at the points of bonding. The resultant fabric is rigid and the stability of the system is enhanced. This characteristic distinguishes a bonded clay from an unbonded or remoulded clay. In an unbonded clay, the cementing bonds are poorly developed, or absent. The mineral framework is not held together rigidly at ee and es points. The fabric is not rigid and is characteristic of unbonded and remoulded clays. The only forces available for bonding in this instance are the electrical forces due to the specific charges on the clay minerals. These are defined as interparticle force bonds. In a flocculated clay, for example, the dominant bonding agent or forces in an unbonded clay would likely be the es forces. In bonded clays, other bonding agents as described above, will be added on to the es forces. In shear displacement, the primary component of shear resistance is derived from destruction or breakdown of the bonding agents and the es bonds. In a bonded clay, this develops a high strength over a small displacement. The typical stress—strain curve for this type of sample is given in Fig.10.14. For comparison, Fig.10.14 also gives the stress—strain curve for a random fabric. This is typical of unbonded and remoulded clays. Because of the weaker ss bonding and absence of bonding agents, there will be a flatter slope in the stress—strain curve. In many instances the curves are characterized by the absence of a peak point. Fabric unit interference during relative displacement of the units also contributes to the shape of the curves. A higher degree of physical interference is found in the unbonded flocculated fabric.

In drained strength tests where no pore pressures are generated, the reorientation of fabric units or the breakdown of bonds and the consequent reorientation of the units together with physical interference will define the stress—strain curves. In an undrained test where pore pressures are generated, restricted fabric unit orientation due to the hydrostatic pressure in the pore fluid

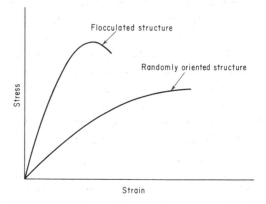

Fig.10.14. Influence of structure on stress-strain relationships.

will influence total interaction and consequently provide a different set of parameters for shear strength. This must be considered in the evaluation of shear strength parameters. It is important to realize that in laboratory testing of soils for evaluation of shear strength the pore pressure generated in the laboratory test sample must be that anticipated in the field. If a drained condition is expected in the field under actual loading conditions, this type of test must be performed in the laboratory. In this case, shear strength development is a function of the bonding agents and the physical arrangement of fabric units without generation of pore pressure. The significant property is that arising from physical breakdown of bonding agents and es bonds. For undrained tests, resistance to fabric unit rearrangement may be provided through total pore-pressure response, which would produce different yield characteristics of the soil sample. If one is able to accurately measure the actual generated pressures in the fluid in the soil sample, it would be possible to compute what pressures might be due to physical breakdown and reorientation of particles. This would provide a more realistic evaluation of soil performance.

Shear strength interpreted from interparticle forces

In postulating a mechanism for shear strength based on interparticle forces of attraction and repulsion, Lambe (1960) lists the following forces acting between particles: F_m = force where contact is mineral–mineral; F_a = force where contact is air–mineral; F_w = force where contact is water–mineral or water–water; R' = electrical repulsion between particles; A' = electrical attraction between particles.

Equating the sum of these forces acting across the potential plane of failure to the combined mineral stress, and expressing the equation in terms of stresses rather than forces, one obtains the equation for the combined normal stress across the failure plane σ_n as:

$$\sigma_n = \bar{\bar{\sigma}} a_m + p_a a_a + u a_w + R - A \qquad (10.13)$$

where: $\bar{\bar{\sigma}}$ = effective stress between minerals derived from and associated with F_m; a_m = unit contact area for mineral–mineral; p_a and a_a = unit pressure and unit area directly associated with F_a; a_w = unit area directly associated with F_w; R and A represent unit electrical forces of repulsion and attraction respectively.

Lambe's equation (eq.10.13) represents an idealized breakdown of the distribution of stresses induced in the soil (and acting across the potential shear plane) as a result of the applied normal stresses. It is not possible to measure all of the component items. As a simplification, we can reduce eq.10.13 to:

$$\sigma_n = \bar{\sigma} + u + (R - A) \qquad (10.14)$$

where u = pore water pressure. Designating $[u + (R - A)]$ as $-u_m$, eq.10.14 becomes similar to eq.10.2.

Measurements of resistance to direct translatory shear on samples of clays of known composition, with known differences in interparticle forces, may be used to relate shear strength to interparticle forces. From this relationship, it is possible to separate the interparticle forces responsible for shear strength and to set up a model for their action (Warkentin and Yong, 1960). Shear strength is not simply related to the interparticle forces in terms of forces of attraction minus forces of repulsion. Swelling and non-swelling clays behave differently. Swelling clays, in which repulsion can be the dominant interparticle force, retain strength even when wet. When non-swelling clays are wetted to the same water content, the samples fall apart. This difference is apparent in the marked tendency for non-swelling clays such as kaolinite to erode under moving water.

Fig.10.15 shows the shear strength values measured for sodium montmorillonite, a high-swelling clay, at different void ratios and at two salt concentrations. In this clay, single particles may be the fabric units. At the higher salt concentration 1.0 M NaCl, interparticle repulsion is lower, attraction is higher and shear strength is lower. In Fig.10.16, calcium montmorillonite with a lower swelling, hence a lower net repulsion, and higher attraction, has the lower shear strength at the same void ratio or same average interparticle distance. For the calcium clay, the fabric unit is a tactoid of several particles (see Chapter 2).

In all of these measurements, shear strength increases as net repulsion increases. This increase in repulsion is accompanied by an increase in soil suction (Chapter 3), so that a greater strength would be predicted from effective stresses.

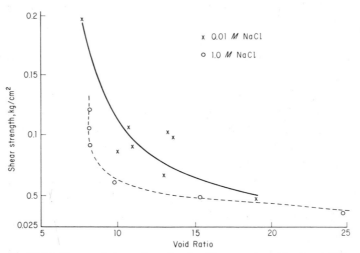

Fig.10.15. Measured shear strength of sodium montmorillonite at different salt concentrations in the pore water. (Warkentin and Yong; 1962.)

However, this does not explain the existence of shear strength in a clay where the swelling shows that a net repulsion between particles exists.

During shear, particles and fabric units in the failure plane become oriented parallel to each other, and any force resisting this orientation should contribute to shear strength of remoulded clays. Fig.10.17 suggests how interparticle repulsion can play this role for high-swelling clays. For low salt concentrations in this kind of clay, the single particle model is applicable.

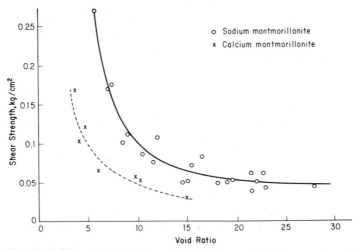

Fig.10.16. Measured shear strength of sodium and calcium montmorillonite. (Warkentin and Yong, 1962.)

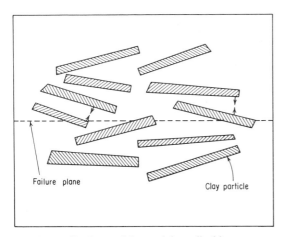

Fig.10.17. Single particle model applicable to montmorillonite, demonstrating interparticle repulsion resisting particle rearrangement during development of failure plane. (Warkentin and Yong, 1962.)

Formation of the failure plane requires particle movement, which in many instances requires one particle to move closer to another. Interparticle repulsion resists this movement, accounting qualitatively for measured strength values. It is difficult to apply this concept quantitatively because the original particle orientation is not known and this generally changes as void-ratio changes. Particle orientation becomes more parallel as the void ratio decreases, so that while the net repulsion can be measured (as the swelling pressure), the amount of movement required in the failure plane cannot be specified.

The configuration at failure of soil particles in a clay soil requires minimal particle interference along the failure plane. Thus an arrangement parallel or near

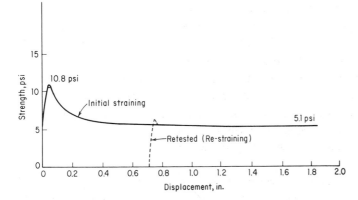

Fig.10.18. Immediate direct shear retesting of a failed clay sample. (Skempton, 1964.)

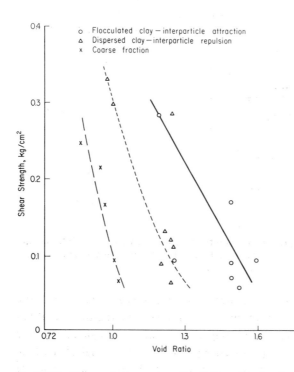

Fig.10.19. Measured shear strength of flocculated and dispersed kaolinite and of clay- plus silt-size fraction of kaolinite. (Warkentin and Yong, 1962.)

parallel to the failure plane is favoured. The resistance to the parallel arrangement accounts for the major component of strength. In Fig.10.18, a failed clay sample tested at strains beyond the failure strain shows little or no measurable change in the stress—strain relationship beyond failure — thus indicating a minimal particle-interference pattern. This is the residual strength concept (Skempton, 1964) and is similar to the concept developed in Fig.9.7 in Chapter 9. In the residual stress range, we note that there is no further change in soil volume, which confirms the postulate of minimal fabric unit or particle interference.

Fig.10.19 shows measured shear strength values for kaolinite, a low-swelling clay. Highest shear strength was measured for the flocculated clay, where attraction is the dominant interparticle force. The random fabric clay, with net repulsion, will have a lower strength at the same void ratio. Interparticle forces are still important in determining shear strength, as shown by the low value for the coarse fraction of the clay.

The forces of attraction between fabric units leading to flocculation and establishment of es and ee bonds are not sufficiently strong to contribute appreciably to shear strength in a direct way, i.e. in resisting separation of fabric units. Interparticle attraction could keep the units in a certain orientation, contributing to strength by resisting the rearrangement required to form a failure plane. If the flocculated fabric is visualized as a random arrangement of fabric units with some es bonding, forces of attraction would resist the rearrangement to parallel orientation because of the es bonds. In the random clay, repulsion keeps the units in an arrangement which is more nearly parallel to begin with, thus utilizing the ss bonding arrangement, and requiring less force and less movement to rearrange units in the failure plane.

Shear strength can thus be interpreted on the basis of interparticle forces for remoulded and unbonded clays. These forces lead to specific fabric unit arrangements and stability and disruption of soil fabric would account for resistance to shear. This depends upon both the force required to move the units and the amount of movement in the failure plane. For bonded soils, cementation bonds and other bonding agents together with es and ee bonding are dominant. Destruction of these bonds would result in subsequent control of behaviour being conditioned by interparticle forces.

The schematic diagram shown in Fig.10.20 interprets the fabric arrangement of clays in general. Also shown is an idealized representation of a typical set of forces and bond mechanisms for the system.

The distinction between the unbonded and bonded soils can be seen in the absence or presence of strong bonds established by cementation, carbonate and/or

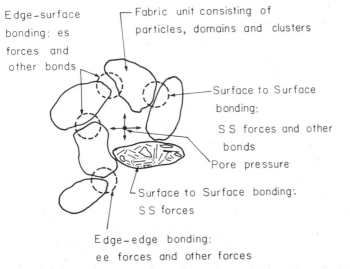

Edge−surface bonding: es forces and other bonds

Fabric unit consisting of particles, domains and clusters

Surface to Surface bonding: S S forces and other bonds

Pore pressure

Surface to Surface bonding: S S forces

Edge−edge bonding: ee forces and other forces

Fig.10.20. Forces and bond action in soil stability. Note that "other" bonds include cementation, carbonate and organic complexing bonds. For unbonded soils, "other" bonds do not influence behaviour.

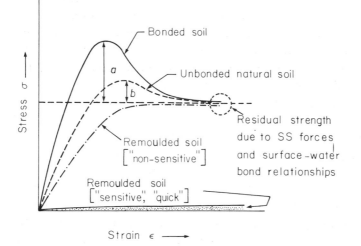

Fig.10.21. Comparison of stress–strain curves for bonded and unbonded clays. Magnitude of *a* is due to structural ("other") bonds plus interparticle force bonds plus mechanical interference. Magnitude of *b* is due to interparticle force bonds plus mechanical interference. Note the remoulded performance for both "non-sensitive" and "sensitive" clays.

organic bonds, as described previously. These are sometimes called "structural" bonds and are rigid if they consist primarily of cementation and carbonate bonds. This is because the bonds will break at very low strains, i.e. the bonds are brittle bonds. The clay organic bonds are generally thought to be semi-rigid or "plastic", i.e. the bonds can allow a certain amount of stretch before breakage. The interparticle force bonds consisting of ee, es and ss electrical forces can be thought to be plastic.

A demonstration of the contribution by structural bonds to strength characterization may be seen in Fig.10.21. The strength increases above residual for the bonded and unbonded clays are shown in Fig.10.21 as *a* and *b*. Residual strength is due to: (1) mechanical interference between particles and fabric units adjacent to the failure plane; (2) ss forces; and (3) surface energy relationships established between particles and fabric units with water.

The magnitude of *b* is due to interparticle force bonds (ee, es, and ss bonds) plus any structural bonding available. By definition, unbonded clays possess little or no structural bonds. The contribution to clay soil strength from structural bonds is shown by the magnitude of *a*. Included in *a* are the resistance mechanisms which typify *b* and the residual strength. In passing, we note that the loss of *a* or *b* characterizes the sensitivity of a soil, which is defined as:

$$\text{sensitivity} = \frac{\text{undisturbed unconfined compression strength}}{\text{remoulded or disturbed strength}}$$

We note that for the "non-sensitive" soil shown in Fig.10.21, a finite value for sensitivity can be obtained[1]. In the case of the "sensitive" clay, it is not unusual to obtain values for sensitivity of from 8 to infinity.

The influence of structural bonds on development of resistance to shear or deformation can be studied by selectively destroying the structural bonds without destroying or disrupting the fabric and all the other interparticle force relationships. An example is the work of Kenney (1967) on removal with EDTA (a disodium salt of ethylene diamine-tetra-acetic acid) of iron compounds which act as cementing agents.

If the initial bonding of natural soils is disrupted or broken due to pretesting procedures, such as application of a cell-confining pressure, the magnitude of a or b may become obscured or lost. La Rochelle and Lefebvre (1971) show that for a strongly bonded clay, the stress–strain curves obtained are conditioned by initial cell-confining pressure. Fig.10.22 shows the stress–strain curves obtained in drained triaxial testing. In Fig.10.22a, for $\sigma_{\text{conf.}} < \sigma_{\text{p.c.}}$ where $\sigma_{\text{p.c.}} =$ preconsolidation pressure, a large portion of the structural bonds are intact and these will contribute to strength, as shown previously in Fig.10.21. At $\sigma_{\text{conf.}} = \sigma_{\text{p.c.}}$, the stress–strain curves in Fig.10.22b show that structural bonding is not an important factor in the characterization of shear resistance beyond the yield point. Resistance to deformation will involve the residual shear mechanisms previously explained for Fig.10.21.

Fig.10.22c and d show that when $\sigma_{\text{conf.}} > \sigma_{\text{p.c.}}$, initial preconditioning due to application of $\sigma_{\text{conf.}}$ not only destroys initial structural bonds, but also precompresses the clay soil such that application of a shearing force would invoke resistance mechanisms due primarily to mechanical interference from adjacent fabric units. At this stage, the stress–strain curve resembles the characteristics of work hardening material. The response behaviour of the material is not dissimilar to that of remoulded clays. Thus stress–strain characterization and yield or failure can be adequately described at this point by the mechanisms for remoulded clays. Residual strength for the material can only be attained at very large deformation since the initial state of the material shows in-situ water contents greater than the liquid limit. The problem with high initial water contents is that the mechanisms previously described for residual shear cannot be invoked until fabric unit

---

[1] The general classification for sensitive clays shows that medium sensitive clays have values of sensitivity of from 2 to 4, and very sensitive clays possess values of from 4 to 8. For values above 8, the clay is defined as "quick" clay.

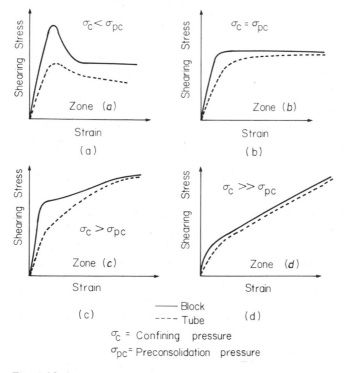

Fig.10.22. Stress–strain curves for bonded clay at various confining pressures. (LaRochelle and Lefebvre, 1971.)

separation distances become smaller i.e. at lower water contents. This instance means that a considerable amount of water will have to be driven out during continued straining before the residual strength mechanisms become operative.

The Mohr-Coulomb diagram for bonded clays described in Fig.10.21 and 22 show that where the preconditioning effects do not destroy the structural bonds, the failure envelope obtained does not describe a linear relationship between $(\bar{\sigma}_1 - \bar{\sigma}_3)/2$ and $(\bar{\sigma}_1 + \bar{\sigma}_3)/2$ as shown in Fig.10.23. The region a shown in Fig.10.23 corresponds to Fig.10.22a. Correspondingly, b, c and d in Fig.10.23 relate to Fig.10.2b, c and d. As noted previously, when $\sigma_{conf.} > \sigma_{p.c.}$, we can treat the material behaviour in terms of a remoulded clay model. This will be discussed in a later section when we deal with yield and failure of cohesive soils.

The results shown in Fig.10.23 are in many respects similar to those obtained for overconsolidated clays (Fig.10.11 and 10.12), where region b represents the preconsolidation pressure region, and the behaviour in region a is typical of overconsolidated clays.

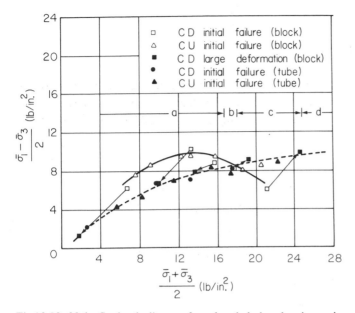

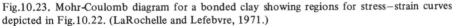

Fig.10.23. Mohr-Coulomb diagram for a bonded clay showing regions for stress—strain curves depicted in Fig.10.22. (LaRochelle and Lefebvre, 1971.)

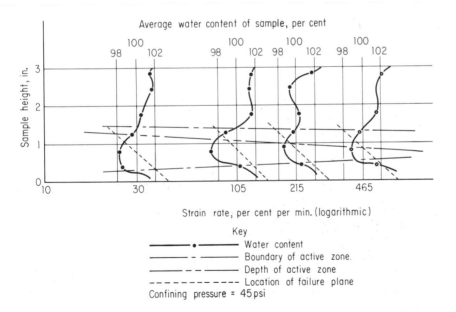

Fig.10.24. Correlation of active-zone depth with strain rate for kaolin. Note that the water content scale gives 100% as the reference for original water content. Thus, for example, 98%, refers to 98% of original water content. (Leitch and Yong, 1967.)

10.5 STRENGTH AND SOIL STRUCTURE

Fabric and soil-water potential

The results from a series of shear tests at varying rates of strain are shown in Fig.10.24. The change in water content due to failure is shown plotted against strain rate. The degree of water content change may be considered indicative of the

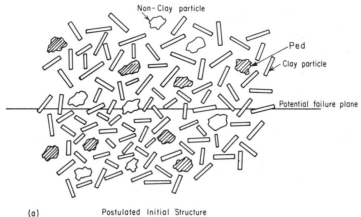

(a)   Postulated Initial Structure
                              (Bell Clay)

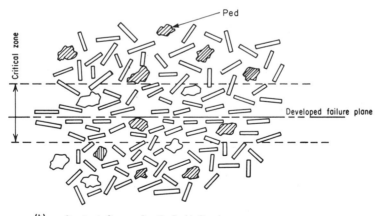

(b)   Structural Changes Due To Rapid Shearing

Fig.10.25. Postulated structure change during shearing of kaolinitic clay. a. Initial structure. b. Structure after rapid shearing. (Leitch and Yong, 1967.) Note that forces acting between fabric units for characterization of structure are not shown in the diagram.

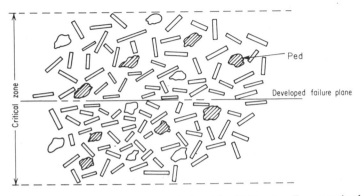

Fig.10.26. Structure after slow shearing of kaolinitic clay. Forces acting between particles and fabric units are not shown int he diagram. (Leitch and Yong, 1967.)

degree of change in original fabric. The probable change in configuration for the critical zones (defined as the zone of water content change) is shown for two cases in Fig.10.25 and 10.26. Under rapid shear, the critical zone (see Fig.10.24) is smaller, but the change in water content is higher. Thus, greater fabric distortion within that zone occurs. The fabric units within this zone are rearranged into parallel array to a greater degree than under slow shear (cf. Fig.10.25 and 10.26). Leitch and Yong (1967) show that computations of the amount of work needed to redistribute the water content for the samples shown in Fig.10.24 can be directly correlated with the input work required to shear the samples to failure. For

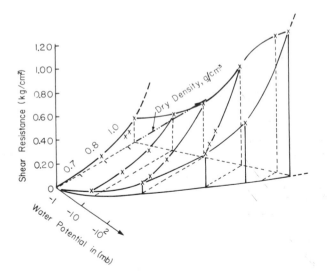

Fig.10.27. Shear strength surface related to dry density and soil-water potential for a remoulded kaolin clay. (Yong et al., 1971.)

unbonded soils, the relationship is a direct measure between input energy and energy stored within the soil fabric which is reflected in the soil-water potential. The application of external work to shear a sample distorts the soil fabric, breaking or changing the bond arrangements shown in Fig.10.20 with the result that changes in the free energy of the soil water will occur. The relationship between the soil-water potential and shear resistance for a kaolin clay is shown as a three-dimensional surface in Fig.10.27. Similar surfaces can be obtained for other soils through laboratory measurements of soil-water potential, volume change, and shear strength. The need for an analytical shear strength parameter such as that described in Section 10.1 does not arise. We take advantage of the fact that the sets of forces resisting deformation in the sample can be characterized by the prevailing energy state. In addition, full saturation of soil is not required since measurement of soil-water potential does not require complete water saturation.

Interpretation of fabric change from Mohr-Coulomb diagram

The stress paths shown in Fig.9.8 (Chapter 9) may be used to interpret fabric charges during shear. In a drained test, since no pore pressures are generated during application of $\Delta\sigma$, $A = 0$ since $\Delta u_1 / (\Delta\sigma_1 - \Delta\sigma_3) = 0$. In an undrained test, if we assume that $\Delta u_1 = (\Delta\sigma_1 - \Delta\sigma_3)$, then $A = 1$. The drained and undrained stress paths shown in Fig.10.28 are interpreted to show the domains for A in relation to the stress paths described.

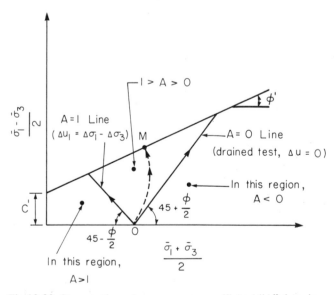

Fig.10.28. Stress paths and pore-pressure coefficient "A" domain.

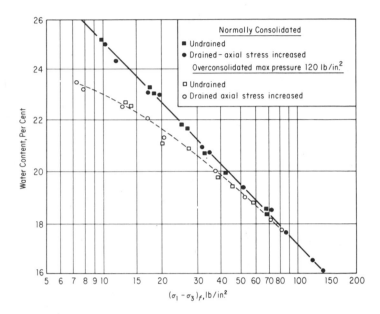

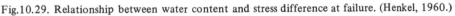

Fig.10.29. Relationship between water content and stress difference at failure. (Henkel, 1960.)

To interpret fabric changes in drained tests, the A lines are examined in relation to Δu_1. For $A < 0$, we will observe that $\Delta u_1 < 0$. Thus dilatation occurs if no constraints on volume change are imposed. The region $A < 0$ indicates that test samples demonstrating a stress path in this region would show expanding fabric changes. Correspondingly, stress paths in the region $A > 1$ would indicate large contracting fabric changes. Parts of any stress path, e.g., *OM* in Fig.10.28, can also be examined, where the slopes of the increments are compared to the A lines for the appropriate interpretation of expanding or contracting fabric change.

The relationship between strength and water content shows that strength increases linearly as water content decreases for normally consolidated clays (Fig.10.29). For overconsolidated clays, this relationship is not linear but lies slightly below the relationship for normally consolidated samples at higher water content. The two lines converge at high strengths.

The general relationship between the ratio of the major to the minor principal effective stress $(\bar{\sigma}_1/\bar{\sigma}_3)$ and the ratio of the maximum consolidation pressure to the average effective stress, (J_m/J_f) is shown in Fig.10.30. The drained and undrained test results of compression and extension tests of Weald clay and London clay all lie on the same curve (Henkel, 1960).

If the results of drained and undrained tests were plotted in the stress plane given by the diagram in Fig.10.31 and Fig.8.14 (p.280), the stress paths for undrained tests on normally consolidated clays and overconsolidated clays will be

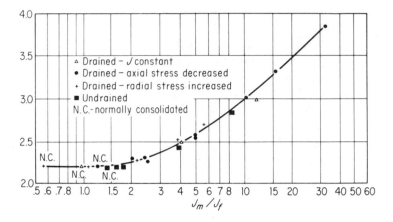

Fig.10.30. $\overline{\sigma}_{1f} / \sigma_{3f}$ vs. J_m/J_f relationship. (Henkel, 1960.)

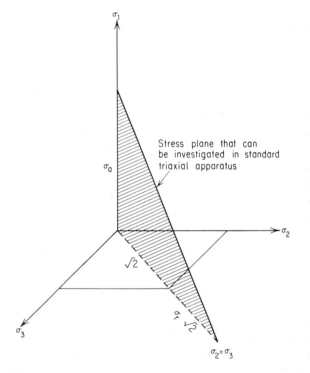

Fig.10.31. Stress plane for triaxial investigation. (Henkel, 1960.)

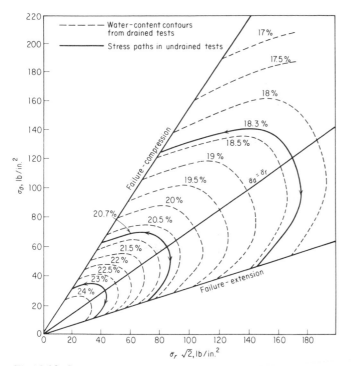

Fig.10.32. Stress vectors for normally consolidated specimens. (Henkel, 1960.)

shown for compression and extension tests as in Fig.10.32 and 10.33. Contours of constant water content similar in form to the stress paths shown in Fig.10.28 may be constructed from drained compression and extension tests. The undrained stress paths fit in very well with the contours from the drained tests for both normally consolidated and overconsolidated clays. Using this diagram, known as the Rendulic diagram (Rendulic, 1937), the unique relationship between the effective stress and the water content becomes evident. However, the main difference between normally consolidated and overconsolidated clays lies in the shape of the stress paths. The similarity in stress path characteristics between Fig.10.32, 10.33 and 10.28 is due to the fact that similar mechanisms are involved, even though the test data are expressed in different ways. Compressional changes in fabric occur in Fig.10.32 whilst Fig.10.33 indicates that at water contents above 19.5%, dilatational effects dominate and fabric expansion occurs. Expansion of fabric leads to higher void ratios and water contents if volume expansion is allowed during shear. As the cell pressure reaches $\sigma_{p.c.}$, the stress paths begin to show fabric compressional effects to the point that where $\sigma_{conf.} > \sigma_{p.c.}$ in Fig.10.33, the stress paths will be similar to those shown in Fig.10.32.

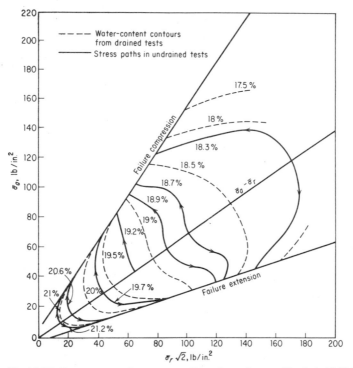

Fig.10.33. Stress vectors for overconsolidated specimens. (Henkel, 1960.)

Anisotropic effects

Anisotropy in general refers to material structure and/or properties which do not exhibit the same characteristics and/or properties in every direction. In soil usage, we must distinguish between material structure and property anisotropy, in addition to external anisotropy constraints such as stress anisotropy. Thus:

(1) *Material structure anisotropy* — relates primarily to anisotropy of fabric which would influence development of interparticle force relationships.

(2) *Property anisotropy* — refers to strength, compressibility, permeability, conductivity and other mechanical properties which are not equal in all directions, i.e. the material property demonstrated is a function of the sample tested.

(3) *External constraint anisotropy* — refers particularly to applied stresses and boundary constraints. This consideration is not within the scope of treatment of this book.

Fabric anisotropy was discussed in Chapter 4, where we noted that two levels of fabric anisotropy can be obtained, namely fabric unit anisotropy which requires only that preferred orientation of fabric units be achieved, and total anisotropy

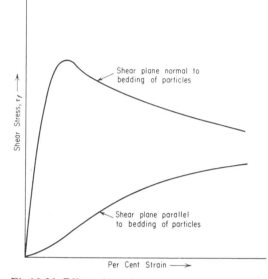

Fig.10.34. Effect of particle orientation on stress–strain relationships for a precompressed clay.

where both fabric units and particles within the units are oriented in the same direction. Because of stress preconditioning, as for example from surcharge loading and consolidation effects, fabric anisotropy occurs which would result in material structure anisotropy. Material property anisotropy will also develop because of structure and fabric anisotropy. This is shown, for example, as a property interrelationship between material structure (soil-water potential) and strength as in Fig.10.27.

As an example of strength derived from orientation in a clay soil, the results of shear tests on a highly-precompressed glacial clay are shown in Fig.10.34. The stress–strain relationship performed such that the potential shear plane was perpendicular to the bedding of the clay fabric units and the second series such that the potential shear plane was parallel to the bedding of the clay sample. The increase in strength for the tests in which the potential shear plane was oriented normal to the bedding is due not only to developed interparticle forces but also to the difficulty in reorientation of the fabric units during shear. With an anisotropic arrangement there is a greater resistance to reorientation of particles for definition of a failure plane, since the units must be rotated through a 90° angle. There will be a sharp rise in shear strength, to a peak point defined by rotation and reorientation of the fabric units, following which the system will possess strength in the form of ss forces for slip along the potential shear plane. This reduced strength is similar to the case of shear parallel to the bedding plane. Resistance due to displacement and

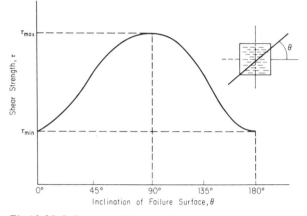

Fig.10.35. Influence of directional shear on shear strength of a well-oriented clay fabric.

reorientation of fabric units and particles is at a minimum, and because of the greater facility for slip along the potential shear plane we will obtain lower shear stress with correspondingly higher shear strain. The variation in strength due to orientation of the failure plane for a near parallel configuration is shown in Fig.10.35.

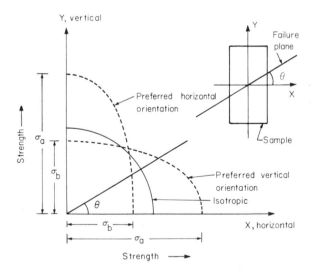

Fig.10.36. $X-Y$ orientation of strength of anisotropic soils relative to preferred X or Y orientation of fabric units in soil samples.

Another way of expressing property anisotropy information is to plot the material property variation in terms of orientation relative to the x- or y-coordinate property as in Fig.10.36. The strength variation for an isotropic material is also shown for comparative purposes. The x or y preferred orientation of fabric units can be deduced directly from the strength trace with θ orientation shown in Fig.10.36. The greater the σ_a / σ_b ratio, the greater is the degree of anisotropy.

In characterizing the strength parameters for anisotropic soils, it is important to bear in mind that both stress and strain path dependencies are reflected in the development of material resistance. Saada and Zamani (1969) show that the analytic stress parameters can vary considerably with θ variation which obviously

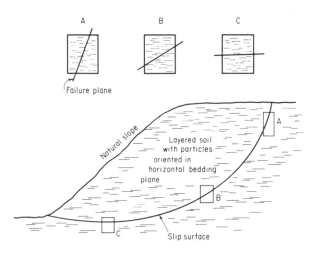

Fig.10.37. Influence of particle bedding on shear strength in analysis of slope stability: note directional shear and bedding of particles.

reflects variations in stress and strain path effects on fabric distortion. Application of anisotropic-strength considerations by Lo (1965) to slope stability computations show that anisotropy influences become increasingly more significant with flatter slopes. This is due to the fact that the failure arc in a slope failure will be dependent on height and steepness of slope. Fig.10.37 shows a medium steep slope of a fully oriented clay deposit. By varying the failure arc above or below the one shown, using the same toe point we observe that the two variations can be either small or large, i.e. large for steep slopes and small for shallow slopes. Thus, if σ_a/σ_b is significant, anisotropy becomes an important consideration for a small θ variation since we can be sampling more of σ_b as opposed to σ_a for shear resistance.

10.6 SOME METHODS AND MECHANISMS FOR LABORATORY EVALUATION OF STRENGTH PARAMETERS

The scope of this book does not allow us to study in detail all the presently available methods and mechanisms for laboratory evaluation of the shear strength parameters. Of the several methods presently in use, the ultimate choice depends on the availability of facilities and the particular type of analysis to which one is partial. It is important to remember, and this cannot be overemphasized, that the *c* and *ϕ* parameters obtained from evaluation of the test results are only *demonstrated mechanical properties*, peculiar to the test system and the method used for analysis of results. The relationship between these mechanical properties and actual microscopic properties is not easily defined since there does not exist common agreement as to what friction or cohesion means in clay soils, and further, test systems developed to date suit those who are biased toward one or another concept of strength parameters.

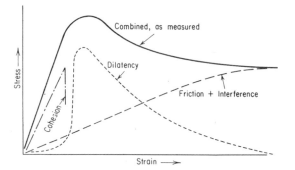

Fig.10.38. Components of shear resistance. (Lambe, 1960.)

Lambe (1960) divides the components of shear resistance into three basic components, namely cohesion, dilatancy, and friction. The separation in terms of contribution among the three components is not too distinct. Qualitatively, this is shown in Fig.10.38. Cohesion is mobilized at very small strains and ceases to be operative at some higher strain. Dilatation increases from zero to a maximum value, but after particle interference is overcome, this component must decrease. When the stress–strain curve approaches an asymptotic value, since cohesion and dilatation are no longer important contributors to shear strength, we must conclude that "friction" is responsible for the continued shear strength of the soil.

Separation of shear strength into cohesion and friction parameters

In the technique proposed by Schmertmann and Osterberg (1960) the

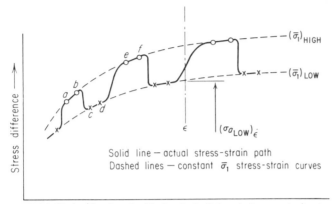

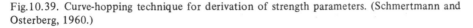

Solid line — actual stress-strain path
Dashed lines — constant $\bar{\sigma}_1$ stress-strain curves

Axial Strain (Developed with Constant Rate of Compression) ⟶

Fig.10.39. Curve-hopping technique for derivation of strength parameters. (Schmertmann and Osterberg, 1960.)

induced pore-water pressure is varied to maintain a constant, preselected value of $\bar{\sigma}_1$. In Fig.10.39 points a and b are obtained during strain at $\bar{\sigma}_1$ (high) after which the induced pore pressure is changed to maintain a new constant value of $\bar{\sigma}_1$ (low). Points c and d may then be recorded for this value of $\bar{\sigma}_1$ (low). This procedure of alternating between two chosen values of $\bar{\sigma}_1$ permits the derivation of two stress—strain curves from one test. Values of $(\sigma_1 - \sigma_3)/2$ and σ_3 are plotted to obtain straight-line relations at various strains (Fig.10.40). Application of this method requires that the assumption of validity of the Mohr-Coulomb failure criterion at any strain. This assumption is not totally valid, as will be shown in a later section. However, if we treat the strength parameters as analytic parameters only, the following treatment can be pursued. Rewriting $\kappa\bar{\sigma}_p$ in eq.10.12 as c', both

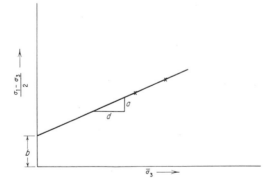

Fig.10.40. Computation for cohesion and friction parameters. (Schmertmann and Osterberg, 1960.)

cohesion c' and friction ϕ' may then be obtained mathematically from the expression:

$$\frac{\sigma_1 - \sigma_3}{2} = \left(\frac{\sin \phi'}{1 - \sin \phi'}\right) \bar{\sigma}_3 + c' \left(\frac{\cos \phi'}{1 - \sin \phi'}\right)$$

where:

$$\text{friction parameter } \phi' = \sin^{-1}\left(\frac{a}{a + d}\right)$$

and:

$$\text{cohesion parameter } c' = b \left(\frac{1 - \sin \phi'}{\cos \phi'}\right)$$

The variation of the analytic cohesion and friction parameters with axial compressive strains is shown in Fig.10.41. With the curve hopping technique, we note that the cohesion component generally develops its maximum value at a very-low compressive strain, while the friction component requires a much greater strain to reach its maximum value.

The stress locus technique used by Yong and Vey (1962), interprets the effective shear strength parameters of normally consolidated clays in terms of the measured stress on the potential failure plane. In this technique, an attempt at

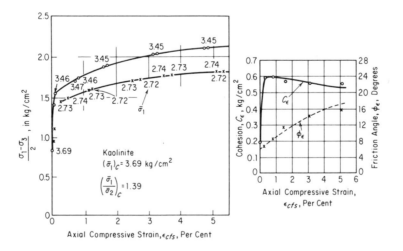

Fig.10.41. Derivation of ϕ and c with relation to strain. (Schmertmann and Osterberg, 1960.)

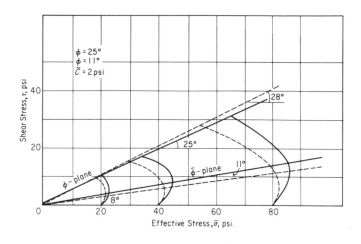

Fig.10.42. Stress loci for Chicago clay. (Yong and Vey, 1962.)

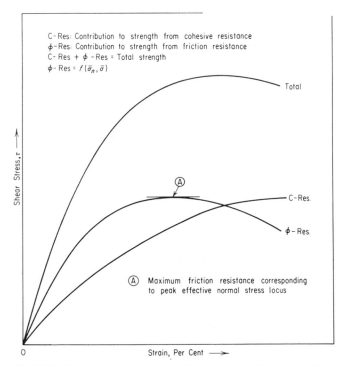

Fig.10.43. Strength parameter contributions to total strength. (Yong and Vey, 1962.)

obtaining a mechanistic interpretation of the analytic parameters is made. Defining friction as that property in the shear stress versus normal stress region wherein the shear stress increases with the effective normal stress and the cohesion parameter $\bar{c}$ as the parameter in the same region where the shear stress increases despite a decrease in the effective normal stress, the relative influence of these parameters will be evident from an examination of the stress locus (Fig.10.42). For any one particular confining pressure, the cohesive component will be given by the vertical distance of the intersection of the stress locus with the ϕ and $\bar{\phi}$ lines.

On a shear stress versus strain plot (Fig.10.43), the ϕ component increases rapidly in the initial stages of load application. It continues to increase until it reaches a maximum while the cohesion component is still increasing. Beyond the peak for the friction component, the total shear strength will be due mainly to the increase in the cohesive component. By this definition and interpretation of the stress locus, the mechanical interference term in residual strength development (Fig.10.21) will be the friction component and the surface force relationships with water and ss forces will provide the "cohesion" component.

Reaction rates and rate process

To apply the theory of specific reaction rates, we consider a cohesive soil to be composed of particles or fabric units in water and behaving as a soil-water solution. The relationships for an ideal solution provide us with the free-energy difference ΔF as:

$$\Delta F = \Delta F^\circ + RT\Sigma n \, \ln a_i \tag{10.15}$$

where: ΔF° = free-energy difference at standard state; R = gas constant; T = absolute temperature; n = number of moles in solution; a_i = activity coefficient of ith ion or particle.

For $\Delta F = 0$, we will obtain:

$$\Delta F^\circ = - RT \ln K_p \tag{10.16}$$

where: K_p = equilibrium constant for the ideal gas at one atmospheric pressure.

By differentiating eq.10.16 with respect to T and subsequently multiplying through by T and making the appropriate substitution:

$$T \left(\frac{\partial (\Delta F^\circ)}{\partial T} \right)_p = \Delta F^\circ - RT^2 \frac{d \ln K_p}{dT} \tag{10.17}$$

For the case where all substances are in their standard state, the Gibbs-Helmholtz relationship gives us:

$$\Delta F^\circ - \Delta H^\circ = T \left(\frac{\partial(\Delta F^\circ)}{\partial T} \right)_p \qquad (10.18)$$

where: ΔH° = standard change in Heat Content, i.e. heat content change at standard state.

Eq.10.17 and 10.18 can be combined, and ΔH substitutes for ΔH° without specifying standard states to give:

$$\frac{d\ln K_p}{dT} = \frac{\Delta H}{RT^2} \qquad (10.19)$$

Eq.10.19 is the Van't Hoff relationship. In terms of energy:

$$\frac{d\ln K_p}{dT} = \frac{\Delta E}{RT^2}$$

where: ΔE = energy change; $K_c = K/K'$ = reaction rate; K and K' = specific reaction rates.

Thus:

$$\frac{d\ln K}{dT} - \frac{d\ln K'}{dT} = \frac{\Delta E}{RT^2} \qquad (10.20)$$

Van't Hoff suggests that eq.10.20 can be split as follows:

$$\frac{d\ln K}{dT} = \frac{E}{RT^2} + \bar{A} \quad \text{and} \quad \frac{d\ln K'}{dT} = \frac{E'}{RT^2} + \bar{A}$$

where: $E - E' \doteq \Delta E$; $\bar{A}$ = constant.

If $\bar{A} = 0$, taking only the first of the above equation:

$$\frac{d\ln K}{dT} = \frac{E}{RT^2} \qquad (10.21)$$

eq.10.21 is the Arrhenius equation. Thus:

$$\ln K = - \frac{E}{RT} + \text{const.}$$

$$K = A \exp\left(- \frac{E}{RT}\right) \tag{10.22}$$

where A = constant describing frequency of reaction.

Eq.10.22 can also be obtained by way of activated energy states. Consider the reaction given as:

$$AB + C \rightarrow ABC \rightarrow A + BC \tag{10.23}$$

Fig.10.44 demonstrates the reaction process. If $E_3 > E_1$, the reaction is called endothermic; if $E_3 < E$, the reaction is exothermic.

In a soil-water system, the application of external stressing will cause the fabric units to move if the soil sample deforms or distorts. The change in energy state brought about by the external stressing system is achieved by the change in positions of the fabric units in the soil sample. The amount of energy needed to achieve the energy change can be examined as in the theory of reaction rates beginning with eq.10.15 or by the rate process mechanism utilizing the concept of activation energy. The probability that a flow unit made up of a representative fabric unit will possess an energy in excess of E (activation energy) at temperature T is related to the Boltzmann factor $\exp(-E/RT)$.

Restricting the energy to translational motion requirements in shear displacements, and introducing a frequency factor A which describes the total frequency of encounter between adjacent flow units $K = A \exp(-E/RT)$ as in eq.10.22.

In the treatment given by Mitchell et al. (1969), the approach which stems from Eyring's rate process theory (Eyring, 1936) has been expanded from initial considerations of Murayama and Shibata (1961) to provide the rate of deformation of a soil as:

$$\dot{\varepsilon} = X \frac{\beta T}{h} \exp - \frac{\Delta F}{RT} \exp \frac{f\lambda}{2\beta T} \tag{10.24}$$

where: $\dot{\varepsilon}$ = rate of deformation = equivalent to k in eq.10.22; X = function of flow units in the direction of deformation; β = Boltzmann constant; h = Planck's constant; f = average shear force acting on each flow unit; λ = separation distance between successive equilibrium positions.

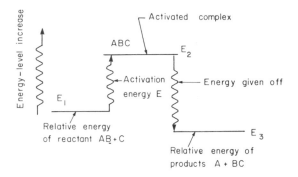

Fig.10.44. Energy of activation and reaction state.

The calculations that can be made using either eq.10.22 or the specialized soil equations from Mitchell et al. (1969) represented in eq.10.24 will lead to the estimation of the number of mathematical bonds and their strength. The distance between successive equilibrium positions λ is presumed to be equal to a representative flow unit, i.e. a fabric unit. The emphasis on the term mathematical bond is due to the fact that the physical presence of a bond and its real identity need not be established. The expanded Eyring model will lead to the calculation of these bonds from available experimental information suitable for analysis using eq.10.24. This treatment is useful in that a measure of compatibility between analytical and physical behaviour models is achieved. The results obtained for example have been given in Fig.10.13.

10.7 YIELD AND FAILURE

In this section, we will continue the examination of yielding and failure begun in Chapter 8 and will evaluate cohesive soil performance in the light of available information of material behaviour and relevant theories. The treatment for this section relies on previous developments by Yong and McKyes (1971) and Yong et al. (1972).

The difficulties surrounding the application of suitable analytical techniques for description of soil yielding and failure, where failure is identified as the terminal stage of yielding, centre around the problem of finding or developing specific analytical models which can accurately describe the actual physical behaviour of the material under stress. We showed in Fig.8.1, Chapter 8, that in ideal elastic-plastic behaviour, a linear elastic-stress–strain relationship is obtained until the point of yield identified as B in Fig.8.1d, following which subsequent irrecoverable shear deformation occurs under no stress increase. The ideal plastic behaviour beyond point B requires that no volume change occurs during the process of deformation after initial yield. One implication from this theory is that the point of yield is defined independently of stress path. Applications of plasticity theory to analysis of soil mechanics problems require certain adaptations — see, for example, Drucker and Prager (1952), and Shield (1955). These are directed towards an attempt to reconcile the consequences of plasticity theory where some account for increases in yield shear stresses with compressive stresses is recognized. The consideration of normality of the plastic strain-increment vector to the yield surface in stress space discussed in the last part of Section 8.5 requires the assumption that the plastic potential surface, to which the strain-increment vectors are always normal, coincides with the yield surface for ideal isotropic plastic behaviour. If a yield surface varies with mean normal stress, as shown in Fig.10.45 the theory shows that volume changes accompanying shear flow are a direct result. The deviatoric planes which are normal to the space diagonal, (identified as π planes in Fig.10.45) will show varying sizes of yield surfaces dependent on the position of the π plane from the origin of axes. To obtain information which would facilitate development of suitable yield and failure criteria, we begin by examining stress–strain relationships from true triaxial tests.

Experimental three-dimensional stress–strain relationships for unbonded and bonded clays may be examined in terms of octahedral shear stress and strain components (Fig.10.46 and 10.47):

$$\tau_{oct} = \frac{1}{3}\left[(\sigma_1 - \sigma_2)^2 + (\sigma_2 - \sigma_3)^2 + (\sigma_3 - \sigma_1)^2\right]^{1/2}$$

$$\gamma_{oct} = \frac{1}{3}\left[(\varepsilon_1 - \varepsilon_2)^2 + (\varepsilon_2 - \varepsilon_3)^2 + (\varepsilon_3 - \varepsilon_1)^2\right]^{1/2} \tag{10.25}$$

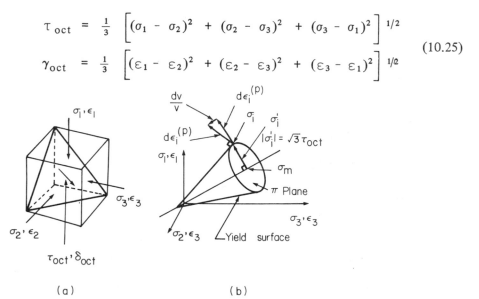

(a)   (b)

Fig.10.45. Principal and octahedral shear stresses and strains. a. On material element in physical space. b. In principal stress and strain space. (Yong and McKyes, 1971.)

where τ_{oct} is octahedral shear stress and γ_{oct} octahedral shear strain, and $b = (\bar{\sigma}_2 - \bar{\sigma}_3)/\bar{\sigma}_1 - \bar{\sigma}_3)$, i.e. b introduces the effect of intermediate principal stresses.

The results shown in Fig.10.46 and 10.47 represent many tests where sample stiffnesses vary according to initial preconditioning due to initial applied stresses

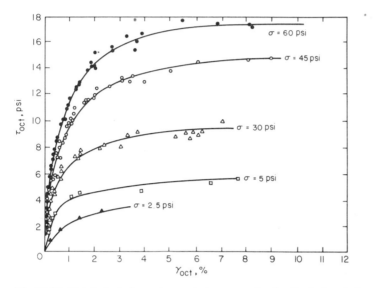

Fig.10.46. Octahedral shear stress—strain curves for kaolin (unbonded and remoulded) clay. (Yong et al., 1972.)

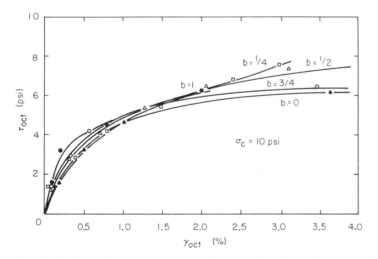

Fig.10.47. Octahedral shear stress—strain curves for Champlain (bonded) clay. (Yong et al., 1972.)

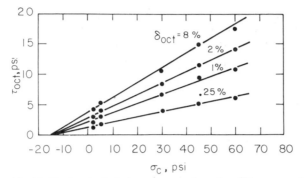

Fig.10.48. Octahedral shear stress values for different strains as influeneed by consolidation pressure for kaolin clay. (Yong and McKyes, 1971.)

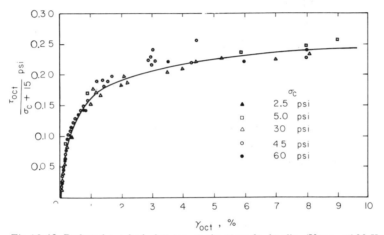

Fig.10.49. Reduced octahedral stress—strain curve for kaolin. (Yong and McKyes, 1971.)

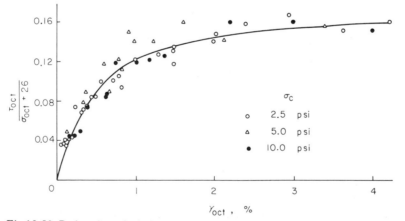

Fig.10.50. Reduced octahedral stress—strain curve for Champlain clay. (Yong et al., 1972.)

prior to shearing. The common relationship which seeks to relate material stiffness to shear resistance must first succeed in reducing the varied stress–strain curves to a single family or even to one relationship as follows:

$$\gamma_{oct} = f(\tau_{oct}, \sigma_{conf}, b,) \tag{10.26}$$

In the technique adopted by Yong and McKyes (1971), τ_{oct} is plotted against σ_c (i.e. $\sigma_{confining\ pressure}$) for various strains. Fig.10.48 shows the typical diagram achieved from interpretation of the results shown in Fig.10.46.

By using the linear stress–strain relation with average consolidation pressure, together with the common point of intercept on the abcissa, we can reduce the various stress–strain curves shown in Fig.10.49 and 10.50 to a common locus of reduced octahedral shearing stress versus octahedral shearing strain. The reduced octahedral stress parameter takes the following form:

$$\bar{\tau} = \frac{\tau_{oct}}{\sigma_{oct} + const.} \tag{10.27}$$

The curves shown in Fig.10.49 and 10.50 demonstrate that a functional relationship for eq.10.26 does exist. The scatter of results for the Champlain clay which is larger than that for the kaolin clay is due to the bonded nature of the natural Champlain clay.

Plasticity analysis

The reduced octahedral shear-stress parameter–strain curves of Fig.10.49 and 10.50 can be analyzed using a modified form of plasticity theory. The yield function would be of the Von Mises type with a required modification to account for the change in stiffness of the clays due to increasing consolidation pressure. The formulation of the modified plastic strain-increment relationship may begin with either (1) the Von Mises work hardening plastic relations given by Hill (1950) or Mendelson (1968), or (2) a particular solution of the generalized work hardening plastic strain-increment relation of Drucker (1960).

The result is the same in both cases for materials exhibiting yield behaviour independent of the average normal stress:

(1) Von Mises plastic yielding

$$d\varepsilon_{ij}^p = \sqrt{\frac{3}{2}}\, \sigma'_{ij}\, \frac{d\gamma_{oct}^p}{\tau_{oct}} \tag{10.28}$$

if:

$$H' = \sqrt{\frac{2}{3}} \frac{d\tau_{oct}}{d\gamma_{oct}}$$

we will obtain:

$$d\varepsilon_{ij}^P = \frac{3}{2} \frac{\sigma'_{ij} d\tau_{oct}}{H' \tau_{oct}} \tag{10.29}$$

where:

$d\varepsilon_{ij}^P$ = plastic strain-increment tensor;

$\sigma'_{ij} = \sigma_{ij} - \frac{1}{3} \sigma_{ii} \delta_{ij}$ = the deviatoric component of the stress tensor;

(a)

(b)

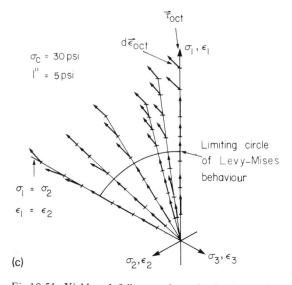

(c)

Fig.10.51. Yield and failure surfaces for kaolin. a. In principal stress space. b. On deviatoric plane in principal stress space. c. Incremental strain vectors and stress paths on a π plane in stress–strain space. (Yong et al., 1972.)

$$d\gamma_{oct}^{p} = \frac{1}{3}\left[(d\varepsilon_1 - d\varepsilon)^2 + (d\varepsilon_2 - d\varepsilon_3)^2 + (d\varepsilon_3 - d\varepsilon_1)^2\right]^{1/2}$$

$$= \text{plastic increment of octahedral shear strain;}$$

$$H' = \sqrt{\frac{2}{3}} \text{ times the slope of the octahedral shear stress–strain curve.}$$

(2) Drucker work hardening plasticity:

$$d\varepsilon_{ij}^{p} = G \frac{\partial f}{\partial \sigma_{ij}} \frac{\partial f}{\partial \sigma_{mn}} d\sigma_{mn} \tag{10.30}$$

If $f = \tau_{oct}$ we will obtain:

$$\frac{\partial f}{\partial \sigma_{ij}} = \frac{1}{3}\frac{\sigma_{ij}}{\tau_{oct}}, \qquad d\varepsilon_{ij}^{p} = \frac{G}{3}\frac{\sigma'_{ij} d\tau_{oct}}{\tau_{oct}} \tag{10.31}$$

where:

$$G = \frac{9/2}{H'} = \frac{9}{2}\sqrt{\frac{3}{2}}\frac{d\gamma_{oct}^{p}}{d\tau_{oct}}$$

If we now denote:

$$H^* = \sqrt{\frac{2}{3}} \frac{d\bar{\tau}}{d\gamma_{oct}^p} = \sqrt{\frac{2}{3}} \frac{d\tau_{oct}}{d\gamma_{oct}^p} \frac{1}{(\sigma_c + \text{const})}$$

$$= \frac{H'}{\sigma_c + \text{const}}$$

eq.10.29 becomes:

$$d\varepsilon_{ij}^p = \frac{3}{2} \frac{\sigma'_{ij} \, d\tau_{oct}}{H^* (\sigma_c + \text{const}) \tau_{oct}} \tag{10.32}$$

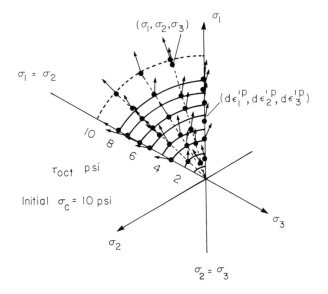

Fig.10.52. Incremental strain vectors and stress paths on a π plane in stress–strain space for Champlain clay.

The validity of eq.10.32 for shear strain increments in the unbonded kaolin remoulded clay has been adequately shown by Yong and McKyes (1971). The deviatoric-stress vectors and increments of strain vectors in the deviatoric plane show that when the limit of Prandtl—Reuss behaviour occurs, isotropic behaviour terminates (Fig.10.51b). According to eq.10.29 and 10.32 the components of these vectors should be proportional, and hence the vectors should be co-linear. In addition, eq.10.30 requires plastic strain-increment vectors to be normal to the yield surface, since the components are proportional to those of the yield function gradient $\partial f/\partial \sigma_{ij}$. As the yield function is τ_{oct} in eq.10.31 this requirement gives the same results as that of eq.10.29 because the yield function is a circle in principal stress space (Fig.10.51b) and vectors normal to the circle are co-linear with radial stress position vectors.

The influence of bonding in the bonded Champlain clay is significant in the control of strain directions. In Fig.10.52 the strain-increment vectors and stress points are plotted for Champlain clay initially consolidated to 10 psi. Referring to Fig.10.23, a σ_c value of 10 psi is less than $\sigma_{p.c.}$ This puts the material behaviour in category a in Fig.10.23 and 10.22. Thus we can confirm from the Figures that in the case of the bonded clay, for $\sigma_c < \sigma_{p.c.}$, yielding at low strains cannot be categorized or analyzed by plasticity theories. Normality of the strain-increment vectors to the yield surfaces and co-linearity with stress vectors are not established, as compared to the unbonded remoulded clay shown in Fig.10.51b. Thus, eq.10.32 will not fit the situation where initial structural bonding exists. The resistance to

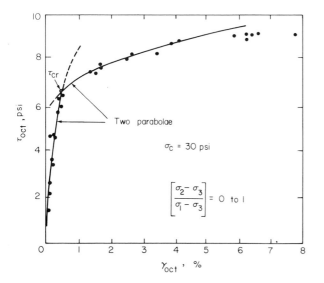

Fig.10.53. Octahedral shear stress—strain curve for kaolin. (Yong et al., 1972.)

shear deformation due to the bonded nature of the clay will provide a cemented character to the material.

Yielding of bonded and unbonded clays

From the typical stress deformation result shown in Fig.10.46 it is apparent that two separate parabolae may be conveniently fitted to the test points (Fig.10.53). The results shown in Fig.10.46 are representative of all the tests performed using various stress paths. Preconditioning of the material through preconsolidation effects will produce different stress–strain characteristics. The corresponding effect of variation in the consolidating cell pressure σ_c will change the nature of the resultant parabolae. The linear relationships shown in Fig.10.54 are obtained by reducing the test data to accommodate the basic properties of parabolae. These demonstrate that a major change in material characteristics occurs in the region identified as the "yield" point. The characteristics of the yield point are related to the physical yielding of the real material. For soft and medium clays, nonlinearity in stress and strain is more the rule than the exception. Since initial irrecoverable deformations will occur under the smallest of loads, the effective yield behaviour of clays must be defined consistent with both physical behaviour and analytical requirements. The critical point, defined by stress τ_{cr} in Fig.10.54 is consistent with that presented in Fig.10.51, for the definition of the limit of Prandtl-Reuss behaviour. This point is significant and demonstrates compatibility between physical behaviour and analytical requirements.

For the bonded clay we recall from Fig.10.52 that since non-normality and non-co-linearity occur, the definition of a workable yield point becomes all the more important. The same technique developed for the remoulded clay (Fig.10.54) can also be used to define τ_{cr}. However, the technique for bonded clays is more empirical since Fig.10.49 shows that in actuality, yielding in the "plasticity" sense cannot be established for tests on bonded clays where confining pressure σ_c is less than the preconsolidation pressure $\sigma_{p.c.}$. We will accept that the technique demonstrates consistency in unbonded clays and empirical realization for bonded clays.

The results shown in Fig.10.54 indicate that as the density of the material increases the nonlinear stiffness K of the material correspondingly increases. Thus:

$$\gamma_{oct} = k\tau_{oct}^2, \qquad \tau_{oct} < \tau_{cr}$$

Whilst the octahedral yield stress τ_{cr} is not influenced by the intermediate principal stress, strain conditions appear to obey the following relationship:

$$\varepsilon_2 - \varepsilon_3 = \frac{\sigma_2 - \sigma_3}{\sigma_1 - \sigma_3}(\varepsilon_1 - \varepsilon_3)$$

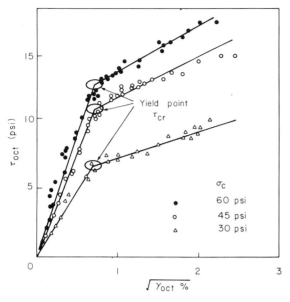

Fig.10.54. Octahedral shear stress-root of strain curves for kaolin showing influence of confining pressure. (Yong et al., 1972.)

In undrained triaxial testing $\varepsilon_1 + \varepsilon_2 + \varepsilon_3 = 0$. Hence:

$$\varepsilon_1 = \frac{1}{3}(2\sigma_1 - \sigma_2 - \sigma_3)k\tau_{oct} = \sigma'_1 k\tau_{oct}$$

$$\varepsilon_2 = \frac{1}{3}(2\sigma_2 - \sigma_1 - \sigma_3)k\tau_{oct} = \sigma'_2 k\tau_{oct} \qquad (10.33)$$

$$\varepsilon_3 = \frac{1}{3}(2\sigma_3 - \sigma_2 - \sigma_1)k\tau_{oct} = \sigma'_3 k\tau_{oct}$$

The values of τ_{cr} and K are obtained from experimentation, and the stress–strain relationships applicable for the region below yield can be found from eq.10.33.

Failure

Using the π representation shown in Fig.8.2 (p.265), we can plot the failure stress envelopes (failure surfaces) for the unbonded and bonded clays tested and shown in Fig.10.46 and 10.47. The failure surfaces are shown in Fig.10.55 and 10.56 together with the stress paths taken. For comparative purposes, we show data obtained from other sources. These confirm that for unbonded clays, of medium

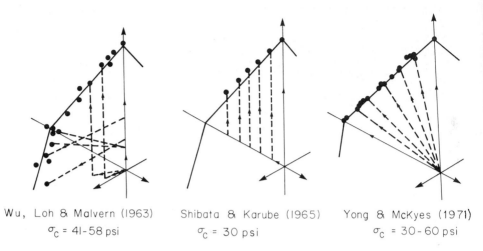

Fig.10.55. Failure stress envelopes for kaolin clay and other insensitive clays.

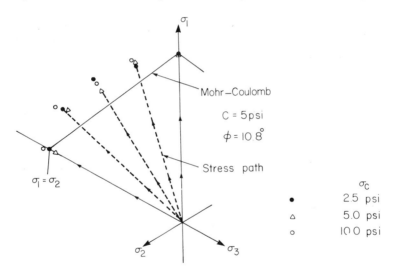

Fig.10.56. Failure stress envelope for Champlain clay with $\sigma_c < \sigma_{p.c.}$ (Yong et al., 1972.)

Fig.10.57. Failure plane in kaolin soil subject to true triaxial testing. A. Through thin section using polarizing microscopy. Failure plane is shown by the light diagonal band running down from right to left in the picture. B. Montage of scanning electron micrographs showing the failure plane. (Yong and McKyes 1971.) Note arrangement of particles adjacent to the failure plane

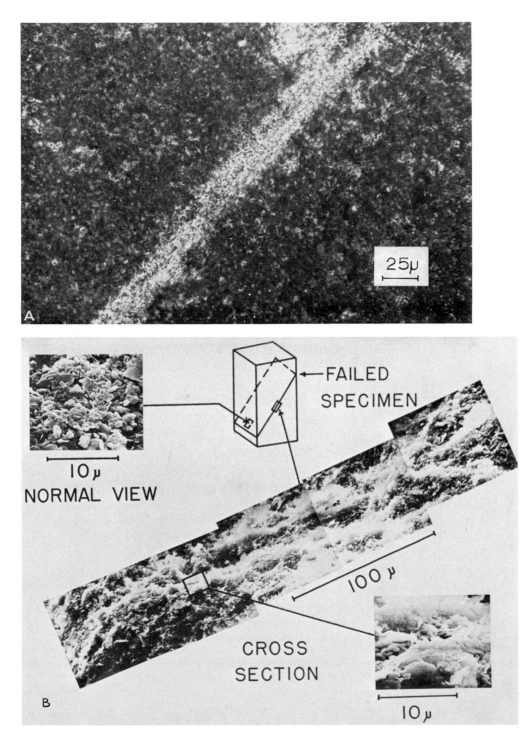

A

B NORMAL VIEW

FAILED SPECIMEN

CROSS SECTION

Fig.10.58. Transmission electron micrograph of failure plane shown in Fig.10.57. This picture shows the orientation of the clay particles in the failure zone. Note the presence of fabric units of various sizes.

plasticity if proper accounting is given to boundary energy and dilatational energy corrections, the Mohr-Coulomb failure theory can adequately describe failure. The conditions for application of the theory are strict and must be considered in careful terms for general application.

In bonded clays, when $\sigma_c < \sigma_{p.c.}$ the presence of structural bonds will deny admissibility of the Mohr-Coulomb failure theory as shown in Fig.10.56. When $\sigma_c > \sigma_{p.c.}$ the stress–strain curves are adequately represented by elastic-plastic and workhardening models and failure conditions are not unlike those obtained for unbonded and remoulded clays. This is due to the fact that at $\sigma_c \geq \sigma_{p.c.}$ the structural bonds initially present are destroyed prior to shear application, thus making the bonded clay similar to an unbonded clay.

In order to establish physical compatibility with the frictional Mohr-Coulomb analytical rupture model, electron microscopy may be used to determine the absence or presence of oriented particles and fabric units which characterize frictional slip lines in failed clay specimens. The montage shown in Fig.10.57 and the transmission picture in Fig.10.58 bear out the correspondence of theory with physical fact and indicate the Mohr-Coulomb model to be quite consistent with the actual mode of clay rupture.

10.8 SUMMARY

The summary given in Section 8.6 can equally apply here. The points of particular interest relate to obtaining a proper insight into the deformation behaviour of cohesion soils to allow for the rational application or development of the corresponding analytical model. The presence of bonding forces creating structural bond effects will provide an initial rigidity to the cohesive soil which will not permit the rigorous application of the Mohr-Coulomb failure theory, Until such bonds are broken, failure of the bonded clay will be conditioned by the rigidity of the bonds. No adequate failure theory exists at the present time for such cases, although Mohr-Coulomb approximations can be, and have been used for test conditions where $\sigma_c < \sigma_{p.c.}$. The strict condition for application shows that the Mohr-Coulomb criterion will be applicable for bonded clays when test conditions allow for $\sigma_c > \sigma_{p.c.}$. We note that application to the field situation requires that the same σ_c conditions be met.

Chapter 11

SOIL FREEZING AND PERMAFROST

11.1 INTRODUCTION

In regions where surface temperatures fall below the freezing point of water, freezing of the pore water in the soil is possible. This occurs in land areas located close to the Arctic and Antarctic. These areas constitute the frozen belt. Subsoils within this belt may be completely frozen through the entire year, or may be frozen only in the winter months. The limits of the frozen belt defined as the frost boundary are not rigidly established because of the fluctuation in winter temperatures. Fig. 11.1 shows the approximate limits of subsoil permanently frozen through the entire year.

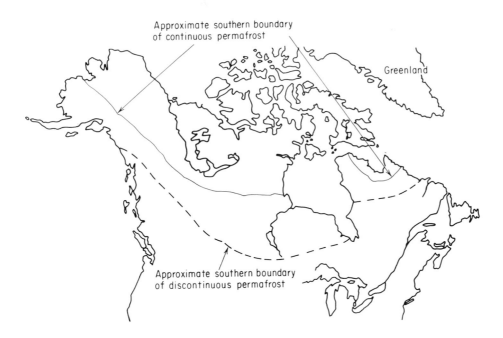

Fig.11.1. Approximate boundary for permafrost for North American continent.

11.2 GEOTHERMAL PROFILE

The variation of soil temperature with depth from the ground surface is generally referred to as the geothermal profile. Its characteristics at a particular location and time are dependent on the net transfer of heat between the ground surface and air, and the thermal properties of the subsoil. The heat-transfer mechanisms include radiation, convection and conduction.

At the ground surface, i.e. air–ground interface, radiation and convection–conduction mechanisms are responsible for the net heat transfer. Fig. 11.2 shows the idealized operative mechanisms at the interface for a condition of bright sun with little cloud cover.

The primary factors and variables involved in characterizing the net heat transfer at the interface are:

Radiation:

(1) Location and view factor: this accounts for geographical location, altitude and exposure view to the sun.

(2) Albedo: ground surface texture and charactistics which determine resultant absorbtivity to radiation.

(3) Greenhouse effect: cloud cover responsible for longwave re-radiation (reflection).

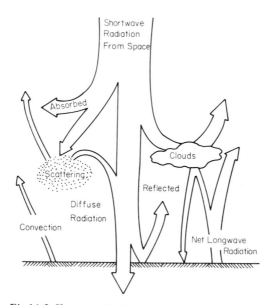

Fig. 11.2. Heat transfer between ground surface and air on a sunny day. (Aldrich, 1956.)

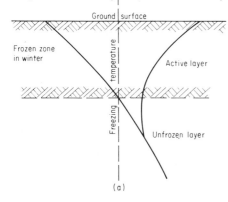

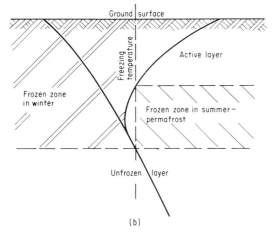

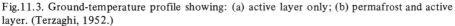

Fig.11.3. Ground-temperature profile showing: (a) active layer only; (b) permafrost and active layer. (Terzaghi, 1952.)

(4) Cloud cover and atmospheric vapour pressure: determination of equivalent sunshine and solar absorption and scattering.

Convection–conduction:

(1) Surface features and inclination: this includes vegetative cover, snow cover, surface roughness and other features which would influence the characteristics of wind or air flow at the surface.

(2) Wind velocity.

(3) Evaporation and condensation.

In view of the above, it is apparent that the ground-surface temperature need not be the same as ambient air temperature. The variation of subsurface temperature with depth may be plotted for any one time to show the limits of penetration of the cold front into the soil at any one location. A cold front which freezes the pore water in the subsoil is generally defined as the *frost front*. In Fig. 11.3 two conditions of the frost front are given to show the effect of deep penetration of the front.

The *active layer* shown in Fig. 11.3 is the layer of subsoil that freezes in winter and thaws in summer. The thickness of this layer depends upon several factors, but chiefly upon the penetration of the frost front into the subsoil. If frost penetration is deep and if suitable conditions prevail, then it is possible to have a layer of soil that will remain frozen in summer. This is shown in Fig. 11.3b. The permanently frozen subsoil layer is defined as the permafrost layer. Although the thickness of this layer depends on the thermal soil properties, the primary factors are duration, intensity and penetration of the frost front. The permafrost layer has been found to vary in thickness from a few inches to hundreds of feet.

About one-fifth of the land area of the world is underlain by permafrost. These areas lie close to or border the Arctic and Antarctic Oceans and the thickness of permafrost will naturally vary according to the location of the area. The closer one gets to the North or South Pole, the greater will be the thickness of the permafrost layer. In contrast to this, the further one is away from the North or South Pole, the thicker will be the active layer.

Permafrost regions can be broken up into two distinct zones identified in terms of lateral continuity of the permafrost layer. Where the permafrost layer is laterally continuous, the region is identified as a continuous permafrost zone. The discontinuous permafrost zone contains non-continuous permafrost which may at times be over one hundred feet. Limited permafrost occurrence (islands or patches of permafrost) in "non-permafrost" regions are identified as "sporadic" permafrost.

Since it is difficult to physically drill into the subsoil to establish continuity of permafrost, various criteria have been used to delineate the division between the various zones (Brown, 1970). In the main, the methods rely on a particular isotherm ($-5°C$) of mean annual ground temperature measured just below the zone of annual variation.

Changes in climate will cause permafrost to thaw, and ultimately disappear if warming conditions prevail. The term *perennially frozen ground* appears to be better suited to describe "frozen ground" in general without connotations of permanency.

Since the active layer freezes and thaws seasonally, *frost action* is a direct function of freezing of the active layer.

11.3 FREEZING INDEX

The variation of temperature with time can be plotted graphically to show the intensity of temperature in terms of duration of existing temperature. Fig. 11.4 shows the idealized typical temperature–time curves for ambient air temperature and various kinds of ground surface influence on measured surface temperatures. As pointed out in Section 11.2 the various transfer mechanisms operative at the interface contribute to the differences in observed surface temperatures. A surface correction factor can be used to render the surface temperature equal to ambient air temperature. Obviously, as seen from Fig. 11.2, there will be many different surface correction factors to account for differences in ground-surface features, stratigraphy, soil type, thickness of ground cover, and all the various other aspects of surface, subsurface and air interactions.

Since there is the local fluctuation in temperature between day and night, the mean daily temperature is used in Fig. 11.4. The freezing-temperature intensity is the area contained by the curve below freezing and is defined as the number of degree days below freezing. This is a time–temperature phenomenon which shows as a finite quantity, the temperature intensity factor vital to the consideration of frost penetration and its associated problems.

The temperature intensity, considered in terms of degree days below freezing, is an indication of the length of the freezing period, which coupled with the magnitude of the surface freezing temperature, contributes directly to the penetration of the frost front. The term *freezing index F* refers specifically to the number of degree days below freezing and is computed with the aid of a graph similar to that shown in Fig. 11.4.

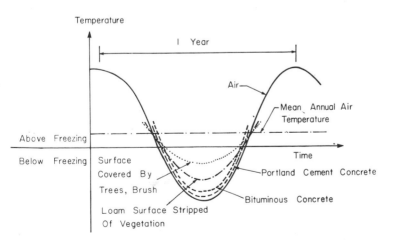

Fig.11.4. Idealized temperature curves for various surfaces. (Aldrich, 1956.)

The mean daily temperature is plotted for a period of one year. The freezing index F is the area defined and bounded by the temperature curve in the below freezing range. If T_f represents the mean surface temperature during the freezing period t, then the freezing index F is $T_f{\cdot}t$ degree days. It is evident that the higher F is, the greater will be the frost penetration. Corresponding considerations for thawing can be made by considering the area contained by the curve shown above freezing in Fig. 11.4. The degree days above freezing indicate the length of the thawing period (pertinent in permafrost areas) and is defined as the thawing index.

11.4 FROST PENETRATION

As mentioned previously frost penetration, or the advance of the frost front into the subsoil, depends upon several factors. There are:

(1) Freezing index and associated temperature and climatic factors.

(2) Ground surface cover and topography.

(3) Soil type and grain size distribution.

(4) Thermal properties of the soil-water system: (a) specific heat of the mineral particles, S_p; (b) volumetric heat of the system, C; (c) latent heat of the pore water, L; (d) thermal conductivity of the soil, k.

(5) Nature of the pore water.

Except for the freezing index, associated temperature and climatic factors, and surficial terrain features, the factors mentioned are intrinsic properties and characteristics of the soil-water system.

The units for specific heat are generally given as cal. g^{-1} $°C^{-1}$ or Btu lb.$^{-1}$ $°F^{-1}$. This represents the quantity of heat required to raise a unit mass of material one degree (Fahrenheit or Centigrade). The volumetric heat is dependent upon the specific heat and can be obtained by multiplying the specific heat by the dry density of the material. Because of the presence or absence of ice it is different for frozen and unfrozen soils. For unfrozen soils, if C_u is the volumetric heat, then

$$C_u = \gamma_d C_{soil\ particles} + \frac{w}{100} \gamma_d C_{water} \tag{11.1}$$

where: γ_d = dry density of the soil; w = water content.

For frozen soils, the volumetric hear C_f is given as:

$$C_f = \gamma_d C_{soil\ particles} + \frac{w}{100} \gamma_d C_{ice} \tag{11.2}$$

The heat content and the change in thermal energy of a soil-water system as it freezes or thaws depends upon C_u, C_f and L. Fig. 11.5 shows the relationship

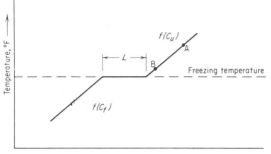

Fig.11.5. Heat content or thermal-energy diagram for an idealized soil-water system.

between these factors. The thermal-energy change is linear with temperature both below and above freezing but is interrupted by the latent heat of fusion L of the pore water at the freezing point. While it is recognized that there could be dissolved foreign matter in the pore water, the latent heat L for the pore water is generally taken to be that of water. The latent heat of fusion L represents the change in thermal energy per unit volume of soil when the soil water freezes or thaws.

Diffusion

Consideration of idealized heat flow through soil in terms of heat transmission is a preliminary step towards estimation of depth of frost penetration. As pointed out previously, heat transfer can be achieved by means of radiation, conduction, or convection. In partially saturated soils, convection and conduction are the more likely mechanisms for heat transfer, with conduction playing the major role. In fully saturated soils, however, radiation is reduced to a negligible quantity and heat transfer may be analyzed solely on the basis of heat conduction without appreciable errors. In this instance, this is the transfer of kinetic energy from the molecules in the heat sink (from the warm portion of subsoil at lower depths) to those in the cooler portion of the subsoil closer to the ground surface. If the temperatures of all bodies are considered in relative terms, all physical bodies can be considered as heat storages. If the temperature surrounding the bodies is much lower than that of the bodies, some of the thermal energy contained within the bodies will be released in order to maintain thermal equilibrium.

Considering temperatures above or below the freezing point, the change in thermal energy between two arbitrary points A and B in Fig. 11.5, is given as follows:

$$u_A - u_B = C(T_A - T_B) \tag{11.3}$$

where: C = volumetric heat of the body; u = thermal energy; T_A, T_B = temperature corresponding to thermal energy states of u_A and u_B respectively.

While the general term for C has been used in eq. 11.3, it is understood from Fig. 11.5 and eq. 11.2 and 11.3 athat either C_u or C_f should be used as the case may be.

For small changes, by letting $u_A - u_B$ approach zero in eq. 11.3, we obtain:

$$du = C\, dT \qquad \text{or} \qquad \frac{\partial u}{\partial T} = C \qquad\qquad (11.4)$$

In the case of heat transfer by conduction, Q is given by the Fourier equation which has a form similar to Darcy's equation (Chap. 5). This is not altogether surprising since these are laminar flow laws relying on the principle of conduction. Hence:

$$Q = kiA = k\,\frac{T_A - T_B}{L}\,A$$

More basically:

$$q = -k\,\frac{\partial T}{\partial x} \qquad\qquad (11.5)$$

where: Q = heat transfer; i = thermal gradient; k = thermal conductivity; A = area; $q = Q/A$, heat flux; x = depth taken as downward from top of ground surface.

To estimate the rate and depth of frost penetration, thermal continuity may be considered. For simplicity, only one dimensional flow of heat, in the x direction, is considered. In the absence of freezing or thawing, i.e. without introduction of the term L as shown in Fig. 11.5, it follows from the conservation of thermal energy that:

$$\frac{\partial u}{\partial t} + \frac{\partial q}{\partial x} = 0$$

where t refers to the particular instant of time. From eq. 11.4,

$$\frac{\partial u}{\partial t} = C\,\frac{\partial T}{\partial t}$$

By making the appropriate substitutions, and since:

$$\frac{\partial q}{\partial x} = -k\,\frac{\partial^2 T}{\partial x^2}$$

therefore:

$$\frac{\partial T}{\partial t} = \frac{k}{C} \frac{\partial^2 T}{\partial x^2}$$

Defining k/C as "a" the thermal diffusivity constant:

$$\frac{\partial T}{\partial t} = a \frac{\partial^2 T}{\partial x^2} \tag{11.6}$$

Eq. 11.6 is the diffusion equation, which represents the temperature profile in the subsoil at any instantaneous time, as for example in Fig. 11.7. Its importance and its role will be demonstrated in the prediction of the depth of frost penetration.

Estimation of depth of frost penetration

The simplest assumption that may be made for estimation of frost penetration into the subsoil is that the variation of temperature from the top of the ground surface to the frost front is linear, and that the temperature remains constant below the frost line. This is the Stefan model. Fig. 11.6 shows these assumptions, which provide for an approximate analysis.

At a point A on the frost front, the equation of continuity must be satisfied. This requires that the latent heat released as the pore water freezes to a depth Δx in

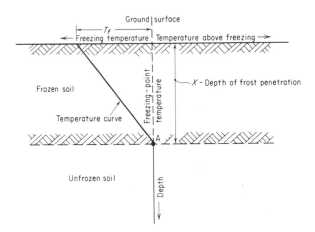

Fig.11.6. Thermal conditions assumed for the Stefan model. (Aldrich, 1956.)

time Δt must be equal to the rate of heat conducted to the ground surface. For small calues of Δx and Δt this may be written as:

$$k_f \frac{T_f}{x} = L \frac{dx}{dt} \tag{11.7}$$

where: T_f = temperature below freezing as shown in Fig. 11.6; L = latent heat of the pore water; $L(dx/dt) = q$; $T_f/x = i$. Thus:

$$\frac{k_f}{L} \int T_f \, dt = \frac{x^2}{2}$$

Therefore:

$$x = \sqrt{\frac{2k_f \int T_f dt}{L}}$$

$\int T_f dt$ in degree hours is equal to the freezing index F reported in degree days with the appropriate correction from Fig. 11.4. Therefore:

$$x = \sqrt{\frac{48 k_f F}{L}} \tag{11.8}$$

Eq. 11.8 is the Stefan equation derived solely from the linear model shown in Fig. 11.6. The limitations other than those contained in the linear variation of temperature from the ground surface to the frost line and the constant temperature below the frost line, include the volumetric heat factors of both the frozen and unfrozen soil. It has been found that prediction of the depth of frost penetration, x, made on this basis tends to overestimate the actual penetration.

The use of the diffusion equation provides a more rigorous analysis for estimation of frost penetration. This method of treatment was developed by Berggren (1943) and later modified by Aldrich and Paynter (1953). In the modified Berggren model, the diffusion equation is used to define the subsoil temperature profile as shown in Fig. 11.7. The soil temperature profile is drawn for any instantaneous time t. The thermal properties for both frozen and unfrozen soil layers are given in Table 11.1.

For any reasonable mathematical analysis, it must be assumed that conditions are analyzed for a particular time, which may then be integrated to cover the interval under consideration.

In the frozen soil layer, the ground-temperature profile is given by the diffusion equation (eq. 11.6) with the appropriate thermal constants as follows:

$$\frac{\partial T_f}{\partial t} = a_f \frac{\partial^2 T_f}{\partial x^2}$$

TABLE 11.1

Notation of thermal properties of frozen and unfrozen soils

	Unfrozen soil	Frozen soil
Thermal conductivity	k_u	k_f
Volumetric heat	C_u	C_f
Diffusivity coefficient	$a_u = k_u / C_u$	$a_f = k_f / C_f$

Similarly, in the unfrozen soil layer:

$$\frac{\partial T_u}{\partial t} = a_u \frac{\partial^2 T_u}{\partial x^2}$$

must be valid.

At a point A on the frost front, the equation for continuity must be satisfied. This requires that the net rate of heat flow from the frost front must be equal to the latent heat supplied by the soil water as it freezes to a depth Δx in time Δt.

Hence for vanishingly small values of Δx and Δt,

$$L \frac{dx}{dt} = \Delta q$$

where Δq is the net rate of heat flow at the frost interface. Therefore:

$$k_f \frac{\partial T_f}{\partial x} - k_u \frac{\partial T_u}{\partial x} = L \frac{dx}{dt} \qquad (11.9)$$

The solution for x, the depth of frost penetration into the subsoil, as defined from the boundary conditions stated has been given by Aldrich and Paynter as:

$$x = \lambda \sqrt{\frac{48kF}{L}} \qquad (11.10)$$

The correction coefficient λ depends on the thermal properties of both the frozen and unfrozen soils. Values for λ can be obtained from Fig. 11.8.

The dimensionless fusion parameter μ is given as:

$$\mu = \frac{C_f}{L} T_f \triangleq \frac{C_f F}{Lt} \qquad (11.11)$$

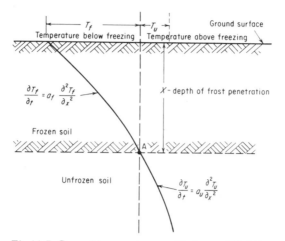

Fig. 11.7. Ground-temperature profile for modified Berggren model. (Aldrich, 1956.)

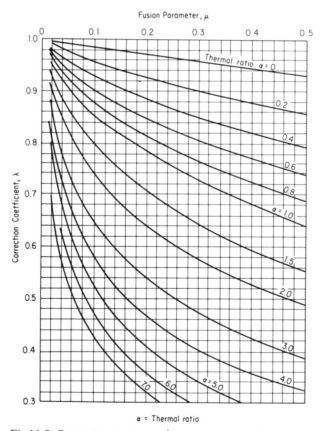

Fig. 11.8. Correction coefficient, λ, for eq. 11.10. (Aldrich, 1956.)

It is obvious from examining the modified Berggren equation that both ideal heat transfer and storage conditions are assumed. The thermal conductivity of the soil mass varies for different soil types and temperatures. (Fig. 11.9 and 11.10).

In order to obtain the correction coefficient λ, both the fusion parameter μ and thermal ratio α are needed. The thermal ratio is defined as the ratio between the temperature above freezing to the temperature below freezing, i.e. $\alpha = T_u/T_f$. As an approximation, T_f may be taken to be equal to F/t. The values for T_u and T_f are absolute values and are measured in terms of temperature difference from the freezing temperature (generally taken as $32°F$). For example:

Given a mean annual surface temperature of $40°F$ and a mean surface temperature during the period of freezing of $10°F$, then: $T_u = 40° - 32° = 8°$ and $T_f = 32° - 10° = 22°$.

The following example illustrates the difference in estimation of depth of frost penetration using the Stefan and modified Berggren equations, i.e. eqs. 11.8 and 11.10:

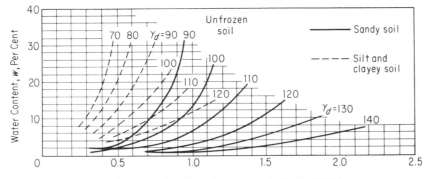

Fig.11.9. Thermal conductivity, k_u, for unfrozen soil. (Aldrich, 1956.)

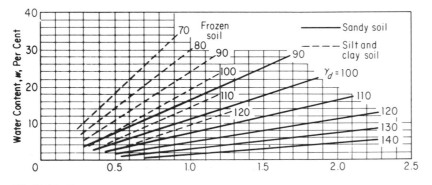

Fig.11.10. Thermal conductivity, k_f, for frozen soil. (Aldrich, 1956.)

Given a uniform homogeneous silty soil deposit; water content $w = 20\%$, dry density $\gamma_d = 100$ pcf. Temperature conditions are as follows: freezing index $F = 2,100$ deg. F days; freezing period $t = 160$ days; mean annual temperature $= 34°F$. Therefore $T_u = 2°F$. From Fig. 11.9 and 11.10, the values for k_u and k_f may be obtained:

$$k_u = 1.02 \text{ and } k_f = 1.30 \text{ (all in btu hr}^{-1} \text{ ft.}^{-1} \text{ °F}^{-1})$$

$$k_{effective} = (k_u + k_f)/2 = 1.16 \text{ btu hr}^{-1} \text{ ft.}^{-1} \text{ °F}^{-1}$$

The latent heat of fusion may now be calculated since both water content and dry density are known.

$$L = 1.434 \, w\gamma_d = 3157.8 \text{ btu/cubic ft.}$$

The Stefan equation may now be used to estimate the depth of frost penetration. From eq. 11.8:

$$x = \sqrt{\frac{48k_f F}{L}} = \sqrt{\frac{48 \times 1.3 \times 2,110}{3157.8}}$$

Therefore:

$$x = 6.46 \text{ ft.}$$

In order to use the modified Berggren equation, it is necessary to calculate both C_u and C_f. From eq. 11.1 and 11.2:

$$C_u = \gamma_d \left(0.17 + \frac{w}{100}\right) = 110(0.17 + 0.2) = 20.9 \text{ btu cu. ft}^{-1} \text{ °F}^{-1}$$

and:

$$C_f = \gamma_d \left(0.17 + \frac{0.5w}{100}\right) = 110(0.17 + 0.1) = 19.8 \text{ btu cu. ft}^{-1} \text{ °F}^{-1}$$

Therefore:

$$C_{effective} = \frac{1}{2}(C_u + C_f) = 20.35 \text{ btu cu. ft.}^{-1} \text{ °F}^{-1}$$

To calculate the fusion parameter μ and the thermal ratio α needed to use Fig. 11.8 to determine the correction coefficient, α is taken to be equal to $T_u t/F$. Therefore:

$$\alpha = \frac{2 \times 160}{2,110} = 0.152$$

From eq. 11.11:

$$\mu = \frac{CF}{Lt} = \frac{20.35 \times 2{,}110}{3157.8 \times 160} = 0.085$$

From Fig. 11.8:

$$\lambda = 0.97$$

Therefore:

$$x = \lambda \sqrt{\frac{48Fk}{L}} \quad \text{from eq. 11.10}$$

which gives $x = 5.92$ ft.

The overestimation of the depth of frost penetration using the Stefan equation can be seen from the preceding example which pays no attention to the volumetric heat of the soil mass, and which further simplifies the estimation for x by assuming linearity in temperature distribution in the soil mass. For multi-layered soil profiles, the effective values for latent heat of fusion and volumetric heat of the subsoil must be computed on a weighted basis since these are dependent upon both water content and dry density of the soil. In the same manner, the thermal conductivity must also be computed for the total subsoil thought to be within the zone of frost penetration. Evidently a first estimate of the depth of frost penetration must be obtained. The Stefan equation can be used most appropriately for this.

11.5 FREEZING IN COARSE-GRAINED SOILS

In coarse-grained (granular) soils, because of the size of the particles, gravity forces are most likely to predominate both in the mineral and liquid phases. As shown previously, the surface forces are, by comparison with the gravity forces, so small that their effect may be neglected. The bulk of the water in a saturated granular soil may then be considered as free water.

When such a soil-water system freezes, migration of water to any incipient bud of ice crystallization is difficult, because of the magnitude of the gravity forces involved in the bulk water. It is easiest for ice to propagate (in the growth process) through the pore spaces than it is for water to be drawn up to the growing ice crystal; i.e. the energy requirements for ice propagation are less than that required to move water to the ice front. In consequence, water in the soil voids will freeze as the ice front propagates or "moves" into the void spaces resulting therefore in little movement of water. Experimental investigations, supported by field studies

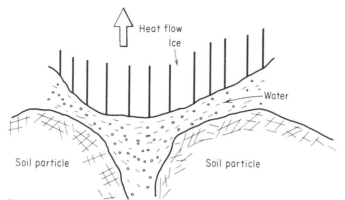

Fig.11.11. Schematic diagram of ice front and curvature of freezing plane.

indicate little or no heaving of soils as a result of freezing of granular soil-water systems. The resultant frozen soil mass may be visualized as an ice matrix studded with granular soil particles. There is experimental evidence to show that a liquid-like film bounds the ice phase. Therefore, the mineral particles may not be in direct contact with the ice phase, but rather would be separated by the liquid-like film.

Initiation of ice growth in a saturated soil-water system composed of coarse-grained particles may be thought of as ordinary freezing of water within the interstices of the porous soil mass.

Following ice nucleation in the pore water, growth and propagation of the ice crystal will progress as long as the energy available due to heat transfer is greater than the ice–water interfacial energy. This is illustrated in Fig. 11.11. The ice–water interfacial relationships are demonstrated in terms of a pressure P_{iw} which we will call the ice–water interfacial pressure. P_{iw} depends on the curvature of the freezing plane (shown in Fig. 11.11) and a constant which is related to the interfacial energy between water and ice. The equation for P_{iw} is given as:

$$P_{iw} = \Delta P = \sigma_{iw}\left(\frac{1}{r_1} + \frac{1}{r_2}\right)$$

where: σ_{iw} = on interfacial energy between water and ice; r_1 and r_2 = principal and r_2 = principal radii of curvature of the curved interface.

If $r_1 = r_2$:

$$P_{iw} = \frac{2\sigma_{iw}}{r_1} \tag{11.12}$$

Eq. 11.12 resembles very closely the form of the surface tension equation given in Chapter 4.

The energy available due to temperature depression or heat transfer may be examined in terms of thermodynamics. If we assume a closed univariant system, then from the second law of thermodynamics:

$$F = Q \frac{T_2 - T_1}{T_2} \qquad (11.13)$$

where: F = free energy or energy available; Q = heat energy transferred from T_2 to T_1 (including the heat of fusion L); $Q \triangleq L$ since the heat transferred from T_2 to T_1 excluding L, is small; T_2 and T_1 = ice and water temperatures, respectively.

Eq. 11.13 may be written in the form of a pressure or suction if F is divided by the volume V with which it is associated. Thus:

$$\frac{F}{V} = H_p = L \left[\frac{T_2 - T_1}{T_2} \right] \qquad (11.14)$$

where: H_p = suction or pressure.

Eq. 11.14 is similar to that given by Schofield (1935) for calculation of freezing-point depression. Thus as long as H_p is greater than P_{iw}, freezing of the pore water in a saturated granular soil mass will continue. In other words, the ice front shown in Fig. 11.11 will "propagate" downward through the neck pore separating the two particles and into the pore space below. In the equation given by Schofield, the following form is used:

$$H_s = \frac{Lt_f}{T_g}$$

where: H_s = suction in centimeters of water; L = latent heat of fusion; t_f = freezing point depression; T = absolute temperature; g = gravitational acceleration.

From Chapter 4, it was established that the soil-water potential could be used to describe the energy state of water in a soil-water system. The value of H_s reflects the matric potential, assuming gravitational potential to be zero, and no salt. Thus, for freezing of water in the pore spaces to continue (i.e. to "propagate"), the energy available due to temperature depression, must exceed the matric potential. It follows that when energy requirements for "propagation" of ice exceed that available, the ice front will not be able to move downward into the pores. Instead, either water will now be drawn up to the ice to continue the growth process, or the ice front will remain stagnant. If water is drawn upward to the ice front to continue growth of the ice in the pore space above, segregation will occur. This constitutes the beginning of the ice-lensing phenomenon which will be discussed in the next section.

The limitations of a closed univariant system are further examined in Section 11.8. Deviation of actual from theoretical freezing is in large measure due to the idealized assumptions utilized.

It is likely that each void space in the soil mass would contain one or more ice crystals. The formation of ice crystals may occur on the basis of both heterogeneous and homogeneous nucleation; heterogeneous nucleation being the initiation of growth or formation of an ice crystal by a foreign substance, and homogeneous nucleation being initial growth on the basis of a bud of crystallization formed within the water phase.

If water within the soil mass is supercooled and subsequently jarred suddenly, spontaneous nucleation would be likely to occur. The soil-water system would freeze instantaneously, thus giving rise to multicrystal formation. If non spontaneous nucleation were to occur, fewer crystals would be formed and the resultant ice mass would be composed of larger crystals growing as a result of propagation of ice fronts. This, however, must depend upon factors such as temperature and duration of freezing.

Heterogeneous nucleation may be explained on the basis of the presence of foreign bodies that act as buds of crystallization for the resultant freezing. Homogeneous nucleation in a pore-water system containing no foreign bodies may occur when a cluster of the water molecules attains a certain critical radius. The cluster radius will be determined by the free surface energy of the molecules, and the critical cluster radius is in turn a function of the temperature.

Energy is required to support the overburden pressure at any level in the subsoil. This places the soil-water system under a state of stress and consequently lowers the critical temperature T_c at which ice and the pore water have equal vapour pressure (Fig. 11.12). For ice to begin to form, i.e. for nucleation to occur, the pore water in the soil-water system must be cooled to the nucleation temperature T_n. Experimental tests have shown that T_n must be at least 7°F less than T_c. Nuclei form slightly below T_c, but because of the thermal energy of the water molecules, the statistical odds are greatly in favour of immediate

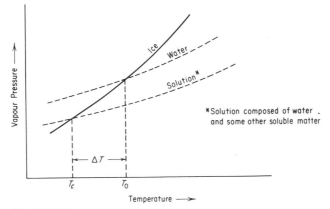

Fig. 11.12. Vapour-pressure curves for soil-water system.

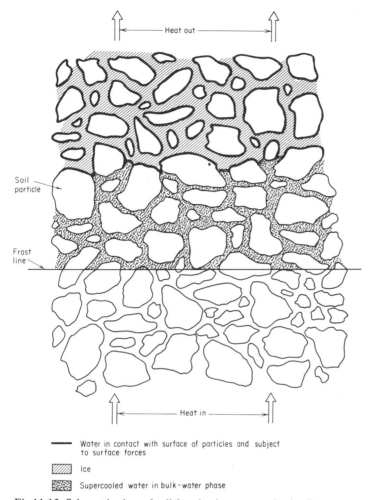

Fig.11.13. Schematic view of soil freezing in coarse-grained soils.

disintegration. Thermal energy is also supplied in the form of water being transported to the nuclei. Obviously the temperature of the system must be lowered in order to decrease the possibility of disintegration of nuclei.

Fig. 11.13 shows a schematic view (enlarged and considerably exaggerated) of soil freezing in a coarse-grained soil-water system. The system is assumed for simplicity of presentation to be completely saturated. A liquid film separates the ice phase from the mineral phase because of the altered structure of the water next to the surface of the particles. This has been designated as super-cooled water in the diagram.

For a semi-infinite subsoil system, a uniaxial direction of freezing as shown in Fig. 11.13 is a valid assumption. The frost line represents the position of the T_c

isotherm. This is not necessarily a straight horizontal line since several other physical properties of the soil mass have to be considered. The water contained in the soil pores immediately behind the frost line is supercooled and will nucleate when the frost line advances further thus lowering the temperature to T_n, the nucleation temperature.

11.6 FREEZING IN FINE-GRAINED SOILS

Measurements of the volumetric expansion of fine-grained soils subjected to freezing temperatures show that the amount of volume increase is greater than the expected $10 - 11\%$ increase in pore volume due to freezing of the water in the soil. The increase in volume may amount to 100% or more, and is the result of formation of lenses of ice in the subsoil. Heaving of the ground surface results and detrimental action to surface structures can arise from this frost heave.

The conditions necessary for frost heaving to occur are:

(1) Frost susceptible soil. Casagrande has shown that for well-graded soils, if 3% or more of the soil particles are less than 0.02 mm in particle size, the possibility of frost action is high. For uniformly graded soils, if 10% or more of the soil particles are less than 0.02 mm in size, then frost action is possible. Variations in the percentage of fines will, in general, be expected in regard to frost susceptibility in view of depth of frost penetration.

(2) Temperature. Subfreezing temperature must penetrate into the subsoil. The rate of freezing must be relatively slow to allow for growth of ice lenses, i.e. consistent with rate of water movement to the growing ice crystal.

(3) Availability of water. There must be a supply of water for the growth of ice lenses. This generally means that the ground-water level should be not less than 6 ft. from the ground surface.

For an explanation of the phenomenon of ice-lens growth, the water surrounding the soil particles and fabric units may be classified into two categories:

(a) Water close to the surfaces of soil grains, held at low value of matric potential, i.e. at high suction.

(b) Water further away from the surface, having higher values of matric potential. This is often called free or bulk water.

Nucleation is more likely to occur in the free water, on the basis of energy requirements. This means that ice nucleation will begin in inter fabric unit pore spaces (i.e. pore spaces established between fabric units) since these pores are comparatively much larger in individual volume than pore spaces between the particles within the fabric units, as shown in Chapter 5. There is more free water available in inter fabric unit pores, thus enhancing or increasing the likelihood for ice nucleation and growth in these pores. After nucleation and subsequent crystallization, growth would continue as long as heat loss exceeds heat gain. Any addition of water to the bud of crystallization would mean an addition of heat. It follows then that unless heat loss from this bud is equal to or greater than the heat gain from water brought in to the bud to foster growth, subsequent melting would occur. If heat loss is equal to heat gain, equilibrium is maintained.

When the free water in the immediate pore space is used up for crystal growth, either one or all of three things can happen if further growth is to occur:

(a) The water held at lower potentials would be used for further growth.

(b) The ice will "propagate" through the neck pores into other pore spaces as in the case of freezing in coarse-grained soils.

(c) Water is drawn to the bud of crystallization from lower unfrozen layers of the soil.

However, the temperature must now be lowered significantly as (a) the water is associated with lower potentials, (b) the neck pores are extremely small and would thus require considerable energy to form the very-small front curvature of the ice to "propagate" through the neck pore, and (c) energy would be required to draw water up to the growing bud. If the heat balance is maintained after all the water in the immediate pore space has been expended, no further growth occurs. If, however, the temperature is further depressed, conditions for subsequent growth are now made available.

As the temperature is decreased, there is an induced pressure deficiency at the ice front. Whether this deficiency in pressure is sufficient to result in movement of water to the growing bud would depend upon the effective radius of the neck pore, and the availability of the water in adjacent pores. By and large, it will be apparent that the effective radius of the neck pore will be too small to permit easy "propagation" of the ice front, thus creating the situation for water to move into the pore space containing the ice crystal. This process initiates ice segregation. The heat balance or unbalance would become greater because the growing crystal requires more water. This process can be carried further and further as long as free water can be brought in from pore spaces not in the immediate vicinity, and water held to the particle surfaces from the immediate and adjacent pore spaces is used to satisfy the demand for water. However, a point is reached when the energy required to sustain growth becomes too large and growth stops. Then another nucleus would be formed further ahead near the ice front, where the temperature may be within the nucleation-temperature range. The pictorial sequence for growth of ice lenses is presented in Fig. 11.14.

In Stage 1, the bud of crystallization is formed in the free water. If the heat balance conditions are satisfied, the ice crystal will grow until all the "available" water (consistent with the energy balance) within this pore space is added to the

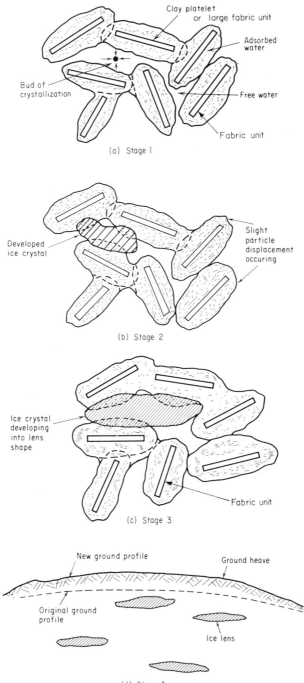

(a) Stage 1

Clay platelet or large fabric unit

Adsorbed water

Bud of crystallization

Free water

Fabric unit

(b) Stage 2

Developed ice crystal

Slight particle displacement occuring

(c) Stage 3

Ice crystal developing into lens shape

Fabric unit

(d) Stage 4

New ground profile

Ground heave

Original ground profile

Ice lens

growing crystal. We note here that all the available water used for ice-crystal growth in the pore space does not necessarily include the entire adsorbed water regime. As a matter of fact, as will be shown later, some of the adsorbed water will almost never freeze. When all the available water in the pore has been used up, as shown in Stage 2, water is drawn from the adjacent pore, and water at lower potentials may also be brought in to add to ice growth. Particle or fabric unit displacement occurs as a result both of water loss and of crystal growth. During this stage, and also as a result of initial ice segregation, some propagation of the ice crystal into the neck pores may occur—especially if particle- or fabric-unit displacement occurs due to initial ice segregation.

With further temperature decreases, more particle- or fabric-unit displacement occurs as water is drawn from sources further away from the bud of crystallization and as the ice crystal grows. This is shown as Stage 3 in Fig. 11.14. Since heat transfer is essentially uni-directional, the crystal begins to assume the shape of a lens perpendicular to the direction of heat loss. If growth of the ice lens is to continue, the temperature must be decreased further and the heat transfer must consequently be much greater. More of the water in the immediate pore space held to the soil particles and fabric units, and more available water from adjacent pore spaces must be drawn up to the ice lens.

Stage 4 of the growth of an ice lens occurs when all the available water within the neighbouring area has been used and a further supply cannot be made available because of the high energy requirements. When this happens, another lens will begin to grow lower down in the same way. The process will be repeated as long as the cold front moves downward, and water is available.

The growth of ice lenses displaces the soil surrounding the lens causing in general a total volume change, and in particular, a resultant upheaval of the soil mass on the surface. This is frost heaving.

Ice lenses may also be formed in closed systems, where water is not readily available, but such lenses would be small since only the original water content provides the source of water for ice-lens formation and growth. As a result, little or no heaving occurs.

Fig. 11.14. Stages in growth of ice lenses. a. Free water being used in immediate pore space for growth of ice crystal. b. Free water being drawn from neighbouring pores and some adsorbed water being used to further crystal growth. c. Free water being drawn in from pore spaces further away, and more adsorbed water in the immediate pore space being used up; soil displacement occurs because of growth of crystals into lens shape. d. macroscopic view of formation of ice lenses in subsoil due to progress of frost line and depletion of water supply around crystal growth.

11.7 HEAVE AND FROST HEAVING PRESSURES

We note from the previous section that volumetric expansion of pore water and ice segregation constitute two separate mechanisms in the soil-freezing phenomenon. In actual fact, it is not likely that either of the two mechanisms will be operative exclusively—except for possibly the limiting cases of coarse sands and very compact clays. However, in certain cases, because one mechanism dominates in the production of the overall freezing mechanism, it is possible to make computations for heave or heaving pressures based exclusively on one mechanism. The situation in regard to saturated silts for example is a case in point. In such an instance, the ice segregation (i.e. ice lens) model can be used successfully to compute frost heaving pressures.

Calculation of frost heaving pressures

To calculate frost heaving pressures due to ice segregation (i.e. ice lensing) the geometrical model which assumes uniform spheres and cubic packing has been used (McKyes, 1966; Yong, 1967). Using Maxwell's thermodynamic relations written for a curved ice—water interface in a pore:

$$\left(\frac{\partial T}{\partial P}\right)_v = \left(\frac{\partial V}{\partial S}\right)_T$$

and using the usual notation:

$$\left(\frac{\partial V}{\partial S}\right)_T = -\left(\frac{V_i T}{L}\right)$$

For small incremental changes:

$$\frac{\Delta T}{\Delta P} = -\frac{V_i T}{L} \qquad\qquad (11.15)$$

considering ΔP as $2\sigma_{iw}/r$ (from eq. 11.12), where: σ_{iw} = ice—water interfacial energy; r = radius of curved surface; V_i = volume of ice = $1/\rho_i$ for unit volume consideration; L = latent heat of fusion; ΔT = freezing point change at curved surface.

Eq. 11.15 may be written as:

$$\Delta T = -\frac{2\sigma_{iw} T}{\rho_i L r} \qquad\qquad (11.16)$$

where ρ_i = density of ice. Eq. 11.16 is the same form used by Sill and Skapski (1956).

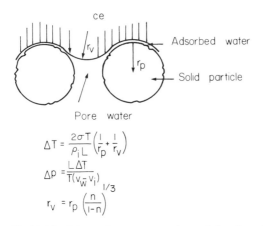

Fig.11.15. Schematic representation of local region at freezing front showing ice-front curvature and temperature – pressure relationships.

For the geometric configuration shown in Fig. 11.15 since the free energy of the ice is equal to that of the water at equilibrium, the reversed curvature of the ice front between pore space and particle must be accounted for. Since the minimum potential energy configuration for the ice front must be a flat plane, both the ice radii in thypore r_v and around the particle r_p must be considered. In this case, if we assume cubicle packing of soil particles:

$$r_v = r_p \left(\frac{n}{1-n} \right)^{1/3} \tag{11.17}$$

$$\Delta T = - \frac{2\sigma_{iw} T}{\rho_i r_v L} \qquad \text{in the pore}$$

$$\Delta T = + \frac{2\sigma_{iw} T}{\rho_i r_p L} \qquad \text{for the particle}$$

Hence in the local region encompassing a particle and a void space, the net ΔT is given as:

$$T = - \frac{2\sigma_{iw} T}{\rho_i L} \left[\frac{1}{r_p} + \frac{1}{r_v} \right] \tag{11.18}$$

which is similar in form to that given by Miller et al. (1960). Since ΔP, the pressure generated between the ice and particle is given by the Clapeyron relation (see, e.g., Penner, 1959), we will obtain:

$$\Delta P = \frac{L \Delta T}{T(V_w - V_i)} \tag{11.19}$$

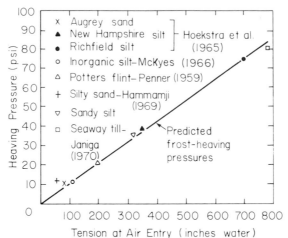

Fig. 11.16. Comparison between measured and predicted frost heaving pressure using radius of curvature computed from moisture potential. (Yong and Osler, 1971, by permission of Nat. Res. Council of Canada.)

Substituting the appropriate values for r_p, r_v and in eq. 11.18 will provide the value for ΔT in eq. 11.19. Since ΔP in eq. 11.19 would act on the soil particles shown in Fig. 11.15 for a physical demonstration of frost heaving pressure, projection for a total frost heaving pressure must be made considering the effective particle surface area contacted. This is approximated by modifying the total cross-sectional area by $(1 - n)^{2/3}$, where n = porosity.

In silts, since adsorbed water (controlled by surface forces) is sufficiently small, r_p can be reasonably represented by the particle radius. A better determination of r_p and r_v may be obtained by using the matric potential at the air-entry value of the soil. Where the total potential is due to the matric potential, an effective r_p denoted by r_p' can be computed, using $\sigma_{iw} = 72$ ergs/cm^2 for air–water interfacial energy. r_v', the effective pore radius, can thus be computed using eq. 11.17. Fig. 11.16 shows a comparison between predicted and measured frost heaving pressures using the r_v and r_p values computed from known air-entry values and eq. 11.17–11.19.

There are other geometrical models used to compute frost heaving pressures, e.g. Everrett and Haynes (1965), Penner (1968). The computational techniques are essentially similar and variations in treatment are primarily due to assumptions dealing with transmission of pressure between ice lens and soil particle system.

Constraints and heaving pressures

The influence of physical constraints in development of frost heaving pressures may be seen in Fig. 11.17. Since the two cases shown in Fig. 11.17

experienced idential frost penetration, the ability for movement of the soil in the face of developing frost heaving becomes very important. By allowing partial movement (i.e. using a low constraint as shown in Fig. 11.17), it is possible to relieve the frost heave pressures and thus ultimately develop lower total frost heaving pressures. Obviously, if no constraints to frost heaving are provided, free movement of the ground surface occurs and no frost heaving pressures will develop.

Since field soils are seldom fully saturated in the upper layers, unconstrained heave due to frost penetration effects requires examination of free heave in terms of initial saturation and availability of groundwater. The latter point must recognize the opposing factors of specific surface area and hydraulic conductivity. Fig. 11.18 shows a schematic representation of several specific cases for freezing of pore water, as influenced by initial conditions and constraints.

Fig. 11.18a shows that with a low degree of saturation, and in the absence of any significant water movement into the pore caused by attraction to the ice nucleus, the resultant pore-ice size is not sufficient to create outward displacement of the surrounding particles. Increased negative pore-water pressures may result which might cause a volumetric reduction of the frozen soil mass. The main feature of the mechanism shown in Fig. 11.18a is that no observable heave will occur. In fact, the possibility of shrinkage exists when the degree of saturation is sufficiently low.

When water availability increases, the situation presented in Fig. 11.18b becomes likely. In this case, one expects conditions associated with an open system

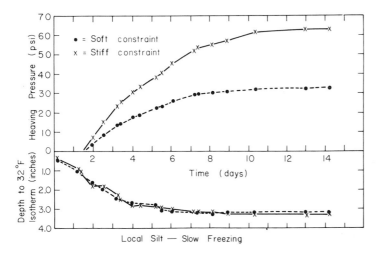

Fig.11.17. Effect of constraints on development of frost-heaving pressures. (Yong and Osler, 1971, by permission of Nat. Res. Council of Canada.)

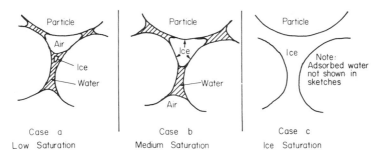

Case a
Low Saturation

Case b
Medium Saturation

Case c
Ice Saturation

Fig.11.18. Schematic representation of ice formation in soil samples with varying degrees of saturation. *Case a.* Ice can expand into air-filled voids. No ice pressure on particles. Possible generation of increased negative pore pressures with volumetric contraction. *Case b.* Ice fills some pores completely, where possible. Partial ice pressures on particles. Start of volumetric increase when quantity of ice is sufficient. *Case c.* Ice in pore expands due to 9% volumetric expansion. Further delivery of water into system will cause macroscopic ice segregation. (Yong and Osler, 1971, by permission of Nat Res. Council of Canada.)

or, as demonstrated by Dirksen and Miller (1966), a closed unconstrained system subjected to large thermal gradients. Since the volumetric expansion of pore water, due to phase change into ice, causes the ice to expand into the diminishing pore air spaces and also exerts partial pressures on the surrounding particles, or their adsorbed water layers, the net result is volumetric expansion. With the addition of small amounts of water from even limited sources, the process continues with observable ground heave. Complete ice saturation within the pores does not necessarily exist at this stage.

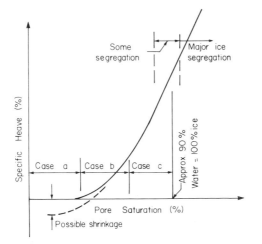

Fig.11.19. Specific heave vs. degree of saturation for an idealized soil system. (Yong and Osler, 1971, by permission of Nat. Res. Council of Canada.)

With continuous water delivery to the frozen zone, we will pass through the stage of 90% water saturation and (or) 100% ice saturation within the pores. At this time, if a stable freezing front exists, ice segregation will occur, resulting in active formation of macroscopic ice lenses. This is demonstrated in Fig. 11.18c. The various mechanisms described above have been combined in Fig. 11.19 to show the range of possible field performance, in terms of specific heave and degree of saturation. Specific heave is defined as total ground heave per depth of frost penetration.

11.8 UNFROZEN WATER IN FROZEN SOILS

The mechanism described for soil freezing implied that not all the water held in the soil is frozen when the soil mass is subjected to subfreezing temperatures. This has been experimentally shown to be true for soils containing a high percentage of fine-grained particles. Surface forces define the nature and amount of the water that surrounds the particles and fabric units. The closer one approaches the particle surface, the higher are the surface forces, the lower is the water potential, and the higher is the force required to move a molecule of water into or out of this area. This is particularly true in intra fabric unit considerations where groups of particles interact to form the particular fabric unit. These forces and potentials are discussed in Chapters 3 and 4. The temperature decrease required for freezing of the water held to the soil particles and fabric units will vary with the degree of interaction or alternatively with the soil-water potential. Because of this, a varying amount of the total water freezes.

The quantity of unfrozen water as a proportion of the total water content, or per cent unfrozen water, varies with: (a) freezing temperature; (b) clay content (per cent of active clay particles in the soil-water system); (c) electrolyte concentration; (d) soil structure; (e) per cent water saturation; (f) charge density of the soil particles; (g) original water content.

Fig. 11.20 shows the variation of unfrozen water with the conditions stated above. In some instances, the freezing point depression caused by the interparticle forces may be as high as 15°C at a distance of about 10 Å from the surface of the particle. The freezing-point depression is the difference between the freezing point of water and the freezing point of the water under the influence of the potentials surrounding the soil particle. From the curves in Fig. 11.20 it is evident that the temperature depression for active clay soils will be more than 20°C.

The freezing history of the soil to be tested must be known since unfrozen water content of the soil will depend upon whether the sample is cooled or warmed to the test temperature. The reasons for the differences in unfrozen water content

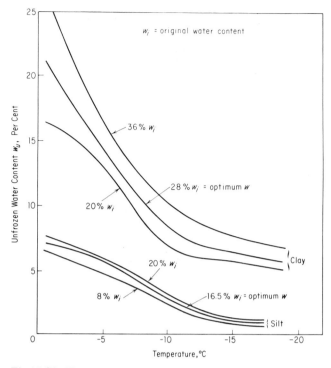

Fig.11.20. Temperature and unfrozen water content relationships for clay and silt. (Yong, 1965.)

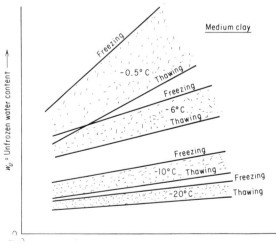

Fig.11.21. Unfrozen water content variation with original water content and freezing history.

with varying freezing histories lie in the fact that the soil fabric may be changed during freezing and thawing, thus causing a change in the overall soil structure. This is demonstrated in Fig. 11.21, where a medium clay was frozen to the desired test temperature and also to a temperature of 2°C lower than the test temperature before warming to the test temperature. The degree of variation in unfrozen water content is seen to be much greater at higher temperatures.

Soil-water potential and unfrozen water

The amount of water remaining unfrozen at any one persisting subfreezing temperature can be related to the soil-water potential. Thus, with measurements of water potential which reflect the degree of interaction arising from surface and curved air—water interface forces, an evaluation of quantity of water remaining unfrozen (designated as unfrozen water content) may be made. The common base used to establish the relationship depends on initial water content. Since little is known of the effect of subfreezing temperatures on water potential relationships, the assumption is made that during the process of freezing, the initial potential of the water must dictate ultimate freezing of the pore water. The assumption is viable

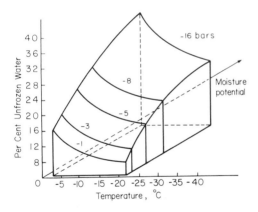

Fig.11.22. Soil-water potential-unfrozen water content surface. (Yong, 1967.)

in that the critical temperature which determines freezing of the pore water will depend on the energy status of the water within the pores. Thus, the soil-water potential-unfrozen water surface shown in Fig. 11.22 can be established. Any part of the surface represents the per cent of original water remaining unfrozen as a function of subfreezing temperature and water potential for the particular clay. Similar surfaces may be obtained for other soils.

Water movement in frozen soils

We have shown in Chapter 5 that water in soil moves in response to potential gradients. When soil freezes, the difference in temperature between the adjacent soil elements will create a temperature gradient which will cause water to move. The magnitude and rate of water movement will obviously depend on the gradient and the thermal and hydraulic properties of the soil. Since unfrozen water exists in frozen soil, water movement in the unfrozen water layers of frozen soil is possible. This is generally identified as part of the phase boundary transport phenomenon. The overall phenomenon also includes transport of salts in the unfrozen water phase – in response to temperature and concentration gradients.

Freezing of water in soil pores will generally tend to exclude the salts during the freeze process. This is sometimes called brine exclusion. Thus one obtains a film of brine surrounding the ice crystal within a pore space. Since pore spaces in soil are not regular or uniform, it follows that concentrations of salt will vary throughout the soil mass, on a local and general basis. Thus, phase boundary transport will provide the means for ionic diffusion in response to increased or decreased activities arising from temperature and concentration differences, and water movement due to temperature gradients.

Considered on a local basis where linear or near linear situations can be assumed, the flow of soil water and salt in unsaturated frozen soil may be written as (Cary and Mayland, 1972):

$$J_w = -k \frac{d\tau}{dz} - \beta \frac{DpH}{R^2 T^3} \frac{dT}{dz} \tag{11.20}$$

and :

$$J_s = -n_s k \frac{d\tau}{dz} - \bar{\gamma} \phi D_s \frac{dn_s}{dz} \tag{11.21}$$

where: J = flux at depth z; subscripts w and s = water and salt respectively; $\bar{\gamma}$ = fraction of unfrozen water at depth z; ϕ = tortuosity; D_s = diffusion coefficient of salt in water; n_w and n_s = mole fraction of water and salt, respectively; β = dimensionless constant $\simeq 2.5$; D = diffusion coefficient of water vapour in air; p = vapour pressure of ice; k = hydraulic conductivity; τ = matric potential; H = latent heat of vaporization.

Fig. 11.23 and 11.24 show field examples of water and salt movement within the frozen soil zone during the freezing period. While eq. 11.20 and 11.21 may be used in controlled quasi steady-state examinations, the field problem of transient temperatures and undefined initial conditions require that the equations be used on a local point basis for analysis or prediction. Bulk analysis involving total freeze

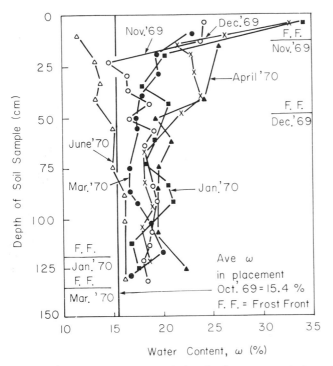

Fig.11.23. Moisture profile variation in frozen zone during winter months. *F.F.* = frost penetration position. (Yong and Janiga, 1972.)

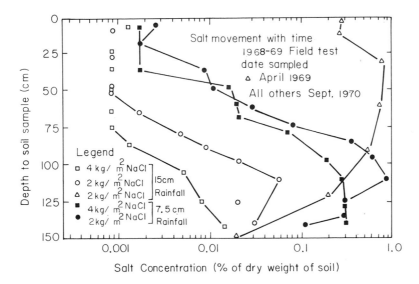

Fig.11.24. Salt movement in field test site with time. (Yong and Sheeran, 1972.)

depth for the transient case becomes difficult or unfeasible. A piecewise increment in time and distance would be the more appropriate form of analytical treatment.

11.9 FIELD FROST HEAVING

The effects of frost heaving are detrimental in two ways: (a) change in the ground surface level due to the formation of ice lenses in the subsoil, and (b) loss of bearing support in the subsoil when the ice lenses thaw leaving pockets of water. The basic conditions giving rise to formation of ice lenses necessary for frost heaving have been detailed previously.

Experimental results have shown that pressures as high as 20 psi were needed to restrict heaving at temperatures of $-5°C$. Surface structures do not provide sufficient overburden pressure to restrain ground heave. Damage to highways is very common. Little if any extra overburden pressure is present, and because of the protective covering of the highway, a greater amount of water is stored in the soil below. The general profile of a highway founded on frost susceptible soil is one with excessive bumps and cracks in winter, and potholes in spring.

Prevention of the detrimental effect of frost heaving requires the elimination or restriction of conditions giving rise to frost heave. The following procedures can be used to change the conditions favouring frost heaving:

(1) Replacement of frost susceptible soil by less susceptible material. It is best to replace frost susceptible soil with granular material. Freezing of the coarse grained soils will not be accompanied by development of ice lenses. The criterion established by Casagrande can be used as a guide for comparison of frost susceptibility, i.e. a wellgraded soil with 3% of its fines less than 0.02 mm in size would be more frost susceptible than the same soil with 2% of its fines less than 0.02 mm in size.

(2) Lowering the freezing point of the pore water. The addition of salts may lower the freezing point of the pore water and thus freezing in the subsoil. Since drainage and groundwater flow during the summer months may carry away the dissolved salts in the pore water, regular additions will be required.

(3) Restricting the water supply. The supply of water can be eliminated or reduced by installing a cut-off blanket in the subsoil. This cut-off blanket may be an impervious clay layer, a bituminous membrane or even a concrete membrane. The idea is to limit the supply of water so that ice lenses may not develop. This method may be feasible, but may not be economical.

(4) Use of chemical additives. Additives used to change the properties of the subsoil may be dispersants, aggregants, or waterproofing agents. When these agents are mixed with the soil and compacted, they will make the subsoil more permeable

or less permeable. If water is fed too fast or too slowly to the bud of crystallization, the bud will either melt or be restricted in size since water supply is one of the prime factors contributing to the growth of ice lenses. Additives which change the permeability characteristics alter the ideal heat balance condition, which in turn either destroys the bud or inhibits its growth. Laboratory tests on the subsoil must be conducted to establish whether dispersants or aggregants can be feasibly used for this purpose.

(5) Adding a surcharge load. Experience has shown that if the water table is more than 6 ft. below the ground surface, the growth of ice lenses is made difficult. Adding a surcharge load raises the ground level and thus effectively lowers the groundwater table. The surcharge load also counteracts, in part, the heaving pressure.

(6) Use of a thermal blanket. If a thermal barrier (such as a foam plastic) is placed at some depth below the surface so that it restricts the frost front to the soil above the barrier, it is possible to reduce detrimental frost heave. The soil above the barrier must however not be susceptible to frost heaving.

11.10 THERMAL EROSION

In permafrost regions, the thicknesses of the active layer and the permafrost layer are established and controlled by surficial environmental factors, temperature regimes and substrate conditions. A delicate dynamic thermal equilibrium exists in undisturbed regions, and the processes contributing towards the continued existence or slow change in the permafrost have been established through long periods of time (Mackay, 1971, 1972). This equilibrium is more stable with lower temperatures, but becomes more unstable at the southern limit of permafrost. The effects of thawing ice-rich permafrost are detrimental in several respects: (a) increased melt conditions leading to formation of local ponding areas; (b) increased thickness of the active layer leading to surface erosion and in the case of slopes, this could produce thaw slides.

Because of the delicate balance of energy established between the above and below ground regimes, any change of the interface properties between the two regimes could contribute to the establishment of thermal erosion effects. Fig. 11.25 shows surface scars and local slumping due to excessive disturbance of surface cover. A change in surface cover can obviously change the properties of the interface. Where the surface cover is removed or reduced, increased penetration of the active layer results. The depth of active layer can be computed in the same manner as that used for determining the depth of frost penetration. Eq. 11.10

which states that:

$$x = \lambda \sqrt{\frac{48kF}{L}}$$

now gives x as the depth of thaw, k refers to the thermal conductivity of the thawed material and F refers to the thawing index (Section 11.3 and Fig. 11.4).

In view of the detrimental effects created by increased thawing of ice-rich permafrost, it is necessary to provide adequate protection of the surface cover and morphology. If alternate cover or replacement procedures are provided for areas where disturbance is excessive, the surficial replacement property should match or improve on the previous properties such that the energy budget is not unduly changed.

11.11 SUMMARY

The primary problems in soil freezing relate to volume change of the pore water due to phase change and to ice segregation. Where overall volume changes are large, detrimental effects are obtained both due to the resultant heave and the subsequent melting of the frozen subsoil. Since large overall volume changes result from ice segregation (i.e. ice lensing), it is apparent that regions of melt would contribute to local instability and failure in the regions of ice lens melt.

Remedial, preventive or corrective measures implemented to combat frost heave effects must change one or more of the following: (a) frost susceptible soil; (b) available water; and (c) temperature regime. In addition, where additives are used for control of frost heave, account must be given for the existence of unfrozen water, due to surface forces and resultant interaction with the pore solution, which provides for mechanisms of water transport in the frozen soil.

In permafrost areas, considerable care and attention must be given to protection of the energy balance between above and below ground regimes to insure that detrimental thawing of the permafrost does not occur. Thermal erosion can contribute to instability, creation of local ponding areas where ice rich permafrost exists, and erosion of slopes.

Fig.11.25. A. Thermal erosion effects on ice-rich permafrost due to surface disturbance. B. Trail erosion due to high disturbance on ground surface.

INTERMOLECULAR ATTRACTION, THE HYDROGEN BOND AND THE STRUCTURE OF WATER

Forces of attraction between atoms or molecules have several sources. The strongest force is the Coulombic or ionic force between a positively charged and a negatively charged atom. This force decreases as the square of the distance separating the atoms. Ionic forces hold together the atoms in a crystal.

Forces of attraction between uncharged molecules are called Van der Waals' forces, and are of several types. Molecules which are neutral but in which the centres of positive and negative charge are separated are called *dipoles*. Certain orientations of adjacent dipoles are statistically preferred and a net attraction results. Even molecules which are not polar can be considered as instantaneous dipoles because the instantaneous positions of the electrons in the atomic shells change. This instantaneous dipole induces an in-phase dipole in an adjacent molecule, with a resultant net attraction. These various Van der Waals' forces decrease as the seventh power of the distance between atoms.

Combinations of these forces are also possible, e.g., ion-dipole attraction. Also, an ion or dipole will induce a dipole in an adjacent molecule to give a force of attraction. With these combinations, the decrease of attraction with distance can be between the second and seventh power of distance.

There is also a repulsion which decreases as the twelfth power of distance, which keeps the molecules from approaching each other beyond a certain distance. It prevents interpenetration of electron shells of adjacent atoms. The combination of attraction and repulsion results in a minimum energy at some finite small distance from the molecule.

The hydrogen atom can attract two electronegative atoms, forming an electrostatic or ionic bond. For example, it can bond the oxygen from a water molecule to the oxygen on the clay particle surface. This is called the *hydrogen bond.* Hydrogen is unique in this property, probably because of its small size. This is a strong bond compared with other bonds between neutral molecules. Hydrogen bonding between two oxygen atoms, which are electro-negative, is important in the structure of water and ice, in bonding layers of clay minerals together, in holding water at the clay surface, and in bonding organic molecules to clay surfaces.

The two hydrogen and one oxygen atoms of the water molecule, H_2O, occur in a V-shape with an angle of $105°$ between the two arms extending to the hydrogen atoms. These hydrogens are 0.97Å from the central oxygen. Water molecules associate in tetrahedral arrangements with each oxygen surrounded by four others, held together by hydrogen bonding. In ice, each oxygen is closely associated with two hydrogens and less closely with two more. When ice melts, only a fraction of the hydrogen bonds are broken and water retains a loose tetrahedral configuration. Breaking some of the bonds allows a closer packing of molecules, and hence water has a higher density than ice. The bonds are not fixed but are continually being broken and reformed. As the temperature increases, more bonds are broken and there is consequently less association between molecules. This accounts for the decrease in viscosity.

THEORETICAL DISTRIBUTION OF EXCHANGEABLE IONS AROUND A SINGLE CLAY PARTICLE

Assume the clay particle is a uniformly charged plate, infinite in two dimensions. The relation between the potential ψ, distance from the plate x, and volume concentration of charges or ions ρ is given by the Poisson equation:

$$\frac{d^2\psi}{dx^2} = -\frac{4\pi\rho}{\varepsilon}$$

where ε = dielectric constant of the solution.

The distribution of ions in the electric field is given by the Boltzmann equation:

$$n_+ = n_0 e^{-(ze\psi/kT)} \quad \text{for cations}$$

and:

$$n_- = n_0 e^{(ze\psi/kT)} \quad \text{for anions}$$

where: k = Boltzmann constant; T = absolute temperature. Also:

$$\rho = \Sigma ez_+ n_+ + \Sigma ez_- n_-$$

where e = electronic charge.

Taking the simplest case of $z_+ = z = 1$ for monovalent ions, the Boltzmann values for n_+ and n_- are substituted into the equation for ρ, and this substituted into the Poisson equation.

By defining $y = -(e\psi/kT)$, a dimensionless potential, the Poisson equation becomes:

$$\frac{d^2y}{dx^2} = \frac{4\pi e^2 n_0}{\varepsilon kT}(e^y - e^{-y}) = \frac{8\pi e^2 n_0}{\varepsilon kT} \sinh y$$

The constants are gathered into one term:

$$K = \sqrt{\frac{8\pi e^2 n_0}{\varepsilon kT}}$$

Integration of the equation with boundary conditions $y = 0$ and $dy/dx = 0$ at $x \to \infty$ gives:

$$\frac{dy}{dx} = \sqrt{2K^2 \cosh y - 2K^2}$$

$$= K\sqrt{4 \sinh^2 (y/2)}$$

$$= 2K \sinh (y/2)$$

The second integration, with boundary conditions, $y \to \infty$ at $x = 0$, gives:

$$\ln \coth (y/4) = Kx$$

This can also be written as:

$$y = 2 \ln \coth (Kx/2)$$

Since $y = \ln (n_+/n_0)$, this results in eq.2.3:

$$n_+ = n_0 \left[\coth \frac{Kx}{2} \right]^2$$

The boundary condition of infinite potential at the particle surface simplifies the solution, but does not describe clay plates which have a finite and measurable charge. This is most easily corrected by calculating the distance between the imaginary plane of infinite potential and the plane at which the potential has decreased to that on the clay surface. This distance x_0 is then subtracted from x to get the actual distance from the clay surface. This calculation is made by integrating the volume charge from x_0 to infinity and setting this equal to the surface charge on the clay. This gives:

$$x_0 = \frac{4zc_0}{K^2 \Gamma}$$

where: c_0 = molar concentration of salt in the outside solution; Γ = surface density of charge on the clay in equivalents per square centimetre.

The values of x_0 are between 1 and 4 Å for clay, so they can be neglected at all but small distances.

THEORETICAL DISTRIBUTION OF CATIONS BETWEEN TWO CHARGED PLATES

This solution is given by Langmuir in *Science,* Vol.88 (1938), p.430, and assumes that the influence of anions can be neglected.

Referring to the previous section, the differential equation neglecting the anion term becomes:

$$\frac{d^2y}{dx^2} = K^2 \frac{e^y}{2}$$

Integration, with the boundary condition that $y = y_c$ at the midplane between particles where $dy/dx = 0$, gives:

$$\frac{dy}{dx} = -K \sqrt{e^y - e^{y_c}}$$

The second integration, with boundary conditions $y \to \infty$ at $x = 0$, and evaluation for $y = y_c$, gives:

$$y_c = 2 \ln \frac{\pi}{Kx}$$

where y_c = potential at midpoint between particles. Since:

$$y_c = \ln \frac{n_c}{n_o} = \ln \frac{c_c}{c_o}$$

The equation can be solved for c_c the concentration of ions at the plane midway between particles:

$$c_c = \frac{\pi^2 c_0}{K^2 x^2}$$

TABLE A.2B.1

Values of y_c

Kx	x at c_o = 0.001 M (Å)	Representative* water content (%)	y_c from elliptic integral	y_c from langmuir equation	y_c from sum of separate potentials
4	400	–	0.15	–	0.15
3	300	–	0.40	0.10	0.40
2	200	150	0.98	1.02	1.10
1.5	150	110	1.50	1.48	1.70
1.0	100	75	2.30	2.28	3.1
0.75	75	56	2.85	2.86	4.1
0.50	50	38	3.60	3.68	5.6

* Calculated for a surface area of 75 m^2/g and parallel arrangement of particles.

The Langmuir approximation, neglecting anions, is valid at small distances between particles, i.e., at low water contents where the anions are effectively excluded from the space between particles. Table A.2B.1 compares the values of y_c from the complete solution of the elliptic integral (including the anions), from the Langmuir approximation, and from summation of the two potentials for single particles. Below y_c values of 1.0, summation of separate potentials can be used, beyond this the Langmuir approximation can be used.

SOIL FLUX AND VOLUME CHANGE IN UNSATURATED-FLOW EQUATIONS

An alternative approach to the examination of the problem of a changing porosity (or void ratio) during flow is to assume that in addition to considerations of motion of water relative to the soil particles (i.e. a modified Darcy relationship as in eq.5.31), the flux of soil particles must satisfy the three physical conditions of: (a) continuity, (b) Newton's law of motion, and (c) rheological equation of state. Since the motion of soil particles in the volume changing soil during fluid flow is slow, the acceleration considerations implicit in Newton's law of motion can be safely ignored. From conservation of mass, the equation of continuity in a fixed coordinate system can be written in general terms as:

$$\text{div} (\rho q_s) + \frac{\partial \rho}{\partial t} = 0 \tag{1}$$

where: q_s = soil flux (vector); ρ = bulk density = $G_s \rho_w /(1 + e)$; ρ_w = density of water; e = void ratio; G_s = specific gravity of solid particle; t = time (2)
From eq.2 and 1:

$$\text{div} \; q_s = \frac{1}{1 + e} \frac{\partial e}{\partial t} + \frac{\vec{q_s}}{1 + e} \cdot \text{grad} \; e \tag{3}$$

Eq.3 which includes the second order term $\vec{q_s}/(1 + e) \cdot$ grad e has been used by Wong (1969). If the second term on the right hand side of eq.3 is ignored, the equation is similar to that used by Zaslavsky (1964).
Defining the soil flux q_s as:

$$q_s = k_s \; \text{grad} \; \Phi \tag{4}$$

where k_s represents the coefficient of particle conductivity, and Φ denotes the pressure potential in the permeating pore fluid which results in particle movement it follows that since Φ depends on the void ratio e:

$$q_s = k_s \frac{\partial u}{\partial e} \text{ grad } e \tag{5}$$

$$= D_s \text{ grad } e$$

|

where D_s = particle diffusivity. Thus eq.3 becomes:

$$\text{div } (D_s \text{ grad } e) = \frac{1}{1+e} \frac{\partial e}{\partial t} + \frac{D_s \text{ grad } e}{1+e} \cdot \text{grad } e \tag{6}$$

writing:

$$q_w = q_{ws} + \theta q_s \tag{7}$$

where q_{ws} = flux of water relative to moving soil particles.
Taking the divergence of both sides, we obtain:

$$\text{div } q_w = \text{div } q_{ws} + \text{div } \theta q_s = -\frac{\partial \theta}{\partial t} \tag{8}$$

By expanding and substituting, the expanded form of the continuity equation can be obtained as:

$$-\frac{\partial \theta}{t} = \text{div } q_{ws} + \theta \left(\frac{1}{1+e} \frac{\partial e}{\partial t} + \frac{q_s}{1+e} \cdot \text{grad } e \right) + q_s \cdot \text{grad } e \tag{9}$$

The above form can be simplified if the second order term is ignored. Thus:

$$-\frac{\partial \theta}{\partial t} = \text{div } q_{ws} + \frac{\theta}{1+e} \left(\frac{\partial e}{\partial t} + q_s \cdot \text{grad } e \right) \tag{10}$$

Eq.10 represents the continuity condition which can now be combined with the modified form of the Darcy relationship as in eq.5.31 to yield the diffusion equation. If q_s is considered small in eq.10 it can be further reduced to (Wong, 1973):

$$-\frac{\partial \theta}{\partial t} = \text{div } q_{ws} + \frac{\partial v}{\partial t} \tag{11}$$

Eq.11 can now be used with the continuity condition shown in Chapter 5.

REFERENCES

Aitchison, G.D. (Editor), 1965. *Moisture Equilibria and Moisture Changes in Soils beneath Covered Areas*. Butterworths, Sydney, 278 pp.

Aldrich, H.P., 1956. Frost penetration below highway and airfield pavements. *Highway Res. Board, Bull.*, 135:124–144.

Aldrich, H.P. and Paynter, H.M., 1953. *First Interim Report, Analytical Studies of Freezing and Thawing in Soils*. Arctic Construction and Frost Effects Laboratory, New England Division Corps of Engineers, Boston, Mass.

Alpan, I., 1957. An apparatus for measuring the swelling pressure in expansive soils. *Proc. Int. Conf. Soil Mech. Found. Eng., 4th*, 1:3–5.

Anonymous, 1954. *Soil Mechanics for Road Engineers*. D.S.I.R., Road Res. Lab., H.M.S.O., London, 541 pp.

Barden, L. and Sides, G.R., 1971. Sample disturbance in the investigation of clay structure. *Geotechnique*, 21:211–222.

Barshad, I., 1965. Thermal analysis techniques for mineral identification and mineralogical composition. In: C.A. Black (Editor), *Methods of Soil Analysis. Am. Soc. Agron., Monogr.*, 9:699–742.

Berggren, W.P., 1943. Prediction of temperature distribution in frozen soils. *Trans. Am. Geophys. Union*, III.

Birrell, K.S., 1962. Physical properties of New Zealand volcanic ash soils. In: R.E. Grim (Editor), *Applied Clay Mineralogy. Conf. Shear Testing of Soils, Melbourne, 1952* – McGraw-Hill, New York, N.Y., pp. 30–34.

Bishop, A.W., 1972. Shear-strength parameters for undisturbed and remoulded soil specimens. *Stress–Strain Behaviour of Soils, Proc. of Roscoe Mem. Symp.*, Foulis, London, pp.3–58.

Bishop, A.W. and Bjerrum, L., 1960. The relevance of the triaxial test to the solution of stability problems. *Norw. Geotech. Inst., Publ.*, 34.

Bishop, A.W. and Blight, G.E., 1963. Some aspects of effective stress in saturated and partly saturated soils. *Geotechnique*, 13:177–197.

Bishop, A.W. and Henkel, D.J., 1962. *The Measurement of Soil Properties in the Triaxial Test*. Arnold, London, 2nd ed., 228 pp.

Bjerrum, L., 1967. Engineering geology of Norwegian normally consolidated marine clays as related to settlements of buildings. *Geotechnique, 7th Rankine Lecture*, 17:81–118.

Bodman, G.B. and Coleman, E.A., 1943. Moisture and energy conditions during downward entry of water into soils. *Soil Sci. Soc. Am. Proc.*, 8:116–122.

Bolt, G.H., 1956. Physico-chemical analysis of the compressibility of pure clays. *Geotechnique*, 8:86.

Bolt, G.H. and Miller, R.D., 1958. Calculation of total and component potentials of water in soil. *Trans. Am. Geophysical Union*, 39:917–928.

Boswell, P.G.H., 1961. *Muddy Sediments: Some Geotechnical Studies for Geologists, Engineers and Soil Scientists*. W. Hoffer and Sons, Cambridge, 140 pp.

Bozozuk, M. and Burn, K.N., 1960. Vertical ground movements near elm trees. *Geotechnique*, 10:19–32.

Brewer, R., 1964. *Fabric and Mineral Analysis of Soils*. Wiley, New York, N.Y., 470 pp.

Brown, R.J.E., 1970. *Permafrost in Canada*. University of Toronto Press, Toronto, Ont., 234 pp.

Buckingham, E., 1907. Studies on the movement of soil moisture. *U.S. Dep. Agr., Bur. Soils, Bull.*, 38:61 pp.

Cary, J.W., 1966. Soil moisture transport due to thermal gradients: practical aspects. *Soil Sci. Soc. Am. Proc.*, 30:428–433.

Cary, J.W. and Mayland, H.F., 1972. Salt and water movement in unsaturated frozen soil. *Soil Sci. Soc. Am. Proc.*, 36:549–555.

Chahal, R.S. and Yong, R.N., 1965. Validity of the soil water characteristics determined with the pressurized apparatus. *Soil Sci.*, 99:98–103.

Chang, R.K. and Warkentin, B.P., 1968. Volume change of compacted clay soil aggregates. *Soil Sci.*, 105:106–111.

Childs, E.C., 1969. *An Introduction to the Physical Basis of Soil Water Phenomena.* Wiley, New York, N.Y., 493 pp.

Christensen, R.M., 1971. *Theory of Viscoelasticity.* Academic Press, New York, N.Y., 245 pp.

Coleman, E.A. and Bodman, G.B., 1944. Moisture and energy conditions during downward entry of water into moist and layered soils. *Soil Sci. Soc. Am. Proc.*, 9:3–11.

Collin, A., 1956. *Landslides in Clay* (transl. by W.R. Schriever). University of Toronto Press, Toronto, Ont., 224 pp.

Conforth, D.H., 1964. Some experiments on the influence of strain conditions on the strength of sand. *Geotechnique*, 14:143–167.

Crawford, C.B., 1964. Interpretation of the consolidation test. *Am. Soc. Civil Engrs., Proc., J. Soil Mech. Found. Div.*, 90, SM5:87–102.

Croney, D. and Coleman, J.D., 1954. Soil structure in relation to soil suction (pF). *J. Soil Sci.*, 5:75–84.

Croney, D., Coleman, J.D. and Bridge, P.M., 1952. The suction of moisture held in soil and other porous materials. *D.S.I.R. Road Res. Lab., Tech. Pap.*, 24:42 pp.

Croney, D., Coleman, J.D. and Black, W.P.M., 1958. Movement and distribution of water in soil in relation to highway design and performance. In: *Water and its Conduction in Soils. Highway Res. Board, Spec. Rep.*, 40:226–252.

De Jong, E. and Warkentin, B.P., 1965. Water retention by clay–glass bead mixtures. *Soil Sci.*, 100:108–111.

Deresiewicz, H., 1958. Mechanics of granular matter. *Adv. Appl. Mech.*, 5:233–306.

Diamond, S., 1970. Pore size distribution in clays. *Clays Clay Miner.*, 18:7–24.

Diamond, S., 1971. Microstructure and pore structure of impact-compacted clays. *Clays Clay Miner.*, 19.239–250.

Dirksen, C. and Miller, R.D., 1966. Closed-system freezing of unsaturated soil. *Soil Sci. Soc. Am. Proc.*, 30:168–173.

Drucker, D.C., 1960. Plasticity. In: J.N. Goodier and N.J. Hoff (Editors), *Proc. Symp. Naval Struct., 1st,* Pergamon, London, pp.407–448.

Drucker, D.C., 1967. *Introduction to Mechanics of Deformable Solids.* McGraw-Hill, New York, N.Y., 445 pp.

Drucker, D.C. and Prager, W., 1952. Soil mechanics and plastic analysis of limit design. *Q. Appl. Math.*, 10:157–175.

Everett, D.H. and Haynes, J.M., 1965. Capillary properties of some model pore systems with special reference to frost damage. *Rilem Bull., New Ser.*, 27:31–38.

Eyring, H., 1936. Viscosity, plasticity and diffusion as examples of absolute reaction rates. *J. Chem. Phys.*, 4:283–291.

Farrar, D.M. and Coleman, J.D., 1967. The correlation of surface area with other properties of nineteen British clay soils. *J. Soil Sci.*, 18:118–124.

Fox, W.E., 1964. A study of bulk density and water in a swelling soil. *Soil Sci.*, 98:307–316.

Gill, W.R., 1959. Soil bulk density changes due to moisture changes in soil. *Trans. Am. Soc. Agric. Engrs.*, 2:104–105.

Gillott, J.E., 1968. *Clay in Engineering Geology.* Elsevier, Amsterdam. 296 pp.

Grim, R.E., 1953. *Clay Mineralogy.* McGraw-Hill, New York, N.Y., 384 pp.

Grossman, R.B., Brasher, B.R., Franzmeier, D.P. and Walker J.I., 1968. Linear extensibility as calculated from natural clod bulk density measurements. *Soil Sci. Soc. Am. Proc.,* 32:570–573.

Haines, W.B., 1923. The volume changes associated with variations of water content in soils. *J. Agr. Sci.,* 13:296.

Hammamji, Y., 1969. *Some Factors affecting Heaving Pressures of Frozen Soils.* Thesis, McGill University, Montreal, Que., 94 pp.

Hansbo, S., 1960. Consolidation of clay, with special reference to influence of vertical sand drains. *Swed. Geotech. Inst., Proc.,* 18:41–61.

Henkel, D.J., 1959. The relationships between the strength, pore water pressure and volume change characteristics of saturated clays. *Geotechnique,* 9:119–135.

Henkel, D.J., 1960. The shear strength of saturated remoulded clays. *Res. Conf. Shear Strength Cohesive Soils, Am. Soc. Civil Engrs.,* pp.533–554.

Highway Research Board, 1969. Effects of temperature and heat on engineering behaviour of soils. *Highway Res. Board., Spec. Rep.,* 103:300 pp.

Hill, R., 1950. *The Mathematical Theory of Plasticity.* Oxford Univ. Press, London, 355 pp.

Hillel, D., 1971. *Soil and Water.* Academic Press, New York, N.Y., 288 pp.

Hoekstra, P., Chamberlain, E. and Frate, T., 1965. Frost heaving pressures. *Cold Reg. Res. Eng. Lab. Res. Rep.,* 176:

Hutcheon, W.L., 1958. Moisture flow induced by thermal gradients within unsaturated soils. *Highway Res. Board, Spec. Rep.,* 40:113–133.

Hvorslev, M.J., 1937. Über die Festigkeitseigenschaften gestörter bindiger Boden. Thesis; Danmarks Naturvidenskabelige Samfund, *Ingeniorvidenskab. Skr., Ser. A,* 45:159 pp.

Hvorslev, M.J., 1960. Physical components of the shear strength of saturated clays. *Res. Conf. Shear Strength Cohesive Soils, Am. Soc. Civil Eng.,* pp.169–273.

Jackson, M.L., 1963. Interlayering of expansible layer silicates in soils by chemical weathering. *Clays Clay Miner.,* 11:29–46.

Jackson, M.L., 1964. Chemical composition of soils. In: F.E. Bear (Editor), *Chemistry of the Soil. Am. Chem. Soc., Monogr.* 2nd ed., 160:71–141.

Jaeger, J.C., 1962. *Elasticity, Fracture and Flow.* Methuen, London, 212 pp.

Janiga, P.V., 1970. *Some Considerations in In-situ Frost Heaving.* Thesis, McGill University, Montreal, Que., 138 pp.

Jennings, J.E., 1965. The theory and practise of construction on partly saturated soils as applied to South African conditions. *Int. Res. Eng. Conf. Expansive Clay Soils, Texas,* pp.345–363.

Kassiff, G., Livneh, M. and Wiseman, G., 1969. *Pavements on Expansive Clays.* Jerusalem Academic Press, Jerusalem, 218 pp.

Keen, B.A., 1931. *Physical Properties of the Soil.* Longmans Green, London, 380 pp.

Kemper, W.D. and Rollins, J.B., 1966. Osmotic efficiency coefficients across compacted clays. *Soil Sci. Soc. Am. Proc.,* 30:529–534.

Kenney, T.C., 1967. The influence of mineral composition on the residual strength of natural soils. *Proc. Geotech. Conf., Oslo,* I:123–129.

Kirkpatrick, W.M., 1957. The condition of failure for sands. *Proc. Int. Conf. Soil Mech. Found. Eng., 4th,* 1:172–178.

Kjellman, W., 1936. Report on an apparatus for consummate investigation of the mechanical properties of soils. *Proc. Int. Conf. Soil Mech. Foud. Eng., 1st.,* 2:16–20.

Ko, H.Y. and Scott, R.F., 1967. Deformation of sand in hydrostatic compression. *Am. Soc. Civil. Engrs., Proc., J. Soil Mech. Found. Div.,* 93, SM3:137–156.

Lafeber, D., 1967. The optical determination of spatial orientation of platey clay minerals in soil thin sections. *Geoderma,* 1:359–369.

Lambe, T.W., 1960a. The structure of compacted clay. *Trans. Am. Soc. Civil Engrs.,* 125:681.

Lambe, T.W., 1960b. A mechanistic picture of shear strength in clay. *Res. Conf. Shear Strength Cohesive Soils, Am. Soc. Civil Engrs.*, pp.555–580.

Lambe, T.W. and Whitman, R.V., 1969. *Soil Mechanics*. Wiley, New York, N.Y., 553 pp.

LaRochelle, P. and Lefebvre, G., 1971. Sampling disturbance in Champlain clays. *Proc. A.S.T.M.*, STP. 483:143–163.

Leitch, H.C. and Yong, R.N., 1967. The rate dependent mechanism of shear failure in clay soils. *Soil Mech. Lab., McGill Univ., Soil Mech. Ser.*, 21:140 pp.

Leonards, G.A. (Editor), 1962. *Foundation Engineering*. McGraw-Hill, New York, N.Y., 1136 pp.

Leonards, G.A. and Ramiah, B.K., 1959. Time effects in the consolidation of clays. *Proc. A.S.T.M.*, STP 254:116–130.

Letey, J., Kemper, W.D. and Noonan, L., 1969. The effect of osmotic pressure gradients on water movement in unsaturated soil. *Soil Sci. Soc. Am. Proc.*, 33:15–18.

Lo, K.Y., 1965. Stability of slopes in anisotropic soils. *Am. Soc. Civil Engrs., J. Soil Mechs. Found. Div. Proc.*, 91, SM4:85–106.

Lowe, J., Jonas, E. and Obrician, V., 1969. Controlled gradient consolidation test. *Am. Soc. Civil Engrs., Proc., J. Soil Mech. Found. Div.*, 95, SM1: 77–98.

Mackay, J.R., 1971. The origin of massive icy beds in permafrost, Western Arctic Coast, Canada. *Can. J. Earth Sci.*, 8:397–422.

Mackay, J.R., 1972. The world of underground ice. *Annals Assoc. Am. Geographers*, 62:1–22.

Mackenzie, R.C. (Editor), 1957. *The Differential Thermal Investigation of Clays*. Mineralogical Society, London, 456 pp.

Maeda, T., 1970. *Water Retention and Movement in Allophane Soils*. Dep. Soil Sci., Macdonald College, Mimeo Rep., 31 pp.

Marshall, T.J., 1959. Relations between water and soil. *Comm. Bur. Soil Sci., Harpenden, Tech. Comm., 50:91 pp.*

Marshall, C.E., 1964. *The Physical Chemistry and Mineralogy of Soils*. Wiley, New York, N.Y., 388 pp.

Martin, R.T., 1966. Quantitative fabric of wet kaolinite. *Clays Clay Miner.*, 9:271–287.

McKyes, E., 1966. *Theoretical and Experimental Determination of Frost Heaving Pressures in a Partially Frozen Silt*. Thesis, McGill University, Montreal, Que., 44 pp.

McKyes, E. and Yong, R.N., 1971. Three techniques for fabric viewing as applied to shear distortion of a clay. *Clays Clay Miner.*, 19:289–293.

Mendelson, A., 1968. *Plasticity. Theory and Application*. MacMillan, New York, N.Y., 353 pp.

Miller, E.E. and Miller, R.D., 1955. Theory of capillary flow. I. Practical implications. *Soil Sci. Soc. Am. Proc.*, 19:267–271.

Miller, R.D. and Richard, F., 1952. Hydraulic gradients during infiltration in soils. *Soil Sci. Soc. Am. Proc.*, 16:33–38.

Miller, R.D., Baker, J.H. and Kolaian, J.H., 1960. Particle size, overburden pressure, pore water pressure and freezing temperature of ice lenses in soil. *Trans. Int. Congr. Soil Sci., 7th*, 1:122–128.

Mitchell, J.K., Singh, A. and Campanella, R.G., 1969. Bonding, effective stresses and strength of soils. *Am. Soc. Civil Engrs., Proc., J. Soil Mech. Found. Div.*, 95, SM5:1219–1246.

Mohr, O., 1900. Die Elastizitätsgrenze und Bruch eines Materials. *Z. Ver. Dtsch. Ing.*, 44:1524.

Moore, C.A., 1968. *Mineralogical and Pure Fluid Influences on Deformation Mechanisms in Clay Soils*. Thesis submitted to University of California.

Murayama, S. and Shibata, T., 1961. Rheological properties of clays. *Int. Conf. I.S.S.M.F.E., 5th, Proc.*, 1:269–273.

Nadai, A., 1950. *Theory of Flow and Fracture of Solids*, 1. McGraw-Hill, New York, N.Y., 2nd ed., 572 pp.

Newmark, N.M., 1960. Failure hypotheses for soils. *Res. Conf. Shear Strength Cohesive Soils, Am. Soc. Civil Engrs.*, pp.17–32.

Olsen, H.W., 1961. *Hydraulic Flow through Saturated Clays.* Thesis, Mass. Inst. Technol., Cambridge, Mass., 195 pp.

Olson, R.E., 1964. Discussion of paper — Effective stress theory of soil compaction. *Am. Soc. Civil Engrs., J. Soil Mech. Found. Div.,* 90, SM2:171–189.

Palit, R.M., 1953. Determination of swelling pressure of black cotton soil — a method. *Proc. Int. Conf. Soil Mech. Found. Eng., 3rd.,* 1:170.

Parr, J. and Bertrand, A.R., 1960. Water infiltration into soils. *Adv. Agron.,* 12:311–363.

Penner, E., 1959. The mechanism of frost heaving in soils. *Highway Res. Board Bull.,* 225:1–13.

Penner, E., 1968. Heaving pressure in soils during unidirectional freezing. *Can. Geotech. J.,* 4:398–408.

Peterson, R. and Peters, N., 1963. Heave of spillway structures on clay shales. *Can. Geotech. J.,* 1:5.

Philip, J.R., 1957. Theory of infiltration. I. The infiltration equation and its solution. *Soil Sci.,* 83:345–357.

Philip, J.R. and De Vries, D.A., 1957. Moisture movement in porous materials under temperature gradients. *Trans. Am. Geophys. Union.,* 38:222–232.

Philip, J.R. and Smiles, D.E., 1969. Kinetics of sorption and volume change in three-component systems. *Aust. J. Soil Res.,* 7:1–19.

Pusch, R., 1966. Quick-clay microstructure. *Eng. Geol.,* 3:433–443.

Quigley, R.M., 1961. *Composition and Engineering Properties of some Vermiculitic Products of Weathering.* Thesis, submitted to Mass. Inst. Technol., Cambridge, Mass.

Quigley, R.M. and Thompson, C.D., 1966. The fabric of anisotropically consolidated sensitive marine clay. *Can. Geotech. J.,* 3(2):61–73.

Rawlins, S.L., 1971. Some new methods for measuring the components of water potential. *Soil Sci.,* 112:8–16.

Rendulic, L., 1937. Ein Grundgesetz der Tonmechanik und sein experimenteller Beweis. *Der Bauingenieur,* 18:459–467.

Richards, L.A., 1949. Methods of measuring soil moisture tension. *Soil Sci.,* 68:95–112.

Richards, L.A., 1950. Laws of soil moisture. *Trans. Am. Geophys. Union,* 31:750–756.

Rose, C.W., Stern, W.R. and Drummond, J.E., 1965. Determination of hydraulic conductivity as a function of depth and water content for soil in situ. *Aust. J. Soil Res.,* 3:1–9.

Rosenqvist, I. Th., 1959. Physico-chemical properties of soils; soil-water systems. *Am. Soc. Civil Engrs., Proc., J. Soil Mech. Found. Div.,* 85, SM2:31–53.

Rosenqvist, I. Th., 1962. The influence of physico-chemical factors upon the mechanical properties of clays. *Clays Clay Miner.,* 9:12–27.

Saada, A.S. and Zamani, K.K., 1969. The mechanical behaviour of cross anisotropic clays. *Proc. Int. Conf. Soil Mech. Found. Eng., 7th, Mexico,* 1:351–360.

Schiffman, R.L., Chen, A.T.F. and Jordan, J.C., 1969. An analysis of consolidation theories. *Am. Soc. Civil Engrs., Proc., J. Soil Mech. Found. Div.,* 95, SM1:285–312.

Schmertmann, J.H. and Osterberg, J.O., 1960. An experimental study of the development of cohesion and friction with axial strain in saturated cohesive soils. *Res. Conf. Shear Strength Cohesive Soils, Am. Soc. Civil Engrs.,* pp.643–694.

Schofield, R.K., 1935. The pF of the water in soil. *Trans. Int. Congr. Soil Sci., 3d,* II:37–48.

Scott, R.F., 1963. *Principles of Soil Mechanics.* Addison-Wesley, Reading, Mass., 550 pp.

Scott, R.F. and Ko, H.Y., 1969. Stress deformation and strength characteristics. *Proc., Int. Conf. Soil Mech. Found. Eng., 7th, State-of-the-art Rep., Mexico,* pp.1–47.

Seed, H.B. and Chan, C.K., 1959. Structure and strength characteristics of compacted clay. *Am. Soc. Civil Engrs., Proc., J. Soil Mech. Found. Div.,* 85, H SM 5,

Sheeran, D.E., 1972. *A Spectrophotometric Technique for the Fabric Analysis of Mono-mineralic Kaolinitic Soils.* Thesis, Northwestern University, Evanston, Ill.

Shibata, T. and Karube, D., 1965. Influence of the variation of the intermediate principal stress on mechanical properties of normally consolidated clays. *Proc., Int. Conf. Soil Mech. Found. Eng., 6th,* I:359–363.

Shield, R.T., 1955. On Coulomb's law of failure in soils. *J. Mech. Phys. Solids,* 4:10–16.

Sill, R.C. and Skapski, A.S., 1956. Method for determination of surface tension of solids from their melting points in thin wedges. *J. Chem. Phys.,* 24:644–651.

Skempton, A.W., 1954. The pore pressure coefficients A and B. *Geotechnique,* 4:143.

Skempton, A.W., 1960. Correspondence. *Geotechnique,* 10:186–187.

Skempton, A.W., 1964. Long-term stability of clay slopes. *4th Rankine Lecture, Geotechnique,* 19:77–101.

Smart, P., 1969. Soil structure in the electron-microscope. *Proc. Int. Conf. Struct., Solid Mech. Eng. Design,* I, Wiley-Interscience, New York, N.Y., 249–255.

Smith, W.O., Foote, P.D. and Busang, P.F., 1929. Packing of homogeneous spheres. *Phys. Rev.,* 34:1271–1274.

Soderman, L.G. and Quigley, R.M., 1965. Geotechnical properties of three Ontario clays. *Can. Geotech. J.,* 2:176–189.

Staple, W.J., 1969. Comparison of computed and measured moisture redistribution following infiltration. *Soil Sci. Soc. Am. Proc.,* 33:840–847.

Stirk, G.B., 1954. Some aspects of soil shrinkage and the effect of cracking upon water entry into the soil. *Aust. J. Agric. Res.,* 5:279.

Swartzendruber, D., 1960. Water flow through a soil profile as affected by the least permeable layer. *J. Geophys. Res.,* 65:4037–4042.

Swartzendruber, D., 1962. Modification of Darcy's Law for the flow of water in soils. *Soil Sci.,* 93:22–29.

Taylor, D.W., 1942. Research on consolidation of clays. *Mass. Inst. Technol., Rep.,* Ser. 82:147 pp.

Taylor, D.W., 1948. *Fundamentals of Soil Mechanics.* Wiley, New York, N.Y., 700 pp.

Terzaghi, K., 1943. *Theoretical Soil Mechanics.* Wiley, New York, N.Y., 510 pp.

Terzaghi, K., 1952. Permafrost. *J. Boston Soc. Civil Eng.,* 39:319–368.

Timoshenko, S., 1956. *Strength of Materials.* II. *Advanced Theory and Problems.* Van Nostrand, New York, N.Y., 510 pp.

Timoshenko, S. and Goodier, J.N., 1951. *Theory of Elasticity.* McGraw-Hill, New York, N.Y., 2nd ed., 506 pp.

Topp, G.C., Klute, A. and Peters, D.B., 1967. Comparison of water content-pressure head data obtained by equilibrium, steady-state, and unsteady-state methods. *Soil Sci. Soc. Am. Proc.,* 31:312–314.

Van Olphen, H., 1963. *An Introduction to Clay Colloid Chemistry.* Interscience, New York, N.Y., 301 pp.

Waldron, L.J., McMurdie, J.L. and Vomocil, J.A., 1961. Water retention by capillary forces in an ideal soil. *Soil Sci. Soc. Am. Proc.,* 25:265–267.

Ward, W.H., Samuels, S.C. and Butler, M.E., 1959. Further studies of the properties of London clay. *Geotechnique,* 9:33.

Warkentin, B.P., 1961. Interpretation of the upper plastic limit of clays. *Nature,* 190:287–288.

Warkentin, B.P. and Bozozuk, M., 1961. Shrinking and swelling properties of two Canadian clays. *Proc. Int. Conf. Soil Mech. Found. Eng., 5th, Paris,* 3A:851.

Warkentin, B.P. and Schofield, R.K., 1962. Swelling pressure of Na-montmorillonite in NaCl solutions. *J. Soil Sci.,* 13:98.

Warkentin, B.P. and Yong, R.N., 1960. Compacted clay. *Trans. Am. Soc. Civil Engrs.,* 125:707–709.

Warkentin, B.P. and Yong, R.N., 1962. Shear strength of montmorillonite and kaolinite related to interparticle forces. *Clays Clay Miner.,* 9:210–218.

White, J.L., 1971. Interpretation of infrared spectra of soil minerals. *Soil Sci.,* 112:22–31.

White, W.A., 1949. Atterberg plastic limits of clay minerals. *Am. Mineralogist,* 34:508–512.

Whittig, L.D., 1965. X-ray diffraction techniques for mineral identification and mineralogical composition. In: C.A. Black (Editor), *Methods of Soil Analysis, Am. Soc. Agron., Monogr.,* 9, Part I: 671–698.

Winterkorn, H.F., 1955. Water movement through porous hydrophilic systems under capillary, electric and thermal potentials. *Proc. A.S.T.M., STP.* 163:27–35.

Winterkorn, H.F., 1958. Mass transport phenomena in moist porous systems as viewed from the thermodynamics of irreversible processes. *Highway Res. Board, Spec. Rep.,* 40:324–338.

Wise, M.E., 1952. Dense random packing of unequal spheres. *Philips Res. Rep.,* 7:321–343.

Wong, C.Y., 1972. *Constitutive Relationships of Granular Materials.* Thesis, McGill Univ., Montreal, Que., 199 pp.

Wong, H.Y., 1969. Preliminary report on unsaturated flow mechanisms in soils. *Soil Mech. Lab., McGill Univ., Soil Mech. Res. Rep.,* 69–1.

Wong, H.Y., 1971. Interim report on unsaturated flow mechanisms in soils. *Soil Mech. Lab., McGill Univ., Soil Mech. Res. Rep.,* 71–1:12 pp.

Wong, H.Y., 1972. Interim report on unsaturated flow mechanisms in soils. *Soil Mech. Lab., McGill Univ., Soil Mech. Res. Rep.,* 72–3.

Wong, H.Y. and Yong, R. N., 1973. A simple solution of some practical engineering problems concerning unsaturated soils. *Civil Eng. Public Works Rev.,* 68:759–765.

Wu, T.H., Loh, A.H. and Malvern, L.E., 1963. Study of the failure envelope of soils. *Am. Soc. Civil Engrs., Proc., J. Soil Mech. Found. Div.,* 89, SM1:145–181.

Yamazaki, F. and Takenaka, H., 1965. On the influence of air-drying on Atterberg's limit. *Trans. Agric. Eng. Soc., Japan,* 14:46–48.

Yong, R.N., 1965. Soil suction effects on partial soil freezing. *Highway Res. Board, Res. Rec.,* 68:31–42.

Yong, R.N., 1967. On the relationship between partial soil freezing and surface forces. In: H. Oura (Editor), *Physics of Snow and Ice.* Inst. Low Temp. Sci., Hokkaido Univ., 1:375–1385.

Yong, R.N., 1971. Soil technology and stabilization. *Gen. Rep. Sess. 5, Asian Reg. Conf. Soil Mech. Found. Eng., 4th, Bangkok,* 2:111–124.

Yong, R.N. and Chen, D.S., 1970. Analysis of creep of clays using retardation time distribution. *Proc., Int. Congr. Rheol., 5th, Kyoto,* Univ. Park Press, 2:309–314.

Yong, R.N. and Chen, D.S., 1972. A probabilistic after-effect analysis of relaxation of clays. *Proc. Int. Congr. Rheol., 6th, Lyon.*

Yong, R.N. and Janiga, P.V., 1971. Field study of moisture movement and ground heave during freeze-up. *McGill Univ., Soil Mech. Lab. Rep.,* 30: 63 pp.

Yong, R.N. and McKyes, E., 1971. Yield and failure of a clay under triaxial stresses. *Am. Soc. Civil Engrs., Proc., J. Soil Mech. Found. Div.,* 97, SM1:159–176.

Yong, R.N. and Osler, J.C., 1971. Heave and heaving pressures in frozen soils. *Can. Geotech. J.,* 8:272–282.

Yong, R.N. and Sheeran, D.E., 1972. Physics of salt intrusion. II. Development of experimental apparatus and pilot testing. *Soil Mech. Lab., McGill Univ., Rep.,* FH-SI-7, 57 pp.

Yong, R.N. and Vey, E., 1962. The use of stress loci for determinations of effective stress parameters. Stress distribution in earth masses. *Highway Res. Board, Bull.,* 342:1038–1051.

Yong, R.N. and Wong, C.Y., 1972. Experimental studies of elastic deformation of sand. *Proc., South East Asian Soils Conf., 3rd., Hong Kong.*

Yong, R.N. and Wong, H.Y., 1973. Water movement in unsaturated swelling soils. *Soil Mech. Lab., McGill Univ., Soil Mech. Res. Rep.,* 73-2, 15 pp.

Yong, R.N., Taylor, L.O. and Warkentin, B.P., 1963. Swelling pressures of sodium montmorillonite at depressed temperatures. *Nat. Conf. Clays Clay Miner. Proc.,* 2:268–281.

Yong, R.N., Japp, R.D. and How, G., 1971. Shear strength of partially saturated clays. *Proc. Asian Reg. Conf. Soil Mech. Found. Eng., 4th, Bangkok*, 1:183–187.

Yong, R.N., McKyes, E. and Silvestri, V., 1972. Yield and failure of clays. *Arch. Mech.*, 24:511–528.

Zaslavsky, D., 1964. Saturated and unsaturated flow equation in an unstable porous medium. *Soil Sci.*, 98:317–321.

Selected reading

A.S.T.M., 1963. Laboratory shear testing of soils. *A.S.T.M., Spec. Tech. Publ.*, 361:505 pp.

Barden, L. and Sides, G.R., 1969. The influence of structure on the collapse of compacted clay. *Int. Conf. Expansive Soils, 2nd, Texas A. & M. Univ.*, pp.317–326.

Barden, L. and Sides, G.R., 1970. Engineering behaviour and structure of compacted clay. *Am. Soc. Civil Engrs., Proc., J. Soil Mech. Found. Div.*, 96, SM4:1171–1200.

Barden, L., Ismail, H. and Tong, P., 1969. Plane strain deformation of granular material at low and high pressures. *Geotechnique*, 19:441–452.

Bear, J. (Editor), 1968. *Physical Principles of Water Percolation and Seepage*. UNESCO, Paris, 465 pp.

Bishop, A.W., 1966. Strength of soils as engineering materials. *6th Rankine Lecture, Geotechnique*, 16:89–130.

Bjerrum, L., 1954. Geotechnical properties of Norwegian marine clays. *Geotechnique*, 4:49–69.

Brewer, R., 1964. *Fabric and Mineral analysis of Soils*. Wiley, New York, N.Y., 470 pp.

Conway, B.E., Bockris, J. and Ammare, I.A., 1951. The dielectric constant of the solution in the diffuse and Helmholtz double layers at a charged interface in aqueous solution. *Faraday Soc. Trans.*, 47:756–767.

Eden, W.J. and Mitchell, R.J., 1970. The mechanics of landslides in Leda clay. *Can. Geotech. J.*, 7:285–296.

Kassiff, G. and Ben Shalom A., 1971. Experimental relationship between swell pressure and suction. *Geotechnique*, 21:245–255.

Kassiff, G., Livneh, M. and Wiseman, G., 1969. *Pavements on Expansive Clays*. Jerusalem Academic Press, Jerusalem, 218 pp.

Komornik, A., 1968. The effect of anisotropy on the swelling of a compacted clay. *Proc. Asian Reg. Conf. Soil Mech. Found. Eng., 3rd, Haifa*, pp.181–185.

Komornik, A. and Zeitlen, J.G., 1965. An apparatus for measuring lateral soil swelling pressures in the laboratory. *Proc. Int. Conf. Soil Mech. Found. Eng., 6th, Montreal*, 1:278–281.

Kruyt, H.R. (Editor), 1952. *Colloid Science. I. Irreversible Systems*. Elsevier, Amsterdam, 389 pp.

Lafeber, D., 1965. The graphical representation of planar pore patterns in soils. *Aust. J. Soil Res.*, 3:143–164.

Lambe, T.W. and Whitman, R.V., 1969. *Soil Mechanics*. Wiley, New York, N.Y., 553 pp.

LaRochelle, P., Chagnon, J.Y. and Lefebvre, G., 1970. Regional geology and landslides in the marine clay deposits of Eastern Canada. *Can. Geotech. J.*, 7:145–156.

Legget, R.F., 1962. *Geology and Engineering*. McGraw-Hill, New York, N.Y., 2nd ed., 884 pp.

Low, P.F., 1961. Physical chemistry of clay–water interaction. *Adv. Agron.*, 13:269–327.

Low, P.F., 1968. Mineralogical data requirements in soil physical investigations. In: *Mineralogy in Soil Science and Engineering. Soil Sci. Soc. Am., Spec. Publ.*, 3:1–34.

Maeda, T., Sasaki, S. and Sasaki T., 1970. Studies on the physical properties of volcanic soils in Hokkaido – properties of pumice with respect to water. *Trans. Japan Soc. Irrig., Drain. Recl. Eng.*, 31:25–28.

Martin, R.T., 1963. Adsorbed water on clay: a review. *Clays Clay Miner.*, 11:28—70.

Meyerhof, G.G., 1961. The mechanism of flow slides in cohesive soils. *Geotechnique*, 7:41—49.

Mitchell, R.J., 1970. On the yielding and mechanical strength of Leda clay. *Can. Geotech. J.*, 7:297—312.

Murayama, S. and Shibata, T., 1964. Flow and stress relaxation of clays. *Proc. Rheol. Soil Mech. Symp., I.U.T.A.M., Grenoble*, pp.99—129.

Nielsen, D.R., Biggar, J.W. and Davidson, J.M., 1962. Experimental consideration in diffusion analysis in unsaturated flow problems. *Soil Sci. Soc. Am. Proc.*, 26:107—111.

Olsen, H.W., 1965. Deviation from Darcy's Law in saturated clays. *Soil Sci. Soc. Am. Proc.*, 29:135—140.

Parlange, J.Y., 1972. Analytical theory of water movement in soils. *Proc. Symp. Fundam. Transp. Phenom. Porous Media, 2nd., Univ. Guelph*, 1:222—236.

Parry, R.H.G. ed., 1972. Stress—strain behaviour of soils. *Proc. Roscoe Mem. Symp.* G.T. Foulis and Co. Ltd., 752 pp.

Pusch, R., 1966. Investigation of clay microstructure by using ultra thin sections. *Swed. Geotech. Inst., Rep.*, 15.

Richards, B.G., 1969. Psychrometric techniques for measuring soil water potential. *C.S.I.R.O., Div. Soil Mech. Tech. Rep.*, 9:1—31.

Rosenqvist, I.Th., 1959b. Mechanical properties of soil-water systems. *Am. Soc. Civil Engrs., Proc., J. Soil Mech. Found Div.*, 85, SM2:31—53.

Sudo, S. and Yasutomi, R., 1965. The rheology of soil as a colloid clay system. *Trans. Agric. Eng. Soc. Japan*, 14:16—20.

Takenaka, H. and Yasutomi, R., 1965. pF concept upon softening and hardening of soil — relation between engineering properties of soil and pF. *Trans. Agric. Eng. Soc., Japan*, 14:54—59.

Taylor, R.M. and Norrish, K., 1966. The measurement of orientation distribution and its application to quantitative X-ray diffraction analysis. *Clay Miner. Bull.*, 6:127—142.

Wooltorton, F.L.D., 1954. The scientific basis of road design. In: *Soil Volume Changes*, IV. Arnold, London, 364 pp.

Wu, T.H., 1966. *Soil Mechanics.* Allyn and Bacon, Inc., 431 pp.

Yamanouchi, T., 1969. Physical chemistry of soils. In: T. Magami (Editor), *Soil Mechanics.* Tokyo Press, Tokyo, pp.1—88.

Yasutomi, R. and Sudo, S., 1968. A concept of softening and hardening of clay based upon the change in soil structure by remoulding. *Soil Sci.*, 105:384—391.

Yong, R.N. and Warkentin, B.P., 1972. Unsaturated flow in expansive soils. *Proc. Symp. Fundam. Transp. Phenomena Porous Media, 2nd, Univ. Guelph*, 1:306—319.

AUTHOR INDEX

SUBJECT INDEX